The BiblioLife Network

This project was made possible in part by the BiblioLife Network (BLN), a project aimed at addressing some of the huge challenges facing book preservationists around the world. The BLN includes libraries, library networks, archives, subject matter experts, online communities and library service providers. We believe every book ever published should be available as a high-quality print reproduction; printed on- demand anywhere in the world. This insures the ongoing accessibility of the content and helps generate sustainable revenue for the libraries and organizations that work to preserve these important materials.

The following book is in the "public domain" and represents an authentic reproduction of the text as printed by the original publisher. While we have attempted to accurately maintain the integrity of the original work, there are sometimes problems with the original book or micro-film from which the books were digitized. This can result in minor errors in reproduction. Possible imperfections include missing and blurred pages, poor pictures, markings and other reproduction issues beyond our control. Because this work is culturally important, we have made it available as part of our commitment to protecting, preserving, and promoting the world's literature.

GUIDE TO FOLD-OUTS, MAPS and OVERSIZED IMAGES

In an online database, page images do not need to conform to the size restrictions found in a printed book. When converting these images back into a printed bound book, the page sizes are standardized in ways that maintain the detail of the original. For large images, such as fold-out maps, the original page image is split into two or more pages.

Guidelines used to determine the split of oversize pages:

• Some images are split vertically; large images require vertical and horizontal splits.
• For horizontal splits, the content is split left to right.
• For vertical splits, the content is split from top to bottom.
• For both vertical and horizontal splits, the image is processed from top left to bottom right.

Aluminium: its history, occurrence, and applications. Illustrated.

Joseph William Richards

ALUMINIUM

ALUMINIUM:

ITS HISTORY, OCCURRENCE, PROPERTIES, METALLURGY AND APPLICATIONS, INCLUDING ITS ALLOYS.

BY

JOSEPH W. RICHARDS, M.A., A.C.,

INSTRUCTOR IN METALLURGY AT THE LEHIGH UNIVERSITY.

SECOND EDITION,

REVISED AND GREATLY ENLARGED.

ILLUSTRATED BY

TWENTY-EIGHT ENGRAVINGS AND TWO DIAGRAMS.

LONDON :

SAMPSON LOW, MARSTON, SEARLE & RIVINGTON, Limited,

ST. DUNSTAN'S HOUSE, FETTER LANE, FLEET STREET.

1890.

PRINTED AT THE COLLINS PRINTING HOUSE,
705 Jayne Street,
PHILADELPHIA, U. S. A.

PREFACE TO THE SECOND EDITION.

IF it was true that no apology was necessary in presenting a work on Aluminium in English, as stated in the preface to the first edition of this book, it is equally true that still less apology is necessary in offering an improvement on that work.

The present volume is designed to be an improvement on the former one in the following respects: Mistakes have been corrected wherever detected by the author or pointed out by his friends; in some instances the order of treatment of different parts has been revised, so as to bring them into strict, logical sequence; the more strictly historical processes are described in greater detail, in order to preserve a complete record of the rise of the aluminium industry; chapters have been added treating on the properties and the preparation of aluminium compounds, on the theoretical aspect of the reduction of aluminium compounds, and on the analysis of commercial aluminium and its common alloys; the original chapters have been in several cases sub-divided, and every part treated more by itself and in greater detail than before; finally, additions have been made throughout, recording and describing the progress achieved in the last three years, with a completeness which it is hoped is up to the standard of the rest of the book.

The method of treatment in the present edition will be found to be more critical, for wherever a reasonable doubt might be expressed as to the correctness of certain claims, or a rational explanation advanced for certain phenomena, the author has not hesitated to put his best thought on the question and to state his conclusions unreservedly.

The friendly criticisms of the scientific press and their suggestions have been kept in view in preparing this new edition. The spelling "aluminium" has been retained, because no sufficient reasons have been advanced for changing it to " aluminum;" and even if each way was equally old and as well-sanctioned by usage and analogy as the other, the author's choice would be the longer spelling, as being more euphonious and agreeable to the ear.

It has been the author's endeavor to make this volume as complete as possible, as accurate as possible, to write it in a manner which will be entertaining to the general reader, and to furnish a treatise which will be of practical value to the practical metallurgist as well as of scientific merit where it touches on matters of theory.

J. W. R.

BETHLEHEM, PA., March 12, 1890.

PREFACE TO THE FIRST EDITION.

No apology is necessary in presenting a work on aluminium in English. In 1858 Tissier Bros. published in France a small book on the subject. H. St. Claire Deville, the originator of the aluminium industry, published a treatise, also in French, in 1859. Deville's book is still the standard on the subject. Until December, 1885, we have an intermission, and then a work by Dr. Mierzinski, forming one of Hartleben's Chemisch-Technische Bibliothek, which is a fair presentation of the industry up to about 1883, this being a German contribution. Probably because the English speaking people have taken comparatively little hand in this subject we find no systematic treatise on aluminium in our language. The present work aims to present the subject in its entirety to the English reader.

Tissier, Deville, Mierzinski, and the German, French, and English scientific periodicals have been freely consulted and extracted from, full credit being given in each case to the author or journal. As this art has of late advanced so rapidly it has been a special aim to give everything that has been printed up to the time of publication.

The different parts of the work are arranged in what seemed their logical order, corresponding closely to that followed by

Deville. The Appendix contains an account of laboratory experiments, etc., several of which, it is trusted, may be of value.

In conclusion, the author wishes to thank the faculty of his "Alma Mater," Lehigh University, for their permission to use his Thesis on Aluminium as the basis of this treatise; also, to acknowledge his indebtedness to Dr. Wm. H. Greene, of Philadelphia, for assistance rendered in the preparation of the work for the press.

J. W. R.

PHILADELPHIA, November 25, 1886.

ABBREVIATIONS USED IN MAKING REFERENCES.

Deville . . . . De l'Aluminium. H. St. Claire Deville.
Paris, 1859.

Fremy . . . . Encyclopédie Chimique. Fremy. Paris,
1883.

Kerl and Stohman . . . . Enclyclopädisches Handbuch der Techni-
schen Chemie. 4th Ed.

Mierzinski . . . . Die Fabrikation des Aluminiums. Dr. Mierzinski. Vienna, 1885.

Tissier . . . . Recherche de l'Aluminium. C. & H. Tissier. Paris, 1858.

Watts . . . . Watts's Dictionary of Chemistry, vol. i.

Ann. de Chim. et de Phys. . Annales de Chimie et de Physique.

Ann. der Chem. und Pharm. } Liebig's Annalen der Chemie und Phar-
  Liebig's Ann.   } macie.

Bull. de la Soc. Chim. . . Bulletin de la Société Chimique de Paris.

Chem. News . . . . . The Chemical News.

Chem. Zeit. . . . . . . Chemiker Zeitung (Cöthen).

Compt. Rend. . . . . . Comptes Rendus de les Séances de l'Acadé-
mie. Paris.

Dingl. Joul. . . . . . . Dingler's Polytechnisches Journal.

E. and M. J. . . . . . The Engineering and Mining Journal.

Jahresb. der Chem. . . . Jahresbericht ueber die Fortschritte der
Chemie.

Jrnl. Chem. Soc. . . . . Journal of the Chemical Society.

Jrnl. der Pharm. . . . . Journal der Pharmacie.

Jrnl. für pr. Chem. . . . Erdmann's Journal für praktische Chemie.

Mon. Scientif. . . . . . Le Moniteur Scientifique. Dr. Quesnes-
ville.

Phil. Mag. . . . . . . The London and Edinburgh Philosophical
Magazine.

Phil. Trans. . . . . . . Transactions of the Royal Philosophical
Society.

Pogg. Ann. . . . . . . Poggendorff's Annalen.

Poly. Centr. Blatt. . . . Polytechnisches Central-Blatt.
Proc. Ac. Nat. Sci. . . . Proceedings of the Academy of Natural Science (Philadelphia).
Quarterly Journal . . . . Quarterly Journal of the Society of Arts.
Rpt. Brit. A. A. S. . . . Report of the British Association for the Advancement of Science.
Sci. Am. (Suppl.). . . . Scientific American (Supplement).
Wagner's Jahresb. . . . . Wagner's Jahresbericht der Chemischen Technologie.
Zeit. für anal. Chem. . . . Zeitschrift für Analytische Chemie.

CONTENTS.

CHAPTER I.

HISTORY OF ALUMINIUM.

CHAPTER II.

OCCURRENCE OF ALUMINIUM IN NATURE.

CHAPTER III.

PHYSICAL PROPERTIES OF ALUMINIUM.

CHAPTER IV.

CHEMICAL PROPERTIES OF ALUMINIUM.

CHAPTER V.

PROPERTIES AND PREPARATION OF ALUMINIUM COMPOUNDS.

CHAPTER VI.

PREPARATION OF ALUMINIUM COMPOUNDS FOR REDUCTION.

B

CHAPTER VII.

THE MANUFACTURE OF SODIUM.

CHAPTER VIII.

THE REDUCTION OF ALUMINIUM COMPOUNDS FROM THE STANDPOINT OF THERMAL CHEMISTRY.

CHAPTER IX.

REDUCTION OF ALUMINIUM COMPOUNDS BY MEANS OF POTASSIUM OR SODIUM.

(Reduction of Chlorine Compounds.)

CHAPTER X.

REDUCTION OF ALUMINIUM COMPOUNDS BY MEANS OF POTASSIUM OR SODIUM (*continued*).

(*Reduction of Fluorine Compounds.*)

CHAPTER XI.

REDUCTION OF ALUMINIUM COMPOUNDS BY THE USE OF ELECTRICITY.

CHAPTER XII.

REDUCTION OF ALUMINIUM COMPOUNDS BY OTHER MEANS THAN SODIUM OR ELECTRICITY.

CHAPTER XIII.

WORKING IN ALUMINIUM.

CHAPTER XIV.

ALLOYS OF ALUMINIUM.

CHAPTER XV.

ALUMINIUM-COPPER ALLOYS.

CHAPTER XVI.

ALUMINIUM-IRON ALLOYS.

CHAPTER XVII.

ANALYSIS OF ALUMINIUM AND ALUMINIUM ALLOYS.

ERRATA.

Page 54, analysis 7 : *Instead of* " silicon 0.80, iron 4.40, *read* " silicon 4.40, iron 0.80."

Page 61, ninth line from foot : *Instead of* Picktet, *read* Pictet.

TEMPERATURES.

Unless otherwise stated, all temperatures are given in Centigrade degrees.

ALUMINIUM.

CHAPTER I.

HISTORY OF ALUMINIUM.

ABOUT 1760, Morveau called the substance obtained by calcining alum-alumina. When, afterwards, Lavoisier first suggested the existence of metallic bases of the earths and alkalies, and alumina was suspected of being the oxide of a metal, the metal was called aluminium. This, long before it was isolated.

The first researches in the preparation of aluminium date back to 1807. Davy tried, but in vain, to decompose alumina by an electric current, or to reduce it by vapor of potassium. Oerstedt, in 1824, believed he had isolated aluminium. He decomposed anhydrous aluminium chloride by potassium amalgam, and obtained, along with some potassium chloride, an amalgam which when decomposed by heat furnished him a metal resembling tin. It is probable that he employed either some moist aluminium chloride or potassium amalgam which contained caustic potash, for it is only when wetted with a solution of caustic potash that aluminium alloys with mercury; for when Wöhler, later, wished to prepare aluminium by this method, he found it impossible to obtain an aluminium amalgam when he employed materials pure and dry. Nevertheless, the method of Oerstedt marks an epoch in the history of aluminium, for, in 1827, Wöhler isolated it by decomposing aluminium chloride by potassium. The metal first isolated by Wöhler was a gray powder, taking under the polisher the brilliancy of tin. It was very easily changed, because of its extreme division, and also because it was mixed

2

with the potassium or aluminium chloride used in excess. At that time no further use was made of these facts. Later, in 1845, on making vapor of aluminium chloride pass over potassium placed in platinum boats, Wöhler obtained the metal in small, malleable globules of metallic appearance, from which he was able to determine the principal properties of aluminium. But the metal thus obtained was scarcely as fusible as cast iron, without doubt because of the platinum with which it had alloyed during its preparation. In addition to this, it decomposed water at 100°, from which we suppose that it was still impregnated with potassium or aluminium chloride. It is to H. St. Claire Deville that the honor belongs of having, in 1854, isolated aluminium in a state of almost perfect purity, determining its true properties.

Thus, while aluminium had been isolated in 1827, for eighteen years its properties *en masse* were unknown, and it was only at the end of twenty-seven years after its discovery that the true properties of the pure metal were established by Deville. The second birth of aluminium, the time at which it stepped from the rank of a curiosity into the number of the useful metals, dates from the labors of Deville in 1854. If Wöhler was the discoverer of aluminium, Deville was the founder of the aluminium industry.

In commencing researches on aluminium, Deville, while he applied the method of Wöhler, was ignorant of the latter's results of 1845. Besides, he was not seeking to produce aluminium that he might turn its valuable properties to practical account, but that it might serve for the production of aluminium protoxide (AlO), which he believed could exist as well as ferrous oxide (FeO). The aluminium he wished to prepare would, he thought, by its further reaction on aluminium chloride, form aluminium proto-chloride (AlCl²) from which he might derive the protoxide and the other proto-salts. But on passing vapor of aluminium chloride over the metallic powder formed by reduction by potassium, this proto-chloride was not thus produced; he obtained, inclosed in a mass of aluminium-potassium chloride (Al²Cl⁶.2KCl), fine globules of a brilliant substance, ductile, malleable, and very light, capable of being melted in a muffle without oxidizing, attacked by nitric acid with difficulty, but dissolved easily by hydrochloric acid or caustic potash with evolution of hydrogen.

Deville troubled himself no more about the proto-salts of aluminium, but, recognizing the importance of his discovery, turned his attention to preparing the metal. He was at this time Professor of Chemistry in the École Normale, Paris, his salary was but 3000 francs, his estate was small, and he was practically without the means of doing anything further.

On Monday, February 6, 1854, Deville read at the *séance* of the Academy a short paper entitled "Aluminium and its Chemical Combinations," in which he explained the results of this experiment as showing the true properties of aluminium and also furnishing a method of purifying it, and declared his intention or commencing immediate search for a process which could be economically applied on a commercial scale. M. Thénard, at the close of the communication, remarked that such experiments ought to be actively pursued, and that, since they were costly, he believed the Academy would hasten the accomplishment of the work by placing at Deville's disposal the necessary funds. As the outcome of this, the Academy appointed Deville one of a committee to experiment on producing aluminium, and 2000 francs were placed at his disposal for the work.

It was on the occasion of the reading of this paper that M. Chenot addressed a note to the Academy on the preparation of aluminium and other earthy and alkaline metals, in which he claimed, in some regards, priority for his inventions. (See further under "Reduction by Carbon.") This note was reserved to be examined by a commission appointed to take notice of all communications relative to the production of aluminium.

With the funds thus placed at Deville's disposal he experimented at the École Normale for several months. As potassium is very dangerous to handle, cost then 900 francs a kilo, and gives comparatively but a small return of aluminium, Deville, in view of the successful work of Bunsen on the electric decomposition of magnesium chloride, tried first the reduction of aluminium chloride by the battery. On March 20, 1854, Deville announced to the Academy in a letter to Dumas that he had produced aluminium without alkaline help, and sent a leaf of the metal thus obtained. At that time Thénard, Boussingault, Pelouze, Peligot, and later, de-la-Rive, Regnault and other well-known scientists shared the honor of assisting in the laboratory experiments. Deville sent, in the following May, a mass of five or six grammes weight to Liebig, making no secret of the fact that it was reduced by the battery; while Balard, at the Sorbonne, and Fremy, at the École Polytechnique, publicly repeated his experiments and explained them in all their details. Although these experiments succeeded quite well, yet, because of the large consumption of zinc in the battery used the process could evidently not be applied industrially, and Deville felt obliged to return to the use of the alkaline metals.

Towards the middle of 1854, Deville turned to sodium, without a knowledge of those properties which render it so preferable to potassium, but solely because of its smaller equivalent (23 to that of potassium 39) and the greater cheapness of soda salts. He studied the manufacture of sodium, with the aid of M. Debray, in his laboratory at the École Normale, and their experiments were repeated at Rousseau Bros.' chemical works at Glaciere, when they were so successful that Rousseau Bros. very soon put metallic sodium on the market at a much reduced price. It is said that while metallic sodium was a chemical curiosity in 1855, costing something like 2000 francs a kilo, its cost in 1859 is put down at 10 francs. Deville carried this process to such perfection that for twenty-five years it remained almost precisely at the status in which he left it in 1859. In order to still further cheapen aluminium, Deville busied himself with the economic production of alumina, which gave later a lively impulse to the cryolite and bauxite industries.

On August 14, 1854, Deville read a paper before the Academy describing his electrolytic methods at length (see under " Reduction by Electricity"), showing several small bars of the metal and also stating some of the results already achieved by the use of sodium but not going into details, since he believed that numerous analyses were necessary to confirm these results—which he was unable to have made with the funds at his disposal. He also stated that the desire to show, in connection with his assertions, interesting masses of the metal, alone prevented the earlier publication of the methods used. Several days before this, Bunsen published in Poggendorff's Annalen a process for obtaining aluminium by the battery, which resembled Deville's method, but of which the latter was ignorant when he read his paper. Thus it is evident that the isolation of aluminium by electrolysis was the simultaneous invention of Deville and Bunsen.

After reading this paper, Deville caused a medal of aluminium to be struck, which he presented to the Emperor Napoleon III. The latter, looking forward to applying such a light metal to the armor and helmets of the French Cuirassiers, immediately authorized experiments to be continued at his own expense on a large scale. This anticipation ultimately proved impracticable, but the ambition in which it was bred was caused for once to minister to the lasting benefit of mankind. Deville, however, about this time accepted, in addition to his duties as professor at the École Normale, a lectureship at the Sorbonne (where he afterwards obtained a full professorship), and it was not until March of the next year that the experiments at the cost of the Emperor were begun.

It was about August, 1854, that two young chemists, Chas. and Alex. Tissier, at the suggestion of Deville, persuaded M. De Sussex, director of a chemical works at Javel, to let them experiment in his laboratory (of which they had charge) on the production of sodium.

Towards the commencement of 1855, Deville took up the industrial question, the Emperor putting at his disposition all the funds necessary for the enterprise, and in March the investigator went to work and installed himself at the chemical works at Javel in a large shed which the director, M. De Sussex, kindly put at his service.

The investigations were carried on here for nearly four months, ending June 29th, and the process elaborated was an application on a large scale of the experiments he had made at the expense of the Academy, which he described in his paper of August 14, 1854, and by which he had been able to obtain a few pencils of metal. In this work such success attended his efforts that on June 18, Deville presented to the Academy through M. Dumas large bars of pure aluminium, sodium and masses of aluminium chloride. The members and large audience were loud in their admiration and surprise at the beauty of the metal. Dumas stated that the experiments at Javel had put beyond a doubt the possibility of extracting aluminium on a large scale by practical processes. Deville's paper was then read, describing all his processes in detail, and concluding with the following words: "After four months of work on a large scale, undertaken without responsibility on my part, and, in consequence, with the tranquillity and repose of mind which are so often wanting to the investigator; without the preoccupation of expense, borne by his Majesty the Emperor, whose generosity had left me entire liberty of action; encouraged each day by distinguished men of science, —I hope to have placed the aluminium industry on a firm basis."

It was the metal made at this time at Javel which was exhibited at the Paris Exposition in 1855. In the Palais de l'Industrie, among the display from the porcelain works at Sevres, were ingots and some manufactured objects. The first article made of aluminium was, in compliment to the Emperor, a baby-rattle for the infant Prince Imperial, for which purpose it must have served well because of the sonorousness of the metal.

After terminating these experiments, Deville continued working at the École Normale, the Emperor defraying his expenses, until April, 1856. The memoir published in the "Ann. de Chim. et de Phys.," April, 1856, contains, besides the results obtained at Javel, the improvements devised in the meantime.

It appears that when Deville first went to Javel, he had for assistants the Tissier Brothers, who were charged by M. de Sussex to give him all the aid they could. Since the previous autumn the Tissiers had been experimenting on sodium furnaces, and now, in concert with Deville, they drew up plans for furnaces, and aided in devising other apparatus. Under these circumstances the furnace for the continuous manufacture of sodium in cylinders was devised, which the Tissiers claim Deville strongly advised them to make their property by patenting, asking only from them the use of it for his experiments. So, immediately after the experiments were ended, in July, the Tissiers patented the furnace in question, and, leaving Paris, took charge of M. Chanu's works at Rouen. On the other hand, Deville always

reproached them for acting in bad faith. He says that after
having assisted for about two months in setting up his apparatus,
being forced to leave the works because of misunderstandings
between them and M. de Sussex, they were admitted to his
laboratory at the École Normale, and initiated by him into the
knowledge of all those processes which they made use of afterwards, then suddenly left, taking drawings of furnaces, details of
processes, etc., which they not only made free use of, but even
patented. However, whichever party was in the right (and those
who comprehend the character of Deville can hardly doubt
which was), the fact stands that in July, 1855, M. Chanu, an
honorable manufacturer of Rouen, founded a works in which
Deville's processes were to be applied, and intrusted the direction
of it to the Tissier Brothers.

The history of the works at Rouen is thus described by the
Tissiers in their book on aluminium, of which we shall speak a
little further on :—

"In July, 1855, Messrs. Malétra, Chanu, and Davey, of
Rouen, formed a company to produce aluminium, and we were
intrusted with the organization and special charge of the industry. The commencement was beset with difficulties, not only
in producing but in using the metal. It then sold at $200 per
kilo, the price being an insurmountable obstacle to its employment in the arts. The small capital at our disposal was not
enough to start the industry, to pay general expenses, and the
losses occasioned by the many experiments necessary. On February 28, 1856, the society was dissolved. In April of the same
year, Mr. William Martin, struck by the results already obtained
and sanguine of greater success, united with us. From that time
daily improvements confirmed M. Martin's hopes, and in 1857
the works at Amfreville-la-mi-Voie, near Rouen, sold the metal
at $60 per kilo ($2 per oz.). The laboratory of this works was
devoted to researches on everything concerning the production
and application of aluminium. M. Martin has our sincere grati-
tude for the kindness with which he so willingly encouraged and
contributed to the progress of the manufacture of this wonderful
metal."

The process ultimately used at Amfreville was the reduction of cryolite by sodium, but the enterprise was not a permanent success, and after running for a few years it was abandoned and the works closed.

Returning to Deville, we find that after leaving Javel one of the first subjects he investigated was the use of cryolite for producing aluminium. The researches made with the aid of MM. Morin and Debray were published in the memoir of April, 1856, and became the basis of the process carried out by the Tissiers at Rouen. Besides this, Deville perfected many of the details of a practicable aluminium plant, with the result that in the spring of 1856 he united with Messrs. Debray, Morin, and Rosseau Bros. (the latter manufacturers of chemicals at Glacière, in whose works aluminium had been made since the middle of 1855) and put up new apparatus in the works at Glacière, the company furthering the work entirely at their own cost. This enterprise lasted for more than a year, during which a number of processes were tried and continued improvements made, so that towards August of the same year aluminium was put on the market in Paris at 300 francs a kilo, being one-third what it cost a year previous.

Finally, in April, 1857, the little works at Glacière, a suburb of Paris, in the midst of gardens and houses, and turning into the air fumes charged with chlorine and salts, was obliged by reason of general complaints to stop making aluminium. The plant was moved to Nanterre, where it remained for some years, under the direction of M. Paul Morin, being on a scale four times as large as the actual demand. Afterwards part of the plant was moved to the works of H. Merle & Co., at Salindres, and later on the whole plant, where the manufacture is now carried on by the firm of Pechiney & Co. The works at Nanterre were really the only "aluminium works" built by Deville, the others were plants installed at general chemical works, but these at Nanterre were built by the united efforts of Deville, his brothers and parents, and a few personal friends. Among those who aided Deville, especially in the problems which the new industry presented, he speaks warmly of Messrs. d'Eichtal, Lechatelier, and Jacquemont.

In 1858 the Tissiers wrote and published a small work entitled

"Recherches sur l'Aluminium," which, in view of what Deville could have written about the subject, was a decided misrepresentation of the results which had been thus far accomplished. Deville thought that the industry was yet too young to merit any sort of publication, yet he näively writes in his work "De l'Aluminium," in 1859, "I will sincerely acknowledge that my writing is a little due to my pride, for I decided to take the pen to speak of my work only to avoid seeing it belittled and disfigured as it has been lately in the book written by the MM. Tissier."

Deville published his book in September, 1859, and he concludes it with these words : " I have tried to show that aluminium may become a useful metal by studying with care its physical and chemical properties, and showing the actual state of its manufacture. As to the place which it may occupy in our daily life, that will depend on the public's estimation of it and its commercial price. The introduction of a new metal into the usages of man's life is an operation of extreme difficulty. At first aluminium was spoken of too highly in some publications, which made it out to be a precious metal ; but later these estimates have depreciated even to the point of considering it attackable by pure water. The cause of this is the desire which many have to see taken out of common field-mud a metal superior to silver itself ; the opposite opinion established itself because of very impure specimens of the metal which were put in circulation. It seems now that the intermediate opinion, that which I have always held and which I express in the first lines of my book, is becoming more public and will stop the illusions and exaggerated beliefs which can only be prejudicial to the adoption of aluminium as a useful metal. Moreover, the industry, established as it now is, can be the cause of loss to no one ; as for myself, I take no account of the large part of my estate which I have devoted, but am only too happy, if my efforts are crowned with definite success, in having made fruitful the work of a man whom I am pleased to call my friend—the illustrious Wöhler."

Contemporary with the early labors of Deville, among the numerous chemists and metallurgists investigating this attractive field we find Dr. Percy in England, and H. Rose in Germany, whose experiments on the reduction of cryolite by sodium were quite successful, and are herein described later on.

As early as 1856 we find an article in an American magazine* showing that there were already chemists in the United States spending time and money on this subject. The following is the substance of the article alluded to: "Within the last two years Deville has extracted 50 to 60 lbs. of aluminium. At the present time, M. Rousseau, the successor of Deville in this manufacture, produces aluminium which he sells at $100 per pound. No one in the United States has undertaken to make the metal until recently Mons. Alfred Monnier, of Camden, N. J., has, according to the statement of Prof. James C. Booth in the 'Penn. Inquirer,' been successful in making sodium by a continuous process, so as to procure it in large bars, and has made aluminium in considerable quantity, specimens of which he has exhibited to the Franklin Institute. Mons. Monnier is desirous of forming a company for the manufacture of aluminium, and is confident that by operating in a large way he can produce it at a much less cost than has heretofore been realized. We would suggest the propriety of giving aid to this manufacturer at the expense of the government, for the introduction of a new metal into the arts is a matter of national importance, and no one can yet realize the various and innumerable uses to which this new metal may be applied. It would be quite proper and constitutional for Congress to appropriate a sum of money to be expended under the direction of the Secretary of the Treasury in the improvement of this branch of metallurgy, and in testing the value of the metal for coinage and other public use."

In the next volume of the "Mining Magazine"† there is a long article by Mr. W. J. Taylor, containing nothing new in regard to the metallurgy of aluminium, but chiefly concerned in calculating theoretically the cost of the metal from the raw materials and labor required by Deville's processes, and concluding that it is quite possible to make it for $1.00 per pound.

In 1859 the first aluminium works in England were started at

* Mining Magazine, 1856, vii. 317.
† Mining Magazine, viii. 167 and 228. Proc. Ac. Nat. Sci., Jan. 1857.

Battersea, near London. No details are attainable respecting the size of these works or the length of time they were in operation. They very probably were merged into the enterprise started the next year, 1860, at Newcastle-on-Tyne, by the Bell Bros.—one of whom was I. Lowthian Bell, so prominent in connection with the metallurgy of iron. In 1862 this company was selling their aluminium at 40 shillings per troy pound, and they continued operations until 1874, when the works were closed.

It was probably shortly after 1874 that the large firm of J. F. Wirtz & Co., Berlin, made an attempt to start an aluminium works. The project drooped before it was well started, and it is only within the last five years that Germany has possessed a flourishing aluminium industry.

The further we get away from an age the better able are we to write the true history of that age. And so, as years pass since the labors of Wöhler, Deville, and Tissier, we are now able to see better the whole connected history of the development of this art. Dr. Clemens Winckler gives us a comprehensive retrospect of the field seen from the standpoint of 1879, from which we condense the following.* "The history of the art of working in aluminium is a very short one, so short that the present generation, with which it is contemporary, is in danger of overlooking it altogether. The three international exhibitions which have been held in Paris since aluminium first began to be made on a commercial scale form so many memorials of its career, giving as they did at almost equal intervals evidence of the progress made in its application. In 1855 we meet for the first time in the Palais de l'Industrie with a large bar of the wonderful metal, docketed with the extravagant name of the 'silver from clay.' In 1867 we meet with it again worked up in various forms, and get a view of the many difficulties which had to be overcome in producing it on a large scale, purifying and moulding it. We find it present as sheets, wire, foil, or worked-up goods, polished, engraved, and soldered, and see for the first time its most important alloy—aluminium bronze. After a lapse of almost another dozen years, we see at the Paris Exhibition of 1878 the maturity

---

* Industrie Blätter, 1879 ; Sci. Am. Suppl., Sept. 6, 1879.

of the industry. We have passed out of the epoch in which the metal was worked up in single specimens, showing only the future capabilities of the metal, and we see it accepted as a current manufacture having a regular supply and demand, and being in some regards commercially complete. The despair which has been indulged in as to the future of the metal is thus seen to have been premature. The manufacture of aluminium and goods made of it has certainly not taken the extension at first hoped for in its behalf; the lowest limit of the cost of manufacture was soon reached, and aluminium remains as a metal won by expensive operations from the cheapest of raw materials.

"There are several reasons why the metal is shown so little favor by mathematical instrument makers and others. First of all, there is the price; then the methods of working it are not everywhere known; and further, no one knows how to cast it. Molten aluminium attacks the common earthen crucible, reduces silicon from it, and becomes gray and brittle. This inconvenience is overcome by using lime crucibles, or by lining an earthen crucible with carbon or strongly burnt cryolite clay. If any one would take up the casting of aluminium and bring it into vogue as a current industrial operation, there is no doubt that the metal would be more freely used in the finer branches of practical mechanics."

At the time of Dr. Winckler's writing, the extraction of aluminium in France was carried on by Merle & Co., at Salindres, while the Société Anonyme de l'Aluminium, at Nanterre, worked up the metal. Both firms were represented at the Exposition of 1878. The prices quoted then were 130 francs a kilo for aluminium, and 18 francs for ten per cent. aluminium bronze. From 1874, when Bell Bros.' works at Newcastle-on-Tyne stopped operations, until 1882, when a new enterprise was started in England by Mr. Webster, the French company were the only producers of aluminium.

Regarding the prospects of the aluminium industry at this period, we can very appropriately quote some remarks of the late Mr. Walter Weldon, F.R.S., who was a personal friend of M. Pechiney (director of the works at Salindres), had given great attention to aluminium, and was considered as a first authority on the subject. Speaking in March 1883, before the London Section of the Society of Chemical Industry, he stated that the only method then practised for the manufacture of aluminium was Deville's classical one; that at Salindres, M. Pechiney had improved and cheapened it, but that was all the progress made in the industry in twenty-five years. Continuing, Mr. Weldon outlined the possible lines on which improvements might be made as :

1st. Cheapening the production of aluminium chloride, or of aluminium-sodium chloride.

2nd. Substituting for these chlorides some other cheaper anhydrous compounds of aluminium not containing oxygen.

3rd. Cheapening sodium.

4th. Replacing sodium by a cheaper reducing agent.

Mr. Weldon exhibited the relative cost of the materials used in making aluminium as then carried on by M. Pechiney as—

    Producing the alumina .   .   .   .   .   . 10 per cent.
        "   "   double chloride   33   "
        "   "   sodium and reducing therewith   . 57   "

"Discussing these figures, it is seen that the cost of the alumina forms but a small item in the cost of the metal, since a saving of 50 per cent. in its cost would only cheapen the metal 5 per cent. A large margin is, however, left in the conversion of the alumina into the chloride, and it is here that a large saving may be expected either in cheaper methods of producing the chloride or by the substitution of some other cheaper salt for the chloride. The only other suitable compounds which might replace the latter are the fluoride, iodide, bromide, sulphide, phosphide, or cyanide. The fluoride has been used to some extent in the form of cryolite, but, from the impurities in the mineral and its corrosive action on the apparatus used for reduction, the metal produced is very much contaminated with iron and silicon. The bromide and iodide, no matter how produced, would always be too costly to replace the chloride. The production of the sulphide in a suitable form from which the metal can be extracted has thus far not proved a success ; and, even if ever it be thus produced in a suitable condition, it is not at all likely to be as cheap a material to use as the chloride. The phosphide and cyanide can thus far only be produced from the metal itself, and are, therefore, totally out of the question. To find a substitute for sodium as a reducing agent has been a favorite object of research among chemists for the past thirty years, and although every element occurring in any abundance or obtainable at a cheaper rate than sodium has been tried under almost all conditions, yet absolutely nothing has been accomplished in this direction that would entitle any one to the belief that aluminium can ever be produced chemically without the use of sodium. So absorbing to those interested in the search for a substitute for sodium has the occupation proved, that the effort to cheapen sodium did not receive anything like its fair share of attention. Since, of the 57 per cent. ascribed to the cost of sodium and reduction, 50 per cent. represents the sodium, which thus costs about 6 shillings a pound, there is seen to be a very large margin for improvements, since the raw materials for a pound of sodium do not cost over 1 or at most 2 shillings."

In 1882, the cost of aluminium was materially cheapened by the application of the inventions of Mr. Webster, which, in accordance with the analysis of the problem made by Mr. Weldon, consisted principally in the cheap production of alumina and its conversion into chloride. Mr. Webster had experimented on this subject many years, and in 1881 and 1882 took out patents for his processes and organized the "Aluminium Crown Metal Company," located at Hollywood near Birmingham, where several thousand pounds were expended in plant. Business was soon commenced on a large scale, the company producing, however, many other alloys besides those of aluminium. The business grew until it soon became the serious competitor of the French company, and practically controlled the English market. However, a radical change of still greater importance in the sodium process was made in 1886 by an invention of Mr. H. Y. Castner, of New York City. This gentleman conceived the plan of reducing sodium compounds in cast-iron pots, from a fused bath of caustic soda, by which the reduction is performed at a much lower temperature and the yield of sodium is very much more than by the Deville method. The application of this process on a large scale, with the use of gas furnaces and other modern improvements, has lowered the cost of sodium from $1 per pound to about 20 or 25 cents. It is but just to say that Mr. Castner's invention was by no means a chance discovery. For four years he worked in a large laboratory fitted up for this special purpose, and after many discouragements in trying to produce aluminium by means other than that of sodium was led finally to consider that the cheapening of this metal was the most promising method for cheapening aluminium, and after much patient, hard work, achieved well-deserved success.

Mr. Castner's patent was taken out in the United States in June, 1886, and, while being the first one granted on that subject in this country, was said also to be the only one taken out in the world since 1808. With the assistance of Messrs. J. H. and Henry Booth, of New York City, Mr. Castner demonstrated the process by building and operating a furnace on a somewhat large scale. This being accomplished, Mr. Castner crossed to England and met the representatives of the Webster process, with whom it was evident a combination would be especially advantageous to both parties; for, with cheap aluminium chloride and cheap sodium, it was clear that a strong process could be built up. Mr. Castner then demonstrated plainly, by erecting a furnace and operating it for several weeks, that his process was all that he claimed for it. As the result of this success, the "Aluminium Company, Limited" was incorporated in June, 1887, with a share capital of £400,000, "to acquire the patents and work and develop the inventions of James Webster for the manufacture of pure alumina and certain metallic alloys and compounds, together with the business at present carried on by the Webster Patent Aluminium Crown Metal Company, Limited, in Birmingham, Sheffield, and London, England; and also to acquire the patents and work and develop the invention of H. Y. Castner for the manufacture of sodium and potassium." Mr. Webster was paid £230,000 for the business, properties, stock, etc., of the Crown Metal Company, while £140,000 was allowed for the sodium patents. The new company appointed Mr. Castner managing director, and the erection of large works was immediately begun at Oldbury, near Birmingham. These works were started in operation at the end of July, 1888. They cover five acres of ground, and have an annual producing capacity of 100,000 lbs. of aluminium. This plant is, at present, the largest aluminium works in the world, and in view of the large part contributed to the establishment of this works by the genius of Mr. Castner, the methods there used are rightly called "The Deville-Castner Process."

We have followed the progress of the Webster and Castner processes up to the date of starting the works at Oldbury because the continuity of the advances made in the old Deville process would hardly allow of a break in order to mention other processes arising meanwhile. However, the five years since 1884 have witnessed not one but several revolutions in the aluminium industry. The great advances made in dynamo-electric machinery in the last decade have led to the revival of the old methods of electrolysis discovered by Deville and Bunsen, and to the invention of new methods of decomposing aluminium compounds electrolytically. It will be recalled that the first small pencils of aluminium made by Deville were obtained by electrolysis, and that he turned back to the use of the alkaline metals solely because the use of the battery to effect the decomposition was far too costly to be followed industrially. This fact still holds true, and we cannot help supposing that if Deville had had dynamos at his command such as we have at present, the time of his death might have seen the aluminium industry far ahead of where it now is.

First in point of time we notice Grätzel's process, patented in Germany in 1883 and used industrially by the "Aluminium and Magnesium Fabrik, Patent Grätzel" at Hemelingen near Bremen. The process was essentially the electrolysis of a bath of fused aluminium salt, such as chloride or fluoride, the improvements on the older experiments being in details of apparatus used, the use especially of anodes of mixed carbon and alumina, and the use of dynamic electricity. Several metallurgists maintained the uselessness of the Grätzel processes, and their position was proved to be not far from the truth, for in October, 1887, the company announced that the addition "Pt. Grätzel" would be dropped from the firm name, since they had abandoned Grätzel's processes and were making aluminium by methods devised by Herr

Saarburger, director of their works. The processes of this latter gentleman not being published, we are unable to state their nature, but they are very probably electrolytic. In October, 1888, Mr. Saarburger reports that their works are producing at the rate of 12000 kilos of aluminium yearly, besides a large quantity of aluminium bronze and ferro-aluminium. The firm also works up the aluminium and its alloys into sheet, wire, tube, etc.

A somewhat similar electrolytic process was patented by Dr. Ed. Kleiner, of Zurich, in 1886. Molten cryolite was decomposed by two carbon poles, the heat generated by the current first melting the cryolite and then electrolyzing it. Since the motive power in this, as in all electric processes, composes one of the chief elements for carrying on the reduction, the Kleiner Gesellschaft, formed to work this method, made an attempt to obtain water rights at the falls of the Rhine, at Schaffhausen, which would furnish 15,000 horse-power. This proposition being refused by the government, an experimental plant was started at the Hope Mills, Tyldesley, Lancashire, England, which is in operation at present; but its commercial success seems still to depend on a more economical application of the electric power and the obtaining of metal of greater purity than is usually made from cryolite.

An electrolytic method, which is probably superior to both the preceding, is the invention of Mr. Chas. M. Hall, of Oberlin, Ohio, which was patented in the United States, April, 1889, but which has already been in successful operation for nearly a year. Mr. Hall is a graduate of Oberlin College, and for several years experimented on a small scale, overcoming many discouragements, at last perfecting the process which is now being operated by the Pittsburgh Reduction Company on Fifth Avenue, Pittsburgh; Mr. Alfred E. Hunt, a well-known metallurgist, being president of the company. The principle involved is different from that in either the Grätzel or the Kleiner process; it is the electric decomposition of alumina suspended or dissolved in a fused bath of the salts of aluminium and other bases, the current reducing the alumina without affecting its solvent. At present the plant is turning out 50 to 75 lbs. of aluminium a day, and is so successful as to first cost that during 1889 they sold aluminium, guaranteed 98 per cent. pure, at $4.50 per pound, the lowest figure the metal

3

had ever touched; but in November, 1889, they captured the aluminium market by cutting the price to $2.00, for which achievement Mr. Hall is to be heartily congratulated.

While the electrolytic processes so far considered use a fluid bath and operate at moderate temperatures with a current of moderate intensity, there have been devised two other prominent processes which operate in a somewhat different manner and attain to very economical results. These primarily depend on the enormous temperature attainable by the use of a powerful electric arc, and secondarily on the reduction of alumina (which at the temperature attained becomes fluid) either by the reducing action of the carbon present or by simple electric decomposition. Which of these two agencies performs the reduction, in either process, is still an unsettled question which we will discuss later on.

Before going further with the history of these two processes, Cowles' and Heroult's, it may not be inappropriate to take note of a few facts antecedent to their appearance. It is well known that Sir W. Siemens devised an electric furnace in which the heat of the arc was utilized for melting steel. In 1882, Mr. Ludwig Grabau, in Hanover, Germany, purchased a Siemens furnace for the express purpose of attempting the reduction of alumina, and after experimenting successfully for some time, modified the apparatus so as to work it continuously, and therewith made aluminium alloys; but on account of the difficulties of the process and the impurity of the alloys produced, Mr. Grabau gave up the experiments, having come to the conclusion that aluminium alloys to be technically valuable should be obtained in a state of almost chemical purity. In the beginning of 1885, Dr. Mierzinski, in his book on aluminium, presented some very striking remarks on the use of the electric furnace, which are so much to the point that they are well worth quoting in this connection: "The application of electricity for producing metals possesses the advantage not to be ignored that a degree of heat may be attained with it such as cannot be reached by a blowpipe or regenerative gas-furnace. The highest furnace temperature attainable is 2500° to 2800° C., but long before this point is reached the combustion becomes so languid that the loss of heat by radiation almost equals the production of heat by combustion, and hinders a further elevation of temperature. But in applying electricity the degree of heat attainable is theoretically unlimited. A further advantage is that the smelting takes place in a perfectly neutral atmosphere, the whole operation going on without much preparation and under the eyes of the operator. Finally, in ordinary furnaces the refractory material of the vessel must stand a higher heat than the substance in it, whereas by smelting in an electric furnace the material to be fused has a higher temperature than the crucible itself. Since the attempt to produce aluminium by the direct reduction of alumina by carbon is considered by metallurgists as impossible, because the temperature requisite is not attainable, the use of the electric current for attaining this end seems to be of so much the more importance."

The Cowles invention was patented August 18, 1885, and was first publicly described before the American Association for the Advancement of Science, at their Ann Harbor meeting, August 28, 1885. The process is due to two Cleveland gentlemen, E. H. and A. H. Cowles, who in the development of their process associated with them Prof. Charles F. Mabery, of the Case School of Applied Science, Cleveland, as consulting chemist. The Cowles Electric Smelting and Aluminium Company, formed to work the process, erected a plant at Lockport, N. Y., where a water power of 1200 horse-power was secured, and where, among other novel apparatus, the largest dynamo in the world, made especially for this purpose by the Brush Electric Company, is in operation. Following the success of this plant in America, the Cowles Syndicate Company, organized to work the patents in England, have put in operation works at Stoke-on-Trent which have a capacity of something like 300 lbs. of alloyed aluminium daily. Springing also from the Cowles process is the "Aluminium Brass and Bronze Company," of Bridgeport, Conn., which was organized in July, 1887, and controls the exclusive rights under the Cowles American patents of manufacturing the alloys of aluminium into sheet, rods, and wire. The extensive plant which this company is starting will employ 300 men, and has been erected at a cost of nearly $300,000.

The principle made use of in the Cowles process is, briefly, that a powerful electric current is interrupted, the terminals being large carbon rods, and the space between having been filled with a mixture of alumina, carbon, and the metal to be alloyed, the intense heat generated in contact with this mixture causes the metal to melt and the alumina to be reduced to aluminium, which combines with the metal, while the oxygen escapes as carbonic oxide.

It is interesting to note as separating the Cowles, as well as the Heroult, process from the previously mentioned electrolytic methods, that while the latter produce almost exclusively pure aluminium in their electric operation, finding it inexpedient, if not, perhaps, impossible to add other metals and form alloys at once—the former experience almost the reverse of these conditions, and as yet are confined exclusively to the direct production of the alloys.

The Heroult process was first put in practical operation on July 30th, 1888, at the works of the Swiss Metallurgic Company (Société Métallurgique Suisse), at Neuhausen, near Schaffhausen. The patents for the process were granted in France and England in April and May, 1887, and in the United States in August, 1888. The company named above is composed of some of the largest metal workers in Switzerland. Previously to their adoption of this process they had experimented with Dr. Kleiner's electrolytic method, but abandoned it, and on becoming the owners of the Heroult process immediately started it up practically on a large scale, and with signal success.

The process consists in electrolyzing molten alumina which has been rendered fluid by the heat of the arc, using as the positive anode a large prism of hard carbon and as the negative a sub-stratum of molten copper or iron, the arrangement of the parts being such that the process seems to proceed, when once well under way, in all respects as the simple electrolysis of a liquid. Using water power for driving the dynamos, the economical production of alloyed aluminium at 4.5 francs per kilo (50 cents per pound), is said to be an assured fact.

The success of this process at Neuhausen was so marked as to attract general attention, and in the latter months of 1888 several large German corporations, prominent among which was the

Allgemeine Electricitäts Gessellschaft of Berlin, sent representatives to arrange for the purchase of the Heroult patents for Germany. The outcome of these examinations and negotiations was the purchase by this German Syndicate of Heroult's continental patents and the founding by them and the former Swiss owners of the Aluminium Industrie Actien-Gesellschaft, with a capital of 10,000,000 francs. In December, 1888, the new company took possession at Neuhausen, and commenced the construction of a plant many times larger than the original one, their plans also including the erection of foundries and mills for casting and manufacturing their alloys. Dr. Kiliani, the well-known writer on electro-metallurgical subjects, is working manager for the new company. The new plant will utilize about 3000 horse-power, and will have a capacity of 20 to 25 tons of 10 per cent. bronze daily.

Besides these works, we learn that a French company, the Société Electro-Metallurgique, has commenced the manufacture of alloys by the Heroult process, their works at Froges (Isere) being equal to a daily output of 3000 kilos of 10 per cent. bronze. Mr. Heroult was also in the United States May–August, 1889, for the purpose of establishing a plant. The works were located at Bridgeport, Conn., and were started in August, but after running a few hours the dynamo was burnt out and operations summarily stopped until the arrival of a dynamo ordered from the Oerlikon works at Zurich. This is hoped to be in place before the end of 1889, and the works in full operation.

Both the Cowles and Heroult processes have been successful in producing aluminium in alloys at a cost far below that at which pure aluminium is made, and they apparently have a good prospect of holding this position for some time to come. Comparing the two processes we see that while on first sight the principle made use of appears similar, yet the different disposition of the parts and the evidently more economical working in the case of the latter seem to point to some deep-seated difference in the reactions made use of in the two cases. However, we shall more minutely discuss these points in their proper place, suffice it to say, in summing up, that while the Cowles process undoubtedly has the merit of having been first in the field, the Heroult has the advantage of more practical and economical application.

Among the many other aluminium processes and companies which have been projected within the last few years, we notice prominently the Alliance Aluminium Company of London, England, organized in the early part of 1888. Having a nominal capital of £500,000, it is said to own the English, German, French, and Belgian patents of Prof. Netto, of Dresden, for the manufacture of sodium and potassium and the reduction of cryolite thereby; the patents of Mr. Cunningham for methods of reduction of the same metals; and methods devised by Prof. Netto and Dr. Saloman, of Essen, for producing aluminium of great purity on a commercial scale. The two latter named gentlemen are said to have invented their processes after long experimenting at Krupp's works at Essen; and, since the apparatus used was mounted on trunnions, many rumors have been spread by the newspapers that aluminium was being made (by tons, of course) in a Bessemer converter by Krupp, of Essen. Prof. Netto reduces sodium by a continuous process, by allowing fused caustic soda to trickle over incandescent charcoal in a vertical retort, the apparatus containing many ingenious details and giving promise of being quite economical. One method of using the sodium in reduction consists in the use of a plunger to which bars of sodium are attached and held at the bottom of a crucible full of molten cryolite; another depends on the use of a revolving cylinder in which the cryolite and sodium react, and appears more chimerical than Netto's other propositions. This latter device, however, is said to be in operation at Essen, though with what success we cannot learn.

In June, 1888, "Engineering" stated that the Alliance Company were located at King's Head Yard, London, E. C., and that several small reduction furnaces were being operated, each producing about 50 lbs. of aluminium a day, estimates of the cost at which it was made giving 6 to 8 shillings per pound. In the early part of 1889 there seems to be a division of the original company. The "Alkali Reduction Syndicate, Limited" have leased ten acres of ground at Hepburn on which to erect a plant for working Cunningham's sodium patents, the sodium produced going to the Alliance Company's reduction works located at Wallsend.

As to the exhibit made by this company at the Paris Exhibition, 1889, we will have some remarks to make later on.

Ludwig Grabau, of Hanover, Germany, has made several patented improvements in producing aluminium, which are in the same direction as Prof. Netto's methods. Mr. Grabau believes that in order that aluminium may possess its most valuable qualities, both for use alone or in alloying, it should be of almost chemical purity; and as the best means of attaining this end economically he has improved the sodium method on these three lines:—

1st. Production of cheap pure aluminium fluoride.

2d. Production of cheap sodium.

3d. Reduction in such a manner that no possible impurities can enter the reduced metal, and that the sodium is completely utilized.

How far Mr. Grabau has succeeded, as regards cheap production, I cannot say, but as for the purity, a sample sent the author contains 99.8 per cent. of aluminium, and is undoubtedly the purest made at present in the world. Mr. Grabau's sodium patents are now pending, but his other processes are described in full in their appropriate places.

"The American Aluminium Company," of Milwaukee, Wis., was organized in July, 1887, with a capital of $1,000,000, to manufacture aluminium by a process of Prof. A. J. Rogers. The process is kept secret, the application made for patents not being yet granted, but the means used are electrolytic and not very different in principle from some others recently granted in England. A small experimental plant put up in the summer of 1888 has given encouraging results as to the purity of metal obtainable and the yield, and it is not improbable that if patents are granted soon the company will have a larger plant in operation and their metal on the market in the early months of 1890.

There is a company hailing from Kentucky, about whose methods no reliable information is to be had, the numerous newspaper articles which it has inspired being glaringly inaccurate and sensational, while our more staid scientific journals seem to treat it on the principle of "the least said the better." The enterprise was first brought to public notice in June, 1888, by an Associated Press dispatch, stating that by reducing common clay and cryolite in steel water-jacketted cupola furnaces, pure aluminium was obtained very cheaply. Two months later the ridiculous statement went the rounds of the press that this concern had exported 150 lbs. of pure metallic aluminium to London, England, selling it at 50 cents per lb. Since then, advertisements have appeared in the newspapers claiming them to be the only manufacturers of pure aluminium in America, offering it at $5 per pound, and also offering for sale various aluminium bronzes, ferro-aluminium, aluminium solders, etc. Later accounts of some methods used seem to point to the utilization of the idea of coating scrap iron with clay and "certain fluxes" and then running it down in a water-jacketted cupola, the castings being said to contain $1\frac{1}{2}$ per cent. of aluminium. This method is identical with one recently patented by other parties in England and on the Continent. Whether this company produces aluminium or not is a question only answered by some very unreliable newspaper statements in the affirmative. The process last referred to is apparently successful in England, and may quite probably give the results claimed by this company, but of the process for making pure aluminium we can only say that it does not appear to be possible.

Colonel William Frishmuth, of Philadelphia, is a German chemist whose name has been often published in connection with aluminium. Before 1860, Col. Frishmuth operated a small chemical works on North Broad Street, Philadelphia, and there followed the production of sodium by Deville's methods, furnishing it to the chemical dealers. It is quite possible that he followed Deville's methods still further, and, by means of sodium, produced aluminium in small quantities. In 1877, Col. Frishmuth was operating a small electro-plating works in the northern part of Philadelphia, claiming to plate an alloy of nickel and aluminium from aqueous solution. While engaged in this, he persuaded some gentlemen in the metal trade to aid him financially in developing a process for making aluminium, but always stipulating that he be allowed to retain the secret of the process. Led on by reports of successes and promises of returns in the near future these gentlemen invested several thousand dollars with the sole result of reports of progress and fresh requests for money. After several years waiting, one, at least, of these gentlemen lost faith in the truth of Frishmuth's statements and withdrew his support, losing all that he had advanced. Other capitalists, however, were induced to step forward and put money into the concern, having nothing but the statements and promises of Frishmuth as their security. Again others, in disgust, threw up their interest in the affair, never regaining a cent of what had been advanced. I have been thus minute in these statements because this is a sample of the whole history of the process. In 1884, Col. Frishmuth obtained a patent for producing aluminium by simultaneously generating sodium vapor in one retort, vapor of a volatile aluminium salt in another, and mixing the vapors in a third retort, where they were to react and form aluminium. The process was never successful, and Frishmuth has since abandoned altogether the use of sodium and has been experimenting of late years with electrolytic methods. On the obtaining of this patent an English syndicate sent Major Ricarde-Seaver, F.R.S.E., to this country to report on the process. Major Seaver was not altogether convinced, from what he was allowed to see, of the practicability of the furnace, and on reporting to the syndicate, a very liberal offer was made to Frishmuth, proposing that he come to England, erect a furnace, and demonstrate its working in a fair, clear manner, the syndicate to pay all the expenses incident to the test, including Col. Frishmuth's personal expenses. This offer was refused, for what reason we need not go far to find. Since this episode other capitalists at home have advanced the funds which were asked for, and aluminium has been sold in moderate quantities, though how it is made, or whether Col. Frishmuth produces it or not, is an unsettled question. The metal sold is unquestionably of good quality, averaging as nearly as can be the same as the best French metal; it is quite probable that with his long experience in handling the metal Col. Frishmuth is quite expert in refining and running down aluminium scrap of all kinds—undoubtedly a difficult thing to do. In 1884–5, the Philadelphia Business Census recorded him as employing ten men and his annual product as valued at $18,000, but since then Col. Frishmuth has grown uncommunicative to the census reporter, so that in 1886 it was stated in the Government Report on the Mineral Resources of the United States that no pure aluminium was made in America in that year—a statement which we may accept as correct. Much public interest was directed to Frishmuth in 1884, when he cast the aluminium cap or apex of the Washington Monument. This casting is of pyramidal form, 10 inches high, 6 inches on a side of its base, and weighs 8½ pounds. An analysis of the metal in this casting is given on p. 54, and shows it to be of a quality equal to the best French aluminium.

Two aluminium companies have come to the author's notice of which I am able to give no more than the bare fact of their existence. "The Aluminium Company of America" was incorporated under the laws of New York, with a capital of $1,500,000; Paul R. Pohl, of Philadelphia, was styled the mineralogist and chemist. I do not think that the company has ever done anything further than to organize, offer stock, and issue a prospectus. It is now extinct. "The United States Aluminium Company," of East St. Louis, was incorporated in March, 1889, with a capital of $1,000,000, for the purpose of manufacturing aluminium and its alloys. The incorporators of the company, process to be used, etc., are unknown to the author.

We will close this historical sketch by referring the reader back to Dr. Winckler's remarks (p. 27), and supplementing them with a notice of the exhibits of aluminium at the Paris Exposition of 1889. If, as Dr. Winckler remarks, the three international exhibitions in 1855, 1867, and 1878 show so many stages in its career, it is quite evident that the exposition of 1889 has shown a more promising view of the industry than any of its predecessors, excepting perhaps the first. It was not unnatural that many hopes were disappointed when the exhibit of 1878 showed so little advance over that of 1867, and when the fact became painfully evident that for twenty years, 1858 to 1878, very little real progress had been made in the industry. I think that if the exposition of 1878 showed anything at all, it showed that, from a metallurgical standpoint, the industry was at a stand-still. Against this we place the exhibit of 1889, and the contrast is striking; this shows not deadness but the most intense and successful activity that the industry has ever known. In place of one exhibitor, five manufacturers compete for honors. In short, this last exposition has shown the aluminium industry re-awakened and rapidly approaching its goal—the placing of aluminium among the common metals. Just now, at the close of 1889, are we not almost inclined to state, in view of recent developments, that it has reached this goal?

As to the exhibits referred to, a detailed account reads as follows:—Société Anonyme pour l'Industrie de l'Aluminium: In a large case, the frame of which was aluminium bronze, samples of aluminium, ferro-aluminium, aluminium bronze, forged and rolled, and numerous articles of the latter alloy. Cowles Electric Smelting and Aluminium Company: Samples of ferro-aluminium, aluminium bronze and aluminium brass of various grades, aluminium silver, and numerous useful articles made of these alloys. Brin Bros.: Samples of aluminium, with thin iron and steel castings made by its use. The Alliance Aluminium Company: Two large blocks of aluminium, cast hollow, weighing possibly 1000 pounds and 500 pounds respectively. The inclosing balustrade and decorations were principally of aluminium or aluminium bronze. The Aluminium Company, Limited: A solid casting of aluminium bronze weighing $\frac{1}{2}$ ton, and on this a solid block of 98 per cent. aluminium weighing the same. In the corners of the case piles of ingots of 99 per cent. aluminium, 10 per cent. bronze, 5 per cent. bronze, 10 per cent. ferro-aluminium, and 20 per cent. aluminium steel. Besides which was a 7 inch bell, springs, statues, aluminium plate, round and square tubes, wire, sheet, etc. Such were the aluminium exhibits which attracted as much interest as the historic ingot of 1855 did at its *début*; and, not taking into account that Mr. Hall's process was not represented and that the German makers were debarred from exhibiting because of international pique, yet the exhibit shown was one which demonstrated the great advances made in the last decade and give cause for the most sanguine hopes for the future. Indeed, it seems more than half true that already "aluminium—the metal of the future, is transformed into aluminium—the metal of the present."

STATISTICAL.

The following table shows the price at which aluminium has been sold since it was first placed on the market.

Date.	Place.					Per kilo.	Per pound.
1856 (Spring)	Paris	.	.	.	.	1000 fr.	$90.90
1856 (August)	"	.	.	.	.	300 "	27.27
1859	"	.	.	.	.	200 "	17.27
1862	"	.	.	.	.	130 "	11.75
1862	Newcastle	.	.	.			11.75
1878	Paris	.	.	.	.	130 "	11.75
1886	"	.	.	.	.		12.00
1887	Bremen	.	.	.			8.00
1888	London	.	.	.			4.84
1889	Pittsburgh	.	.	.			2.00

The selling price of aluminium bronze has until recently depended directly on the price of pure aluminium, since the bronze was made by simply uniting the two metals, but since electrical methods of obtaining the bronze directly have been used the alloy has been sold at a price for the contained aluminium much below what pure aluminium could be bought for. The ten per cent. bronze has been sold as follows :—

Date.	Place.					Per kilo.	Per pound.
1878	Paris	.	.	.	.	18.00 fr.	$1.64
1885	Cowles Bros.	.	.	.		4.50 "	0.40
1888	" "	.	.	.		3.85 "	0.35
1888	Heroult process, Neuhausen	.				3.30 "	0.30

It is almost impossible to estimate how much aluminium has been made since Deville first started the industry. The following figures of annual outputs are gleaned from various sources, some of them being of doubtful accuracy.

		kg.		lbs.
1854–56	Deville	25	=	55
1859	Nanterre (Deville)	720	=	1584
1859	Rouen (Tissier Bros.)	960	=	2112
1865	France	1090	=	2400
1869	France	455	=	1000
1872	Salindres (H. Merle & Co.)	1800	=	4000
1872	England (Bell Bros.)	750	=	1650
1882	Salindres	2350	=	5170
1884	Salindres	2400	=	5280
1883	Philadelphia (Frishmuth)			70
1884	" "			125
1885	" "			230
1885	Cowles Bros. (in alloys)			450
1886	" "			6500
1887	" "			17800

A very approximate estimate of the whole amount of aluminium that had been produced up to 1886, made from a careful comparison and study of the above reports, gives a total of 115,000 lbs. (52,000 kilos). Since then the Cowles Bros. are reported as having turned out 50,000–60,000 lbs. in alloys, the Aluminium Company, Limited, have probably made as much, and the Hemelingen Fabrik, which has been in operation since 1885, is now producing 10,000–15,000 kilos yearly (22,000–33,000 lbs.).

The amount of aluminium imported and entered for consumption in the United States from 1870 to 1887 is as follows:—

Year ending June 30.	Quantity (pounds).	Value.
1870	—	$ 98
1871	—	341
1872	—	—
1873	2	22
1874	183	2125
1875	134	1355
1876	139	1412
1877	131	1551
1878	251	2978
1879	284	3423
1880	341	4042
1881	517	6071
1882	566	6459
1883	426	5079
1884	590	8416
1885	439	4736
1886	464	5297
1887	797	9458
1888	1772	16764

CHAPTER II.

OCCURRENCE OF ALUMINIUM IN NATURE.

THERE is no other metal on the earth which is so widely scattered and occurs in such abundance.

Aluminium is not found metallic. Stocker* made the statement that aluminium occurred as shining scales in an alumina formation at St. Austel, near Cornwall, but he was in error. But the combinations of aluminium with oxygen, the alkalies, fluorine, silicon, and the acids, etc., are so numerous and occur so abundantly as not only to form mountain masses, but to be also the bases of soils and clays. Especially numerous are the combinations with silicon and other bases, which, in the form of felspar and mica, mixed with quartz, form granite.

These combinations, by the influence of the atmosphere, air, and water, are decomposed, the alkali is replaced or carried away, and the residues form clays. The clays form soils, and thus the surface of the earth becomes porous to water and fruitful. It is a curious fact that aluminium has never been found in animals or plants, which would seem to show that it is not necessary to their growth, and perhaps would act injuriously, if it were present, by its influence on the other materials. Most of the aluminium compounds appear dull and disagreeable, such as felspar, mica, pigments, gneiss, amphibole, porphyry, eurite, trachyte, etc.; yet there are others possessing extraordinary lustre, and so beautiful as to be classed as precious stones. Some of these, with their formulæ, are :—

Ruby	Al^2O^3
Sapphire	Al^2O^3
Garnet	$(Ca.Mg.Fe.Mn)^3Al^2Si^3O^{12}$
Cyanite	Al^2SiO^5

---

* Journ. fr. prakt. Chem., 66, p. 470.

Some other compounds occurring frequently are :—

Turquoise	$Al^2P^2O^8.H^6Al^2O^6.2H^2O$
Lazulite	$(MgFe)Al^2P^2O^9 + Aq$
Wavellite	$2Al^2P^2O^8.H^6Al^2O^6.9H^2O$
Topaz	$5Al^2SiO^5.Al^2SiF^{10}$
Cryolite	$Al^2F^6.6NaF$
Diaspore	$H^2Al^2O^4$
Beauxite	$H^6Al^2O^6$
Aluminite	$Al^2SO^6.9H^2O$
Alunite	$K^2SO^4.Al^2S^3O^{12}.2H^2Al^2O^6$

One would suppose that since aluminium occurs in such abundance over the whole earth that we literally tread it under foot, it would be extracted and applied to numberless uses, being made as abundant and useful as iron ; but such is not the case.

Beauxite and cryolite are the minerals most used for producing aluminium, and their preference lies mainly in their purity. Native alums generally contain iron, which must be removed by expensive processes.

BEAUXITE.

Beauxite is a combination between diaspor, $Al^2O^3.3H^2O$, and brown hematite, $Fe^2O^3.3H^2O$; or, it is diaspor with aluminium replaced more or less by iron ; the larger the amount of iron, the more its color changes from white to brown. It was first found in France, near the town of Beaux, large deposits occurring in the departments of Var and Bouches du Rhon, extending from Tarascon to Antibes. Several of these beds are a dozen yards thick, and 160 kilometers in length. Deposits are also found in the departments of l'Herault and l'Arriège. Very important beds are found in Styria, at Wochein, and at Freisstritz, in Austria, a newly discovered locality where the mineral is called Wocheinite. Here it has a dense, earthy structure, while that of France is conglomerate or oölitic. Deposits similar to those of France are found in Ireland at Irish Hill, Straid, and Glenravel. Further deposits are found in Hadamar in Hesse, at Klein Steinheim, Langsdorff, and in French Guiana.

The following analyses give an idea of the peculiar composition of this mineral ; besides the ingredients given there are also traces of lime, magnesia, sulphuric, phosphoric, titanic and vanadic acids.

	a.	b.	c.	d.	e.	f.
Al^2O^3	60.0	75.0	63.16	72 87	44.4	54.1
Fe^2O^3	25.0	12.0	23.55	13.49	30.3	10.4
SiO^2	3.0	1.0	4.15	4.25	15.0	12.0
K^2O and Na^2O . .	—	—	0.79	0.78	—	—
H^2O	12.0	12.0	8.34	8.50	9.7	29.9

	g.	h.	i.	k.	l.	m.
Al^2O^3	64.6	29.80	48.12	43.44	61.89	45.76
Fe^2O^3	2.0	3.67	2.36	2.11	1.96	18.96
SiO^2	7.5	44.76	7.95	15.05	6.01	6.41
K^2O and Na^2O . .	—	—	—	—	—	0.38
H^2O	24.7	13.86	40.33	35.70	27.82	27.61

	n.	o.	p.	q.	r.
Al^2O^3	55.61	76.3	50.85	49.02	73.00
Fe^2O^3	7.17	6.2	14.36	12.90	4.26
SiO^2	4.41	11.0	5.14	10.27	2.15
K^2O and Na^2O . .	—	—	0.26	0.31	—
H^2O	32.33	26.4	28.38	25.91	18.66

Index :—

a and b. from Beaux (Deville).

c.　　dark } Wocheinite (Drechsler).
d.　　light }

e.　　red brown]
f.　　yellow } Beauxite from Feisstritz (Schnitzer).
g.　　white]

h.　　white Wocheinite (L. Mayer and O. Wagner).

i.　　Beauxite from Irish Hill.

k.　　"　　" Co. Antrine (Spruce).

l.　　"　　" Glenravel (F. Hodges).

m and n.　　"　　" Hadamar (Hesse) (Retzlaff).

o.　　from Klein-Steinheim (Bischof).

p and q. from Langsdorff (I. Lang).

r.　　Beauxite from Dublin, Ireland, brought to the Laurel Hill Chemical
　　　Works, Brooklyn, L. I., and there used for making alums. It is
　　　dirty white, hard, dense, compact, and in addition to the ingre-
　　　dients given above contains 0.59 per cent. of lime, and some titanic
　　　acid. It costs $6 per ton laid down in the works. The above
　　　analysis, made by Mr. Joüet, is furnished me by the kindness of
　　　the superintendent of the works, Mr. Herreshoff.

As is seen from the above analyses, the percentage of alumina is very variable, and cannot be determined at all simply by inspection, but only by an analysis, for often the best-looking specimens are the lowest in this base. For instance, a beauxite containing 62.10 Al^2O^3, 6.11 Fe^2O^3, 5.06 SiO^2, and 20.83 H^2O was much darker and more impure-looking than that from Wochein (h), which contained only 29.8 per cent. of alumina.

Beauxite has until recently not been found in the United States, but in 1887 a deposit was discovered in Floyd County, Ga., which is described as follows in a paper read by Mr. Edward Nichols before the American Institute of Mining Engineers at their Duluth meeting, July, 1887.

" Numerous float specimens, covering an area of about one-half an acre, indicate the location of the main deposit, which has thus far been opened to an inconsiderable extent only. The excavations show the beauxite to exist apparently as large masses in clay. The formation is determined by the Geological Survey of Georgia to be Lower Silurian. The surface in the immediate vicinity is covered with numerous fragments of chert, a characteristic rock throughout this formation in Georgia. An examination of the mineral shows it to have the oölitic structure common to several beauxites. It varies in color from light-salmon to dark-red, according to the content of iron sesquioxide. The light-colored specimens are comparatively soft, while the dark-colored are much harder, spots in them being harder than quartz. The chemical composition is interesting because of the presence of titanic acid, in which it resembles the mineral found in Asia Minor. It dissolves with difficulty in acids, but fuses easily with potassium acid-sulphate. Owing to the purity of the deposit, it seems likely to have a value before long for use in some aluminium reduction process, or as a refractory material."

Analyses					Dark specimen.	Light specimens.	
SiO^2	.	.	.	.	2.800	—	2.300
Al^2O^3	.	.	.	.	52.211	57.248	56.883
Fe^2O^3	.	.	.	.	13.504	3.212	1.490
TiO^2	.	.	.	.	3.520	3.600	3.551
H^2O	.	.	.	.	27.721	—	—
CaO	.	.	.	.	0.—	—	—
MgO	.	.	.	.	0.—	—	—
P^2O^5	.	.	.	.	—	—	0.065

4

CRYOLITE.

Cryolite was first found at Ivigtuk in Arksut-fiord, west coast of Greenland, where it constitutes a large bed or vein in gneiss. It was very rare even in mineralogical collections until 1855, when several tons were carried to Copenhagen and sold under the name of "soda mineral." It is a semi-transparent, snow-white mineral. When impure it is yellowish or reddish, even sometimes almost black. It is shining, sp. gr. 2.95, and hardness 2.5 to 3. It is brittle, not infrequently contains ferrous carbonate, sulphide of lead, silica, and sometimes columbite. It is fusible in the flame of a candle, and on treatment with sulphuric acid yields hydrofluoric acid. As will be seen further on, cryolite was first used by the soap-makers for its soda; it is still used for making soda and alumina salts, and to make a white glass which is a very good imitation of porcelain. The Pennsylvania Salt Company in Philadelphia import it from Ivigtuk by the shipload for these purposes; lately they have discontinued making the glass. Cyrolite is in general use as a flux. A very complete description of the deposit at Ivigtuk can be found in Hoffman's "Chemische Industrie."

Pure cryolite contains—

Aluminium	. . . .	13.0
Fluorine	. . . .	54.5
Sodium	. . . .	32.5
		100.00

Or otherwise stated—

Aluminium fluoride	. . . . . .	40.25
Sodium fluoride	. . . . . .	59.75
		100.00

From the reports in the Mineral Resources of the United States we find that there was imported by the Pennsylvania Salt Company in 1887, 11,732 tons, which was valued at nearly $15 a ton. The importers say this value is too low; they sell what they call pure prepared cryolite at $125 a ton. This so called pure article was found by Prof. Rogers, of Milwaukee, to contain 2 per cent. of silica and 1 per cent. of iron.

The only known deposit of cryolite in the United States is that found near Pike's Peak, Colorado, and described by W. Cross and W. F. Hillebrand in the "American Journal of Science," October, 1883. It is purely of mineralogical importance and interest, occurring in small masses as a subordinate constituent in certain quartz and feldspar veins in a country rock of coarse reddish granite. Zircon, astrophyllite, and columbite are the primary associated minerals, the first only being abundant.

CORUNDUM.

Until 1869, the sole sources of corundum were a few river washings in India and elsewhere, where it was found in scattered crystals. Its cost was twelve to twenty-five cents a pound. Within the last twenty years numerous mines have been opened in the eastern United States, the first discovery of which was due to Mr. W. P. Thompson, and is thus described by him :*

"In 1869, in riding over a spur of the Alleghenies in Northern Georgia, I found what has proven to be an almost inexhaustible mine of corundum in the crysolite serpentine, the first instance on record of the mineral being found *in situ*. Previously it had been washed out of débris at Cripp's Hill, N. C., and at a mine in West Chester, Pa., both on the slopes of the crysolite serpentine. The clue being thus obtained accidentally, about thirty mines were shortly afterwards discovered in the same formation ; but of the thousands of tons thus far dug out the larger portion has come from the mines I discovered.

"At present it can be bought at about ten dollars per ton at the mines. It is nearly pure alumina. Disapore, a hydrated aluminia, is also found in the same region and locality. Corundum will probably always be the principal source in America of material from which to manufacture pure aluminiun ; but in Great Britain, in all probability, manufacturers must look to alumina prepared artificially from cryolite or from sulphate of alumina."

In 1887, the production of corundum in the United States was

* Journal of the Society of Chemical Industry, April, 1886.

practically limited to the mines of the Hampden Emery Company
at Laurel Creek, Ga., and at Corundum Hill, Macon County,
N. C., these mines furnishing somewhat over 600 tons. .The
Unionville Corundum Mines Company operate a mine at Unionville, Chester County, Pa., but the extent of their output is
not given.

NATIVE SULPHATE OF ALUMINA.

In the summer of 1884, a large deposit of rock called "native
alum" was discovered on the Gila River, Sorocco County, New
Mexico, about two miles below the fork of the Little Gila, and
four miles below the Gila Hot Springs. The deposit is said to
extend over an area one mile square and to be very thick in places.
The greater part of the mineral is impure, as is usual with native
occurrences, but it is thought that large quantities are available.
A company formed in Sorocco has taken up the alum-bearing
ground. Through the kindness of Mr. W. B. Spear, of Philadelphia, the author was enabled to get a specimen of the mineral.

It is white, with a yellowish tinge. On examining closely it
is seen to consist of layers of white, pure-looking material arranged with a fibrous appearance at right angles to the lamination.
These layers are about one-quarter of an inch thick. Separating
them are thin layers of a material which is deeper yellow, harder
and more compact. The whole lump breaks easily, and has a
strong alum taste. On investigation, the fibrous material was
found to be hydrated sulphate of alumina, the harder material
sulphate of lime.

It is probable that this deposit was the bed of a shallow lake
in which the alum-bearing water from the hot springs concentrated
and deposited the sulphate of alumina. Periodically, or during
freshets, the Little Gila, flowing through a limestone country,
bore into this lake water containing lime, which, meeting the
aluminium sulphate solution, immediately caused a deposit of
calcium sulphate. When the dry season came, the Little Gila
dried up, the deposit of alum was made, and thus were formed
the succession of layers through the deposit.

Analysis showed 7 to 8 per cent. insoluble material, and the remainder corresponded to the formula $Al^2(SO^4)^3.18H^2O$. A small amount of iron was present.

[Further information about some of the native aluminous minerals has unavoidably fallen into the chapter describing aluminium compounds.]

CHAPTER III.

PHYSICAL PROPERTIES OF ALUMINIUM.

COMMERCIAL aluminium is never chemically pure, and therefore displays properties varying more or less from those of the pure metal according to the character and amount of impurities present. In this treatise, whenever the properties of aluminium are mentioned they must be understood to refer to the chemically pure metal, and not to the commercial article unless specifically stated.

The impurities most frequently present in commercial aluminium are iron and silicon. These are found in all brands, varying in amount from 1 per cent. in the purest to 6 and even 8 per cent. in the worst. Besides these, various other impurities are found coming from accidental sources in the manufacture; thus, some of the first metal made by Deville contained a large amount of copper (analysis 1), coming from boats of that metal which he used in his experiments. Metal made later by Deville contained zinc, coming from zinc muffles which he had borrowed and used for retorts, old retorts broken up having been used in the composition of the new ones. More recently, aluminium has been produced by the agency of sodium in the presence of lead, which latter it takes up in small amount. Sodium is liable to remain alloyed in very small proportion, yet it is an element so easily attacked that it destroys some of the most valuable qualities of the aluminium. The distinct effect, however, of each of these usual impurities in modifying the physical properties of aluminium has not yet been investigated in a thoroughly satisfactory manner.

A few years more, however, of increasing familiarity with and handling of the metal on a large commercial scale will, I believe, cause the effect of foreign elements on aluminium to be as plainly recognized as is now the case with carbon and the metalloids in iron. In general, we may say that silicon seems to play a rôle in aluminium closely analogous to that of carbon in iron; the purest aluminium is fibrous and tough, but a small percentage of silicon makes it crystalline and brittle. Carbon, moreover, is said to be dissolved by molten aluminium and to modify its properties quite materially; yet, if so, almost nothing more is known about its influence than this unsatisfactory statement. Here is excellent room for work for some investigator who, as Hampe has done with copper, will prepare the purest aluminium, and by adding to it known impurities tell us precisely, beyond doubt, how these various foreign elements affect its properties.

The following analyses will show the amount of impurities present in commercial aluminium, and also, incidentally, the improvement which has been achieved since the beginning of the industry in 1854 :—

	Aluminium.	Silicon.	Iron.
1. Deville Process	88.350	2.87	2.40
2. " "	92.500	0.70	6.80
3. " "	92.000	0.45	7.55
4. " "	92.969	2.149	4.88
5. " "	94.700	3.70	1.60
6. " "	96.160	0.47	3.37
7. Tissier Bros.	94.800	0.80	4.40
8. Morin & Co., Nanterre	97.200	0.25	2.40
9. " "	97.000	2.70	0.30
10. " "	98.290	0.04	1.67
11. " "	97.680	0.12	2.20
12. Merle & Co., Salindres	96.253	0.454	3.293
13. " "	96.890	1.270	1.840
14. " "	97.400	1.00	1.30
15. " "	97.600	0.40	1.40
16. Frishmuth	97.49	1.90	0.61
17. "	97.75	1.70	0.55
18. Hall's Process	98.34	1.34	0.32
19. Deville-Castner	99.20	0.50	0.30
20. Grabau Process	99.62	0.23	0.15
21. " "	99.80	0.12	0.08

Notes on the above analyses :—

1. Analyzed by Salvètat. Contained also 6.38 per cent. of copper and a
   trace of lead.
2 and 6. Analyzed by Dumas.
3. Parisian aluminium bought in La Haag.
4. Analyzed by Salvètat. Contained also a trace of sodium.
5. Parisian aluminium bought in Bonn and analysed by Dr. Kraut.
7. Made at the works near Rouen, in 1858, from cryolite. Analyzed by
   Demondèur.
8. Analyzed by Sauerwein. Contained also traces of lead and sodium.
9. Analyzed by Morin. Average of several months' work.
10, 11. Analyzed by Kraut. Represents the best product of the French works
   sent to the London Exhibition in 1862.
12, 13. Analyzed by Mallet. The best metal which could be bought in 1880.
   Purchased in Berlin by Mallet and used by him as the material which
   he purified and used for determining the atomic weight of aluminium.
14, 15. Analyzed by Hampe. This was the purest metal which could be
   bought in 1876. No. 14 contained also 0.10 per cent. of copper and
   0.20 per cent. of lead. No. 15 contained 0.40 per cent. of copper and
   0.20 per cent. of lead.
16. Bought in Philadelphia as Frishmuth's aluminium, in 1885, and analyzed
   by the author.
17. Specimen of the metal composing the tip of the Washington Monument,
   cast by Frishmuth. This analysis is reported by R. L. Packard in the
   Mineral Resources of the United States, 1883–4.
18. The best grade of metal made by this process, analyzed by Hunt & Clapp,
   Pittsburgh. For average analyses, etc., see description of process.
19. The best grade made by this process, exhibited at the Paris Exposition,
   1889. Analyzed by Cullen.
20. Analysis by Dr. Kraut of metal being made on a commercial scale.
21. Analysis by Grabau of the purest metal yet obtained by his process.

According to Rammelsberg (Kerl's Handbuch) the silicon
which is always found in aluminium is in part combined with it,
and this combined silicon changes by treatment with hydrochloric
acid into either silica, which remains, or into silicon hydride,
SiH_4, which escapes; while another part of it is combined with
the aluminium just as graphite is with iron; and this part remains on treatment with acid as a black mass not oxidized by
ignition in the air. Two analyses of aluminium reduced from
cryolite by sodium in a porcelain crucible gave—

	1.	2.
Silicon obtained as silica	9.55	1.85
Free silicon	0.17	0.12
Silicon escaping in SiH_4	0.74	0.58

One sample of aluminium analyzed by Professor Rammelsberg contained as much as 10.46 per cent. of silicon, and another sample even 13.9 per cent. The quantity of iron varied from 2.9 to 7.5 per cent.

M. Dumas found that aluminium usually contains gases, about which he makes the following statements :* On submitting aluminium in a vacuum to the action of a gradually increasing temperature up to the softening point of porcelain, and letting the mercury pump continue acting on the retort until it was completely exhausted, considerable quantities of gas were withdrawn. The liberation of the gas from the metal seems to take place suddenly towards a red-white heat. 200 grammes of aluminium, occupying 80 c.c., gave 89.5 c.c. of gas, measured at 17° and 755 mm. pressure. The gas consisted of 1.5 c.c. carbonic acid, and 88 c.c. hydrogen. Carbonic oxide, nitrogen and oxygen were absent.

The author has observed that molten aluminium will absorb large quantities of gas. On passing sulphuretted hydrogen into the melted metal for about twenty minutes some aluminium sulphide was formed while the metal appeared to absorb the gas. On pouring, the metal ran very sluggishly with a thick edge, but when just on the point of setting gas was disengaged so actively that the crackling sound could be heard several feet away, and the thick metal became suddenly quite fluid and spread over the plate in a thin sheet. The gas disengaged seemed by its odor to contain a good proportion of sulphuretted hydrogen, although free hydrogen may have been present in it.

COLOR.

Deville : The color of aluminium is a beautiful white with a slight blue tint, especially when it has been strongly worked. Being put alongside silver, their color is sensibly the same. However, common silver, and especially that alloyed with copper, has a yellow tinge, making the aluminium look whiter by comparison. Tin is still yellower than silver, so that aluminium possesses a color unlike any other useful metal.

* Comptes Rendue xc. 1027 (1880).

Mallet: Absolutely pure aluminium is perceptibly whiter than the commercial metal; on a cut surface very nearly pure tin-white, without bluish tinge, as far as could be judged from the small pieces examined.

The purest aluminium examined by the author is that made by Grabau. On a fresh fracture it is absolutely white, but on long exposure to the air it takes a faint, almost imperceptible bluish tint. On a cut surface it has the faintest suspicion of a yellow tint, not so decided as the yellowish color of pure tin. Ordinary commercial aluminium is bluish on a fresh fracture, the tint being deeper the greater the amount of impurities it contains. A specimen with 10 per cent. of silicon and 5 per cent. of iron was almost as blue as lead. It is my belief that a very small percentage of copper closes the grain and whitens the fracture a little; I have also found that chilling suddenly from a high temperature has the same effect. When ingots of aluminium are exposed a long time to damp air the thin film of oxide forming on them gives a more decided bluish cast to the metal, since the coating is perfectly snow-white and hence, by contrast, heightens the bluish tint of the metallic back-ground. Mourey recommended removing this discoloration by placing the articles first in dilute hydro-fluoric acid, 1000 parts of water to 2 of acid, and afterwards dipping in nitric acid. The oxide would thus be dissolved and the original color restored. Pure aluminium possesses to the highest degree that property expressed best by the French term "éclat." It is rather difficult to see why the blue tint should be more prominent after the metal has been worked, yet I think two reasons will explain this phenomenon; first, aluminium is not a hard metal, and on polishing or burnishing particles of dirt or foreign substances are driven into the pores of the metal, thereby altering its color slightly; second, any metal looks whiter when its surface is slightly rough than when highly polished, in the latter case it being as much the reflected color of the general surroundings as the color of the metal itself which is seen. I have never seen any highly polished white metal which did not look bluish especially when reflecting out-door light. I think this explains why opera glasses, rings, jewelry, etc., generally look bluer than the bar or ingot-metal from which they are made.

Aluminium takes a very beautiful mat which keeps almost indefinitely in the air, the surface thus slightly roughened appearing much whiter than the original polished surface. Aluminium can be polished and burnished without much difficulty if attention is given to a few particulars which it is necessary to observe. (For methods of polishing, etc., see Chapter XIII.)

FRACTURE.

A cast ingot of purest aluminium has a slightly fibrous structure, a section $\frac{1}{2}$ inch thick bending twenty degrees or so from a straight line when sharply bent before showing cracks at the outside of the turn. The fracture of such an ingot is uneven, rough, and very close, often showing a curious semi-fused appearance, as if it had been already exposed to heat and the sharpest points melted down. However, only the purest varieties show these peculiarities. Metal containing 96 to 97 per cent. of aluminium begins to show a crystalline structure, breaks short, and with a tolerably level surface. Metal less than 95 per cent. pure shows large shining crystal surfaces on the fracture, the smaller crystals being on the outside of the ingot where it has been cooled most quickly, while in the centre the crystalline surfaces may be as large as $\frac{1}{16}$ inch in diameter. A specimen containing only 85 per cent. of aluminium broke as short as a bar of antimonial lead, with a large granular, crystalline surface.

Working the metal increases its fibrousness greatly, the section of a square rolled bar of good metal looking very much like that of a low-carbon steel.

HARDNESS.

The purest aluminium is distinctly softer than the commercial, estimated on the scale of hardness proposed by Mohs it would be written as about 2.5, that is, a little harder than can be scratched by the nail. It is not so soft as pure tin. The presence of impurities, however, rapidly increases the hardness. While 99 per cent. aluminium can be cut smoothly with the knife and shavings turned up almost as with pure tin, yet 95 per cent. metal can hardly be cut at all, the shavings break off short and a fine grating is felt through the blade.

Experience in testing various specimens of commercial aluminium with the knife will, I am sure, enable a person to become quite skilful in determining the purity and in separating different grades from each other. Taking this test in connection with the breaking and surface of fracture, it appears to me that these indications are as significant and can be made of as much use as the corresponding tests for iron, steel, and other metals. Mr. Joseph Richards, the author's father, having had many years' experience in testing lead, tin, zinc, and similar metals, in which the knife blade has been put to good service, has been able with very little practice to arrange a number of specimens of aluminium correctly according to their purity simply by noting carefully the way they cut and the color of the cut surface. These tests will in the future, I am sure, be of great use to those handling aluminium on a large scale, especially in the works where it is produced.

Aluminium becomes sensibly harder after being worked, probably owing to the closing of the grain, since we know that its density is also increased.

SPECIFIC GRAVITY.

Mallet: The specific gravity of absolutely pure aluminium was carefully determined at 4° C., and the mean of three closely agreeing observations gave 2.583.

• Commercial aluminium is almost always heavier than this, but the increase is not in direct proportion to the amount of impurities present. There are two reasons why this last statement is correct; first, we cannot say what expansion or contraction may take place in forming the alloy; second, while most of the impurities which occur are much heavier than aluminium, yet silicon, the most frequent of all, has a specific gravity of only 2.34 (Deville's determination), and therefore acts in the opposite direction to the other impurities, though not to as great an extent. The following analyses and specific gravities may give some information on this point :—

ANALYSIS.			SPECIFIC GRAVITY.	
Aluminium.	Silicon.	Iron.	Observed.	Calculated.
97.60	0.60	1.80	2.735	2.61 (2.64)
95.93	2.01	2.06	2.800	2.61 (2.69)
94.16	4.36	1.48	2.754	2.59 (2.74)
78.—	16.—	4.—	2.85	2.66

It is seen in each case that the calculated specific gravity is much less than the observed, which would show contraction in volume by alloying. Indeed, this is a prominent characteristic of aluminium alloys, aluminium often taking up several per cent. of its weight of another metal without its volume being increased, the particles of the other metal seeming to pass between those of the aluminium; thus probably accounting for the extraordinary strength and closeness of many of the aluminium alloys. This subject is treated more at length in the chapter on alloys. We can see the large contraction taking place by inspecting the numbers in parentheses under the heading "Calculated." These are computed on the supposition that the volume of the impure aluminium is equal to that of the pure aluminium entering into it. As these numbers are also less than the observed specific gravities, the extraordinary fact is shown that aluminium can absorb several per cent. of iron and silicon and yet will decrease in volume in doing so.

The remarks thus far made are based on the gravity of cast metal. Aluminium increases in density by being worked; Deville states that metal with a specific gravity of 2.56 had this increased to 2.67 by rolling, which, he says, may explain the differences existing in its properties after being annealed or worked. He remarked further that heating this rolled metal to 100°, and cooling quickly changed its specific gravity very little, lowering it to 2.65. I have observed that on heating a piece of aluminium almost to its fusing point and suddenly chilling it in water, its specific gravity was lowered from 2.73 to 2.69.

The low specific gravity of aluminium, when compared to those of the other metals, is (in the words of a recent lecturer) "the physical property on which our hopes of the future usefulness of aluminium chiefly rest." The following table will facilitate this comparison :—

| | SPECIFIC GRAVITY. | | | |
	Water = 1.	Alumin-ium = 1.	Pounds in a cubic foot.	Kilos in a cubic meter.
Platinum . . .	21.5	8.3	1344	21,500
Gold	19.3	7.4	1206	19,300
Lead	11.4	4.6	712	11,400
Silver	10.5	4.0	656	10,500
Copper	8.9	3.5	557	8,900
Iron and steel . .	7.8	2.8	487	7,800
Tin	7.3	2.7	456	7,300
Zinc	7.1.	2.7	444	7,100
Aluminium . . .	2.6	1.0	163	2,600

In comparing the price of aluminium with that of the metal it is to replace, for such purposes where the bulk of the article is fixed, such as tableware, jewelry, engineering instruments, and a large proportion of all its uses, it is important to take its low specific gravity into the account. Thus, for making spoons, aluminium at $4 per lb. would be as cheap as silver at $1 per lb., since the silver spoons would be four times as heavy. So for such purposes, at the prices prevailing to-day, aluminium is practically only one-twenty-fifth as costly as silver.

FUSIBILITY.

Deville: Aluminium melts at a temperature higher than that of zinc, lower than that of silver, but approaching nearer to that of zinc than silver. It is, therefore, quite a fusible metal.

Mallet: It seems that pure aluminium is a little less fusible than the commercial metal.

Picktet determined the melting point to be 600°, Heeren about 700°, while Van der Weyde placed it as high as 850°.* Prof. Carnelley has lately determined this point himself, and found that a sample containing $\frac{1}{2}$ per cent. of iron melted at 700°, while one with 5 per cent. of iron did not fuse completely until above 730°. These numbers can be accepted as the best determinations yet made, and it results from them that iron raises the melting point and hinders fluid fusion. Since it is already conceded that silicon raises the melting point until, with

* Carnelley's Tables of Melting Points.

a large percentage, the metal can hardly be made fluid at any heat, it is rather puzzling to see why the absolutely pure metal should be less fusible than the commercial metal, as is remarked by Mallet. It may be that the small percentages of iron and silicon present in a high grade of commercial metal act in a manner contrary to the effect of larger percentages, as is known to be true in a few instances with the impurities present in other metals, but we have no definite information to bring forward on this point.

VOLATILIZATION.

Deville: Aluminium is absolutely fixed, and loses no part of its weight when it is violently heated in a forge fire in a carbon crucible.

This statement was made in 1859, and can still be accepted as true as far as ordinary furnace temperatures are concerned. But, with the use of the electric furnace, temperatures have been attained at which aluminium does sensibly volatilize. In Cowles Bros. electric furnace it is stated that the aluminium is almost all produced as vapor and as such is absorbed by the copper or iron present, when these are not present it is found condensed in the cooler upper-part of the furnace. A similar experience has been met in other electric furnace processes, so that the volatilization of aluminium at these extreme temperatures may be accepted as a fact.

ODOR.

Deville: The odor of pure aluminium is sensibly nothing, but the metal strongly charged with silicon will exhale the odor of silicuretted hydrogen, exactly represented by the odor of cast iron. But even under these unfavorable circumstances, the smell of the metal is only appreciable to persons experienced in judging very slight sensations of this kind.

TASTE.

Deville: Pure aluminium has no taste, but the impure and odorous metal may have a taste like iron, in any case only very slight.

MAGNETISM.

Deville: I have found, as also MM. Poggendorff and Reiss, that aluminium is very feebly magnetic.

SONOROUSNESS.

Deville: A very curious property, which aluminium shows the more the purer it is, is its excessive sonorousness, so that a bar of it suspended by a fine wire and struck sounds like a crystal bell. M. Lissajous, who with me observed this property, has taken advantage of it to construct tuning forks of aluminium, which vibrate very well. I also tried to cast a bell, which has been sent to the Royal Institution at London at the request of my friend Rev. J. Barlow, vice-president and secretary of the institution. This bell, cast on a model not well adapted to the qualities of the metal, gives a sharp sound of considerable intensity, but which is not prolonged, as if the clapper or support hindered the sound, which, thus hindered, becomes far from agreeable. The sound produced by the ingots is, on the contrary, very pure and prolonged. In the experiments made in Mr. Faraday's laboratory, this celebrated physicist has remarked that the sound produced by an ingot of aluminium is not simple. One can distinguish, by turning the vibrating ingot, two sounds very near together and succeeding each other rapidly, according as one or the other face of the ingot faces the observer.

The bell referred to above was 20 kilos in weight and 50 centimetres in diameter, but as Deville admits, its sound was not pleasing, and a contemporary writer, evidently not very enthusiastic in sounding the praises of aluminium, said that while the bell was highly sonorous yet it "gave a sound like a cracked pot."

I have not heard that any large bell has since been cast, but it is certain that the metal in bars has a highly musical ring. Faraday's observation has also been verified, for a recent lecturer suspended by one end a bar 6 feet long, $3\frac{1}{2}$ inches wide, and $1\frac{1}{2}$ inches thick, and on striking it a prolonged vibration ensued, two notes being recognized, A sharp and D sharp, the latter more subdued.

CRYSTALLINE FORM.

Deville: Aluminium often presents a crystalline appearance when it has been cooled slowly. When it is not pure the little crystals which form are needles, and cross each other in all directions. When it is almost pure it still crystallizes by fusion but with difficulty, and one may observe on the surface of the ingots hexagons which appear regularly parallel along lines which centre in the middle of the polygon. It is an error to conclude from this observation that the metal crystallizes in the rhombohedral system. It is evident that a crystal of the regular system may present a hexagonal section; while on the other hand, in preparing aluminium by the battery at a low temperature, I have observed complete octahedrons which were impossible of measurement it is true, but their angles appeared equal.

ELASTICITY.

Deville: M. Wertheim has found that the elasticity of aluminium just cast is sensibly the same as that of silver; but when worked it resembles that of soft iron, becoming more rigid and elastic, and giving the sound of steel when dropped on a hard body.

Mallet remarked that absolutely pure aluminium seemed to be less hardened by hammering than ordinary commercial metal. A German firm engaged in making aluminium state that by long, gradual cooling from a red heat aluminium can be made so elastic that it can even be used for hair springs for watches. Annealing by cooling quickly from a red heat makes the metal soft. Aluminium stiffens up very quickly in rolling; the author's father has found the best means of removing this is to heat the metal red hot and plunge into water. Metal thus treated becomes very soft. Fine wire quickly becomes hard in drawing, but can be annealed in the heat over an argand burner.

TENACITY.

W. H. Barlow :* A bar of aluminium three feet long and one-quarter inch square was obtained and different parts of it subjected to tests for tension, compression, and transverse strain, elasticity, elastic range, and ductility. It will be seen on reference to the results that the weight of a cubic inch was 0.0275 pound, showing a specific gravity of 2.688, and its ultimate tensile strength was about twelve tons per square inch. The range of elasticity is large, the extreme to the yielding point being one-two hundredths of the length. The modulus of elasticity is 1,000,000, the extension in samples two inches long being 2.5 per cent. Taking the tensile strength of the metal in relation to its weight, it shows a high mechanical value. These results are thus tabulated :—

	Weight of 1 cubic foot in pounds.	Tensile strength per sq. in. in pounds.	Length of a bar able to support its weight, in feet.
Cast iron	444	16.500	5,351
Bronze	525	36,000	9,893
Wrought iron . . .	480	50,000	15,000
Steel	490	78,000	23,040
Aluminium . . .	168	26,800	23,040

It thus appears that taking the strength of aluminium in relation to its weight, it possesses a mechanical value about equal to that of steel of 35 tons per square inch tensile strength.

Kamarsch† obtained the following results as to the strength of aluminium wire :—

DIAMETER.	TENSILE STRENGTH, GRAMMES.			TENACITY.
Millimetres.	1st trial.	2d trial.	Mean.	Kilos per sq. millimetre.
0.225	661	653	657	12.975
0.205	524	506	515	12.255
0.160	307	311	309	12.700
0.145	246	252	249	11.845

These results are far below that obtained by Barlow, which is equal to 18.92 kilos per square millimetre. The latter figure is,

---

* Rpt. Brit. A. A. S., 1882, p. 668.
† Dingler, 172, p. 55.

however, undoubtedly nearer the truth for good aluminium, since tests of the metal made by the Deville-Castner process average 25,000 to 30,000 lbs. per square inch, being in general higher than the figure given by Barlow.

MALLEABILITY.

Deville: Aluminium may be forged or rolled with as much perfection as gold or silver. It is beaten into leaves as easily as they, and a very experienced gold-beater, M. Rousseau, has made leaves as fine as those of gold or silver, which are put up in books. I know of no other useful metal able to stand this treatment.

Mallet: With absolutely pure aluminium the malleability was undoubtedly improved, the metal yielding easily to the hammer, bearing distortion well, and flattening in two or three directions without cracking. It seemed to be sensibly less hardened by hammering than the ordinary metal of commerce.

Commercial aluminium is now to be had rolled into sheets of almost any size or thickness, and at only a small advance on the price of ingot metal. The only particulars in which it differs much from other metals being that it must be annealed much oftener, and requires an extraordinarily large power to roll it. Mr. J. Richards compares the cold rolling of aluminium to the hot rolling of steel in regard to the power required; he also finds that unless the sheet is rolled until quite hard it does not polish in the rolls.

The aluminium leaf is now in regular use with gilders and decorators. It was first made by M. Degousse, of Paris, and afterwards for several years by C. Falk & Co., of Vienna. The manufacture is rather more difficult than beating out gold or silver, and requires also a pure metal to stand the working. A specimen such as is sold commercially was measured by the author. He found its thickness to be 0.000,638 millimetres or one-forty thousandth of an inch, which compares favorably with that of ordinary gold leaf. It is quite possible that if a test were made with extra pure metal, this result would easily be exceeded.

This leaf was not thin enough to show any color by transmitted light.

Deville has stated that aluminium can be forged with as much perfection as gold or silver, but at what heat it works best is not stated. It can readily be hammered and shaped cold, like silver or copper, but it soon stiffens up, and must be kept soft by frequent annealing.

Aluminium probably stands third in the order of malleability of the metals, gold and silver exceeding it; while it is probably sixth in the order of ductility, being preceded by gold, silver, platinum, iron, and copper.*

DUCTILITY.

Deville: Aluminium behaves very well at the drawing plate. M. Vangeois obtained, in 1855, with a metal far from being pure, wires of extreme tenuity, which were used to make aluminium passementerie. However, the metal deteriorates much in the operation, and the threads become flexible again only after an annealing very delicately performed, because of the fineness of the threads and the fusibility of the metal. The heat of the air coming from the top of the chimney over an Argand burner is sufficient to anneal them.

Aluminium wire is being made at present by numbers of manufacturers, the difficulties being very few when pure metal can be procured to work with. Quite a large amount of power is required for drawing when compared with other metals. Wire as fine as 0.1 millimetre in diameter can be made without very much trouble, and the use of aluminium in this form promises large development in the near future.

EXPANSION BY HEAT.

Fizeau is quoted as authority for the following coefficients of linear expansion of aluminium by heat:—

	For 1° F.	For 1° C.
Cast aluminium	0.00001234	0.00002221
Crystallized aluminium . .	0.00000627	0.00001129

---

* Thurston's Materials of Engineering.

SPECIFIC HEAT.

Deville: According to the experiments of M. Regnault, the specific heat of aluminium corresponds to its equivalent 13.75, from which we may conclude that it must be very large when compared with all the other useful metals. One can easily perceive this curious property by the considerable time which it takes an ingot of the metal to get cold. We might even suggest that a plate of aluminium would make a good chafing-dish. Another experiment makes this conclusion very evident. M. Paul Morin had the idea of using aluminium for a plate on which to cook eggs, the sulphur of which attacked silver so easily; and he obtained excellent results. He noticed, also, that the plate kept its heat a much longer time than the silver one.

The value of this quantity has been quoted differently by different authorities. Regnault obtained 0.2143 as the mean between 0° and 100°, while Kopp obtained 0.2020. In the first case Deville remarks that the metal he gave Regnault was unfortunately contaminated with copper, which would lead to the supposition that the value obtained was somewhat below the truth; we cannot account for the lower value obtained by Kopp. However these may be, more recent and probably more accurate determinations have indicated a higher value. Mallet determined the specific heat of absolutely pure aluminium to be 0.2253, which, he remarked, made its atomic heat 0.2253×27.02 or 6.09. Naccari* observed the specific heat at different temperatures to be—

18°	50°	100°	200°	300°
0.2135	02.164	0.2211	0.2306	0.2401

The author has determined the mean specific heat from 0° to the melting point to be 02.85, and the latent heat of fusion 29.5 calories.

ELECTRIC CONDUCTIVITY.

Deville: Aluminium conducts electricity with great facility, so that it may be considered as one of the best conductors known,

---

* Transactions "Accademia di Torino," Dec. 1887.

and perhaps equal to silver. I found by Wheatstone's Bridge that it conducts eight times better than iron. M. Buff has arrived at results evidently different from mine because we have not taken the same ground of comparison. The difference is due, without doubt, to the metal which he employed containing, as is easily found in many specimens, a little cryolite and fusible materials the density of which is near that of the metal, and which were employed in producing it. The complete separation of the metal and flux is a difficult mechanical operation, but which is altogether avoided by using a volatile flux. This is a condition which must be submitted to in order to get the metal absolutely pure.

The exact value expressing the electric conductivity of aluminium is not beyond dispute. In one place we find the following relative numbers given :—*

	At 0°	At 100°
Copper	45.74	33.82
Magnesium	24.47	17.50
Aluminium	22.46	17.31

M. Margottet states it as being 51.5 if copper is 100; or 33.74 silver being 100. Professor Mattheisen determined the values as follows :—

Pure silver	100
Commercial copper	77
Commercial aluminium	33.76

Watts states that the electric conductivity of aluminium is 56.1 silver being 100. Finally, Benoit† gives the mean electric resistance and conductivity at 0° as follows, the resistance being for a wire 1 metre long, and with a cross section of 0.2 square centimetres (a column of mercury of those dimensions giving resistances of 0.9564 Ohms or 1.0 Siemens).

	Ohms.	Siemens.	Conductivity.
Silver, annealed . . .	0.0154	0.0161	100
Copper, " . . .	0.0171	0.0179	90
Gold, " . . .	0.0217	0.0227	71
Aluminium, annealed . .	0.0309	0.0324	49.7
Magnesium, hard . .	0.0423	0.0443	36.4

* Jahresb. der Chemie, 1881, p. 94.
† Thurston's Materials of Engineering.

If we compare these various results we find the values given to vary as follows :—

	Silver = 100.	Copper = 100.
Jahresb. d. Chemie . . .		{ 49.10 at 0° { 51.18 at 100°
Margottet	33.74	51.5
Mattheisen	33.76	43.8
Watts	56.1	
Benoit	49.7	55.2

THERMAL CONDUCTIVITY.

Deville: It is generally admitted that conductivity for heat and electricity correspond exactly in the different metals. A very simple experiment made by Mr. Faraday in his laboratory seems to place aluminium very high among metallic conductors. He found that it conducted heat better than silver or copper.

It is altogether probable that there was some mistake made in Faraday's experiment, since, as we have seen, aluminium is inferior to silver and copper as a conductor of electricity, and recent investigations also place it inferior to them in thermal conductivity. The writer before quoted (Jahresb. d. Chemie), gives these values :—

	At 0°	At 100°
Copper	0.7198	0.7226
Magnesium	0.3760	0.3760
Aluminium	0.3435	0.3619

or if the conductivity of copper is 100, that of aluminium is 47.72 at 0° and 50.0 at 100°, which it may be observed agree very closely with the values found for electric conductivity by the same investigator. Calvert and Johnson determined the ratio of its conducting power for heat with that of silver to be as 665 is to 1000, which is considerably higher than the values given for electric conductivity. The fact that these values agree in general better when referred to copper, would seem to show that the variable quantity is probably the standard silver used for comparison, although we should have expected to meet with more trouble from the copper in this respect.

CHAPTER IV.

CHEMICAL PROPERTIES OF ALUMINIUM.

WE would here repeat the remark made with regard to the physical properties, that the properties to be recorded are those of the purest metal unless specifically stated otherwise. However, the high grade of commercial metal differs very little in most of its chemical properties from the absolutely pure, so that not many reservations are necessary in applying the following properties to good, commercial metal :

ACTION OF AIR.

Deville: Air, wet or dry, has absolutely no action on aluminium. No observation which has come to my knowledge is contrary to this assertion, which may easily be proved by any one. I have known of beams of balances, weights, plaques, polished leaf, reflectors, etc., of the metal exposed for months to moist air and sulphur vapors and showing no trace of alteration. We know that aluminium may be melted in the air with impunity, therefore air and also oxygen cannot sensibly affect it. It resisted oxidation in the air at the highest heat I could produce in a cupel furnace, a heat much higher than that required for the assay of gold. This experiment is interesting, especially when the metallic button is covered with a layer of oxide which tarnishes it, the expansion of the metal causing small branches to shoot from its surface, which are very brilliant and do not lose their lustre in spite of the oxidizing atmosphere. M. Wöhler has also observed this property on trying to melt the metal with a blowpipe. M. Peligot has profited by it to cupel aluminium. I have seen buttons of impure metal cupelled with lead and become very malleable.

With pure aluminium the resistance of the metal to direct oxidation is so considerable that at the melting point of platinum it is hardly appreciably touched, and does not lose its lustre. It is well known that the more oxidizable metals take this property away from it. But silicon itself, which is much less oxidizable, when alloyed with it makes it burn with great brilliancy, because there is formed a silicate of aluminium.

While the above observations are in the main true, yet it is now well known that objects made of commercial aluminium do after a long exposure become coated with a very thin film, which gives the surface a "dead" appearance. The coating is very similar in appearance to that forming on zinc under the same circumstances. The oxidation, however, does not continue, for the film seems to be absolutely continuous and to protect the metal underneath from further oxidation. This coating can best be removed by very dilute acid (see Mouréy's receipt, p. 57), after which the surface can be burnished to its former brilliancy·

It has also been found that at a high white-heat, especially at the heat of an electric furnace, aluminium burns with a strong light to alumina. It is quite probable that in this case it volatilizes first, and it is the vapor which burns. During the operation of an electric furnace a white smoke formed of invisible particles of alumina is thus formed and evolved from the furnace. Also, in melting aluminium, even the purest, it will be found that the surface seems bound and the aluminium restrained from flowing freely by a minute "skin" which may probably be a mixture of oxide with metal, or perhaps of oxides of foreign metals, but, nevertheless, it is always present and is therefore indicative of oxidation taking place. It seems to protect the metal beneath it perfectly, so that, once formed, it gets no thicker by continued heating.

Wöhler first discovered that when aluminium was in the extremely attenuated form of leaf it would burn brightly in air, and burn in oxygen with a brilliant bluish-light. It is also said that thin foil will burn in oxygen, being heated by wrapping it around a splinter of wood, and fine wire also burns like iron wire, but the combustion is not continuous because the wire fuses too quickly. The alumina resulting is quite insoluble in acids, and as hard as corundum.

ACTION OF WATER.

Deville: Water has no action on aluminium, either at ordinary temperatures or at 100°, or at a red heat bordering on the fusing point of the metal. I boiled a fine wire in water for half an hour and it lost not a particle in weight. The same wire was put in a glass-tube heated to redness by an alcohol lamp and traversed by a current of steam, but after several hours it had not lost its polish, and had the same weight. To obtain any sensible action it is necessary to operate at the highest heat of a reverberatory furnace—a white heat. Even then the oxidation is so feeble that it develops only in spots, producing almost inappreciable quantities of alumina. This slight alteration and the analogies of the metal allow us to admit that it decomposes water, but very feebly. If, however, metal produced by M. Rose's method is used, which is almost unavoidably contaminated with slag composed of chlorides of aluminium and sodium, the former, in presence of water, plays the part of an acid towards aluminium, disengaging hydrogen with the formation of a subchlorhydrate of alumina, whose composition is not known, and which is soluble in water. When the metal thus tarnishes in water one may be sure to find chlorine in the water on testing it with nitrate of silver.

Aluminium leaf, however, will slowly decompose water at 100°. Hydrogen is slowly evolved, the leaf loses its brilliancy, becomes discolored, and after some hours translucent. It is eventually entirely converted into gelatinous hydrated alumina.

ACTION OF HYDROGEN SULPHIDE AND SULPHUR.

Deville: Sulphuretted hydrogen exercises no action on aluminium, as may be proved by leaving the metal in an aqueous solution of the gas. In these circumstances almost all the metals, and especially silver, blacken with great rapidity. Sulph-hydrate of ammonia may be evaporated on an aluminium leaf, leaving on the metal only a deposit of sulphur, which the least heat drives away.

Aluminium may be heated in a glass tube to a red heat in

vapor of sulphur without altering the metal. This resistance is such that in melting together polysulphide of potassium and some aluminium containing copper or iron, the latter are attacked without the aluminium being sensibly affected. Unhappily, this method of purification may not be employed because of the protection which aluminium exercises over foreign metals. Under the same circumstances gold and silver dissolve up very rapidly. However, at a high temperature I have observed that it combines directly with sulphur to give aluminium sulphide. These properties varying so much with the temperature form one of the special characteristics of the metal and its alloys.

Margottet states that hydrogen sulphide is without action on aluminium, as also are the sulphides of iron, copper, or zinc. Aluminium is said to decompose silver sulphide, Ag^2S, setting the sulphur, however, at liberty and alloying with the silver. In regard to its indifference to the first mentioned sulphides, this would give inferential evidence that the reverse operation, $i.e.$, the action of iron, copper, or zinc on aluminium sulphide, would be possible, as will be seen later to be apparently established by direct experiment. As to the action of sulphuretted hydrogen, the author has a different experience to quote. On passing a stream of that gas into commercial aluminium melted at a red heat, little explosive puffs were heard accompanied by a yellow light, while the dross formed on the surface, when cooled, evolved sulphuretted hydrogen briskly when dropped into water, and gave every indication of containing aluminium sulphide. It could not have been silicon sulphide, for the metal contained as large a percentage of silicon after treatment as before. Hydrogen sulphide is also absorbed in large quantity by molten aluminium, and mostly evolved just as the metal is about to set. Some of the gas is entangled in the solidifying metal, forming and filling numerous cavities or blow-holes.

SULPHURIC ACID.

Deville: Sulphuric acid, diluted in the proportion most suitable for attacking the metals which decompose water, has no action on aluminium; and contact with a foreign metal does not help, as with zinc, the solution of the metal, according to M. de la Rive. This singular fact tends to remove aluminium considerably from those metals. To establish it better, I left for several months some globules weighing only a few milligrammes in contact with the weak acid, and they showed no visible alteration; however, the acid gave a faint precipitate when neutralized with aqua ammonia.

Margottet: Sulphuric acid, dilute or concentrated, exercises in the cold only a very slight sensible action on aluminium, the pure metal is attacked more slowly than when it contains foreign metals. The presence of silicon gives rise to a disengagement of silicon hydride (SiH^4), which communicates to the hydrogen set free a tainted odor. Concentrated acid dissolves it rapidly with the aid of heat, disengaging sulphurous acid gas (SO^2).

NITRIC ACID.

Deville: Nitric acid, weak or concentrated, does not act on aluminium at the ordinary temperature. In boiling acid solution takes place, but with such slowness that I had to give up this mode of dissolving the metal in my analyses. By cooling the solution all action ceases. On account of this property, M. Hulot obtained good results on substituting aluminium for platinum in the Grove battery.

HYDROCHLORIC ACID.

Deville: The true solvent of aluminium is hydrochloric acid, weak or concentrated; but, when the metal is perfectly pure, the reaction takes place so slowly that M. Favre, of Marseilles, had to give up this way of attack in determining the heat of a combination of the metal. But, impure aluminium is dissolved very rapidly. At a very low temperature gaseous hydrochloric acid attacks the metal and changes it into chloride. Under these circumstances iron does not seem to alter; able, no doubt, to resist by covering itself with a very thin protecting layer of ferrous chloride. This experiment would lead me to admit that it is the acid and not the water which is decomposed by aluminium; and,

in fact, the metal is attacked more easily as the acid is more concentrated.   This explains the difference of the action of solutions
of hydrochloric and sulphuric acids, the latter being almost inac-
tive.   This reasoning applies also to tin.

When the metal contains silicon it disengages hydrogen of a
more disagreeable smell than that given out by iron under similar
circumstances.   The reason of this is the production of that
remarkable body recently discovered by MM. Wöhler and Buff
—silicuretted hydrogen.   When the proportion of silicon is
small, the whole is evolved as gas; when increased a little,
some remains in solution with the aluminium, and then it requires great care to separate the metal exactly even when the
solution is evaporated to dryness.   If 3 to 5 per cent. of silicon
is present, it remains insoluble mixed with a little silica, as has
been cleverly proven by Wöhler and Buff, by the action of hydrofluoric acid, which dissolves the silica with evolution of hydrogen
without attacking the silicon itself.   On dissolving commercial
aluminium there is sometimes obtained a black crystalline residue, which separated on a filter and dried at 200° to 300° takes
fire in places; this residue is silicon mixed with some silica.   The
presence of silicon augments very much the facility with which
aluminium is attacked by hydrochloric acid.

If hydrochloric acid is present in a mixture of acids, it begins
the destruction of the metal.   Hydrobromic, hydriodic, hydrofluoric acids are said to act very similarly to hydrochloric.

Organic Acids, Vinegar, etc.

Deville: Weak acetic acid acts on aluminium in the same way
as sulphuric acid, i. e., in an inappreciable degree or with extreme
slowness.   I used for the experiment acid diluted to the strength
of strongest vinegar.   M. Paul Morin left a plaque of the metal
a long time in wine which contained tartaric acid in excess and
acetic acid, and found the action on it quite inappreciable.   The
action of a mixture of acetic acid and common salt in solution
in pure water on pure aluminium is very different, for the acetic
acid replaces a portion of the chlorine existing in the sodium

chloride, rendering it free.   However, this action is very slow, especially if the aluminium is pure.

The practical results flowing from these observations deserve to be clearly defined, because of the applications which may be made of aluminium to culinary vessels.   I have observed that the tin so often used and which each day is put in contact with common salt and vinegar, is attacked much more rapidly than aluminium under the same circumstances.   Although the salts of tin are very poisonous, and their action on the economy far from being negligible, the presence of tin in our food passes unperceived because of its minute quantity.   Under the same circumstances aluminium dissolves in less quantity; the acetate of aluminium formed resolves itself on boiling into insoluble aluminia or an insoluble sub-acetate, having no more taste or action on the body than clay itself.   It is for that reason and because it is known that the salts of the metal have no appreciable action on the body, that aluminium may be considered as an absolutely harmless metal.

It may be appropriately remarked here that the rapid tarnishing of polished aluminium articles is more frequently due to the effect of handling than to any other cause.   The perspiration contains about 2 per cent. of sodium chloride and about an equal quantity of organic acids; its action on aluminium is not very great, yet almost always sufficient to spoil a high polish and give a visible tarnish.

Ammonia.

Aqua ammonia acts slowly on aluminium, producing a little alumina, part of which remains dissolved.   Ammonia gas does not appear to act on the metal.

Caustic Alkalies.

Deville: Alkaline solutions act with great energy on the metal, transforming it into aluminate of potash or soda, setting free hydrogen.   However, it is not attacked by caustic potash or soda in fusion; one may, in fact, drop a globule of the pure metal into melted caustic soda raised almost to red heat in a silver vessel, without observing the least disengagement of hydrogen. Silicon, on the contrary, dissolves with great energy under the same circumstances. I have employed melted caustic soda to clean siliceous aluminium. The piece is dipped into the bath kept almost at red heat. At the moment of immersion several bubbles of hydrogen disengage from the metallic surface, and when they have disappeared all the silicon of the superficial layer of aluminium has been dissolved. It only remains to wash well with water and dip it into nitric acid, when the aluminium takes a beautiful mat. Alkaline organic materials, as the saliva, have a tendency to oxidize it, but the whole effect produced is insignificant. M. Charrière has made for a patient on whom he practised tracheotomy a small tube of the metal, which remained almost unaltered although in contact with purulent matter. After a long time a little alumina was formed on it, hardly enough to be visible.

Mallet: The pure metal presents greater resistance to the prolonged action of alkalies than the impure.

Aluminium leaf dissolves with extraordinary quickness in caustic alkali, leaving the iron, which is always present, undissolved. The chemical reaction occurring indicates that aluminium acts the part of a strong acid, forming aluminates of the alkaline metals which stay in solution.

Lime water attacks aluminium in a similar manner, but the resulting calcium aluminate is insoluble in water and is therefore precipitated.

SOLUTIONS OF METALLIC SALTS.

Deville: The action of any salt whatever on aluminium may be easily deduced from the action of its acids on that metal. We may, therefore, predict that in acid solutions of sulphates and nitrates aluminium will precipitate no metal, not even silver, as Wöhler has observed. But the hydrochloric solutions of the same metals will be precipitated, as MM. Tissier have shown. Likewise, in alkaline solutions, silver, lead, and metals high in the classification of the elements are precipitated. It may be concluded from this that to deposit aluminium on other metals by means of the battery, it is always necessary to use acid solutions in which hydrochloric acid, free or combined, should be absent. For similar reasons the alkaline solutions of the same metals cannot be employed, although they give such good results in plating common metals with gold and silver. It is because of these curious properties that gilding and silvering aluminium are so difficult.

These conclusions by Deville are confirmed only when using pure aluminium ; the impure metal, containing iron, silicon, or perhaps sodium, may produce very slight precipitates in cases where pure aluminium would produce none. Some observers have noted different results in some cases even when using aluminium free from these impurities. We will therefore take up these cases and consider them separately.

Mercury.—*Aluminium decomposes solutions of mercuric chloride, cyanide or nitrate, mercury separating out first then forming an amalgam with the aluminium which is immediately decomposed by the water, the result being alumina and mercury. From an alcoholic solution of mercurous chloride the mercury is precipitated more quickly at a gentle heat. A solution of mercurous iodide with potassium iodide is also reduced in like manner.

Copper.—*From solution of copper sulphate or nitrate aluminium separates out copper only after two days' standing, as either dendrites or octahedra ; from the nitrate it also precipitates a green, insoluble basic salt. Copper is precipitated immediately from a solution of cupric chloride ; but slower from the solution of copper acetate. The sulphate or nitrate solutions behave similarly if potassium chloride is also present, and the precipitation is complete in presence of excess of aluminium.

Silver.—*From a nitrate solution, feebly acid or neutral, aluminium precipitates silver in dendrites, the separation only beginning after six hours' standing. From an ammoniacal solution of silver chloride or chromate, aluminium precipitates the silver immediately as a crystalline powder. †Cossa confirms the statement as to the nitrate solution.

Lead.—*From nitrate or acetate solution the lead is slowly pre-

* Dr. Mierzinski.
† A. Cossa, Bull. de la Soc. Chim. 1870, p. 199.

cipitated in crystals; an alkaline solution of lead chromate gives precipitates of lead and chromic oxide.

Zinc.—*An alkaline solution of zinc salts is readily decomposed and zinc precipitated.

Margottet states that all metallic chlorides excepting those of potassium or sodium are reduced from solution. This statement can hardly include chlorides of magnesium or lithium, since magnesium precipitates alumina from solutions of aluminium salts. Alkaline or ammoniacal solutions are more easily decomposed than acid solutions; in alkaline solutions the cause being the facility with which aluminates of the alkaline are formed.

Alkaline chlorides.—A solution of sodium or potassium chloride is not affected by pure aluminium, either cold or warm. However, aluminium which was packed in a case with saw-dust and kept wet with sea-water for two weeks was deeply corroded; whether the result would have been the same without the presence of the saw-dust, I cannot say.

Aluminium salts.—It is a curious fact that a solution of aluminium chloride will attack aluminium, forming sub-chlorhydrate, with evolution of hydrogen. A solution of alum does not attack aluminium, but if sodium chloride is added it is dissolved with evolution of hydrogen. It is interesting to note that while neither of these salts alone attacks aluminium, the mixture of the two does.

SODIUM CHLORIDE.

Fused common salt is used as a flux for aluminium. It does not possess the property, like fluorspar, of dissolving alumina, but it is apparently without any corroding effect on the molten aluminium; neither is it probable that it is capable of reacting alone with any aluminous material to form aluminium chloride, which might volatilize and thus cause loss of material.

FLUORSPAR.

This compound is said to be without action on molten aluminium. It makes a good flux for the metal, especially in connection

* A. Cossa, Bull. de la Soc. Chim. 1870, p. 199.

with cryolite or common salt, and possesses the property of dissolving the alumina with which the metal may be contaminated and which, by encrusting small globules, hinders their reunion to a button.

CRYOLITE.

This salt is largely used as a flux for aluminium and also as a source of the metal.   It is commonly supposed to have the property of dissolving alumina, like fluorspar, but to be without action on the metal itself.   Prof. W. Hampe, however, has recently stated that at a temperature about the melting point of copper, finely-divided aluminium is rapidly dissolved, a sub-fluoride being probably formed; but the metal " en masse" is not sensibly attacked.

SILICATES AND BORATES.

Neither of these classes of compounds can be used as fluxes or slags in working aluminium, since they both rapidly corrode the metal.   Deville had little difficulty in decomposing these salts so completely with metallic aluminium that he isolated silicon and boron.   If aluminium is melted in an ordinary glass vessel it attacks it, setting free silicon from silica, forming an aluminate with the alkali present and an alloy with the silicon set free.   Aluminium melted under borax is rapidly dissolved, an aluminium borate being formed.   It is thus seen that the common metallurgic slags are altogether excluded from the manufacture of aluminium.

NITRE.

Deville: Aluminium may be melted in nitre without undergoing the least alteration, the two materials rest in contact without reacting even at a red heat, at which temperature the salt is plainly decomposed, disengaging oxygen actively.   But if the heat is pushed to the point where nitrogen itself is disengaged, there the nitre becomes potassa, a new affinity becomes manifest, and the phenomena change.   The metal then combines rapidly with the potassa to give aluminate of potash.   The accompanying phenomenon of flagration often indicates a very energetic reaction. Aluminium is continually melted with nitre at a red heat to purify it by the oxygen disengaged, without any fear of loss. But it is necessary to be very careful in doing it in an earthen crucible. The silica of the crucible is dissolved by the nitre, the glass thus formed is decomposed by the aluminium, and the silicide of aluminium formed is then very oxidizable, especially in the presence of alkalies. The purification by nitre ought to be made in an iron crucible well oxidized by nitre inside.

If finely divided aluminium is mixed with nitre and brought to a red heat, the metal is oxidized with the production of a fine blue flame. (Mierzinski.)

ALKALINE SULPHATES AND CARBONATES.

Tissier: Only 2.65 grammes of aluminium introduced into melted red-hot sodium sulphate (Na^2SO^4) decomposed that salt with such intensity that the crucible was broken into a thousand pieces, and the door of the furnace blown to a distance. Heated to redness with alkaline carbonate, the aluminium was slowly oxidized at the expense of the carbonic acid, carbon was set free, and an aluminate formed. The reaction takes place without deflagration.

METALLIC OXIDES.

Tissier Brothers made a series of experiments on the action of aluminium on metallic oxides. Aluminium leaf was carefully mixed with the oxide, the mixture placed in a small porcelain capsule and heated in a small earthen crucible, which served as a muffle. The results were as follows :—

Manganese dioxide.—No reaction.

Zinc oxide.—No reaction even at white heat.

Ferric oxide.—By heating to white heat 1 equivalent of ferric oxide and 3 of aluminium the reaction took place with detonation, and by heating sufficiently we obtained a metallic button, well melted, containing 69.3 per cent. of iron and 30.7 per cent. of aluminium. Its composition corresponds very nearly to the formula AlFe. It would thus appear that the decomposition of ferric oxide will not pass the limit where the quantity of iron reduced is sufficient to form with the aluminium the alloy AlFe.

Lead oxide.—We mixed 2 equivalents of litharge with 1 of aluminium, and heated the mixture slowly up to white heat, when the latter reacted on the litharge with such intensity as to produce a strong detonation. We made an experiment with 50 grammes of litharge and 2.9 grammes of aluminium leaf, when the crucible was broken to pieces and the doors of the furnace blown off.

Copper oxide.—Three grammes of black oxide of copper mixed with 1.03 grammes of aluminium detonated, producing a strong explosion, when the heat reached whiteness.

Bekétoff* reduced baryta (BaO) with metallic aluminium in excess, and obtained alloys of aluminium and barium containing in one case 24 per cent. in another 33 per cent. of barium.

MISCELLANEOUS AGENTS.

Phosphate of lime.—Tissier Brothers heated to whiteness a mixture of calcium phosphate with aluminium leaf, without the metal losing its metallic appearance or any reaction being noted.

Hydrogen.—This gas appears to have no action on aluminium, except to be dissolved in it in a moderately large quantity.

Chlorine.—Gaseous chlorine attacks the metal rapidly. Aluminium foil heated in an atmosphere of chlorine takes fire and burns with a vivid light.

Bromine, iodine, fluorine act similarly to chlorine.

Silver chloride.—Fused silver chloride is decomposed by aluminium, the liberated silver as well as the excess of aluminium being melted by the heat of the reaction.

Mercurous chloride.—If vapors of mercurous chloride are passed through a tube in which some hot aluminium is placed, mercury is separated out, aluminium chloride deposits in the cooler part of the tube, and the aluminium is melted by the heat developed.

* Bull. de la Soc. Chimique, 1887, p. 22.

GENERAL OBSERVATIONS ON THE PROPERTIES OF ALUMINIUM.

Deville : "Aluminium at a low temperature conducts itself as
a metal which can give a very weak base ; in consequence, its resistance to acids, hydrochloric excepted, is very great. It conducts itself with the alkalies as a metal capable of giving a quite
energetic acid, it being attacked by alkaline oxides dissolved in
water. But this affinity is still insufficient to determine the
decomposition of melted caustic potash. For a stronger reason
it does not decompose metallic oxides at a red heat. This is why
in the muffle the alloy of aluminium and copper gives black
CuO, and this also accounts for the alloy of aluminium and lead
being capable of being cupelled. But by a strange exception,
and which does not appertain solely, I believe, to aluminium, as
soon as the heat is above redness the affinities are quickly inverted and the metal takes all the properties of silicon, decomposing the oxides of lead and copper with the production of the
aluminates.

"From all the experiments which have been reported and from
all the observations which have been made, we can conclude that
aluminium is a metal which has complete analogies with no one
of the simple bodies which we consider metals. In 1855, I proposed to place it alongside of chromium and iron, leaving zinc
out of the group with which aluminium had been until then
classed. Zinc is placed very well beside magnesium, there being
intimate analogies between these two volatile metals. There may
be found at the end of a memoir which M. Wöhler and I published in the 'Compt. Rendue' and the 'Ann. de Chem. et de
Phys.,' the reasons why we are tempted to place aluminium near
to silicon and boron in the carbon series, on grounds analogous
to those on which antimony and arsenic are placed in the nitrogen series."

CHAPTER V.

PROPERTIES AND PREPARATION OF ALUMINIUM COMPOUNDS.

In this chapter we propose to note in rather condensed form the prominent characteristics of the various aluminium compounds, with an outline of the methods by which they can be produced, reserving for another chapter however, the preparation of those salts which are now being manufactured on a commercial scale for purposes of further treatment for aluminium. I do not propose this as a substitute for the various chemical treatises on this subject, but simply to add to the completeness of this work in order that a fair understanding of the other parts of the book may not be missed because data of this nature are not immediately at hand. Parts of this chapter are taken from M. Margottet's treatise on aluminium, in Fremy's Enclycopédie Chimique.

GENERAL CONSIDERATIONS.

Structure of aluminium compounds.*—Aluminium is a quadrivalent element, but in its compounds always acts as a double hexad atom $(Al-Al)^{ri}$, one bond or affinity thus serving to bind the two atoms together. The double atom Al^2 can thus unite with six monatomic elements or atomic groups, or their equivalent. Thus we have—

Al^2O^3 Aluminium oxide.
$Al^2(OH)^6$ "   oxyhydrate.
Al^2Cl^6 "   chloride.

The salts of aluminium usually called aluminious salts, are chemically considered as derivatives of the oxyhydrate, the hydrogen atoms being replaced by acid radicals. Thus—

---

* R. Biedermann. Kerl and Stohmans, Handbuch, 4th ed.

$Al^2.O^6.(NO^2)^6$　Aluminium nitrate.
$Al^2.O^6.(C^2H^3O)^6$　　"　　　acetate.
$Al^2.O^6.(SO^2)^3$　　　"　　　sulphate.
$Al^2.O^6.(PO)^2$　　　"　　　phosphate.

The above are normal or neutral salts, all the hydrogen atoms having been replaced. Basic salts result if only part of the hydrogen is replaced.

$Al^2.(OH)^4O^2.(C^2H^3O)^2$ Basic aluminium acetate.
$Al^2.(OH^4)O^2.(SO^2)$　　"　　　"　　sulphate.

Aluminium is very apt to form these basic compounds and others of even greater complexity.

Aluminium oxyhydrate is distinguished from most of the other basic oxides in that its hydrogen atoms are not alone replaced by acid radicals, but by metals forming aluminates. Thus—

$Al^2.O^6.Na^6$　Sodium aluminate.
$Al^2.O^6.Ba^3$　Barium　"

If we consider aluminium oxyhydrate to act in these compounds as an acid, these are its neutral salts. Besides aluminates of this form there are others, natural and artificial, having the general formula Al^2RO^4, R being diatomic. These were written on the old dualistic theory $Al^2O^3.RO$, but they are now considered as derivatives of aluminium anhydro-hydrate. Thus—

$Al^2O^2.(OH)^2$　Aluminium anhydrohydrate.
$Al^2O^2.(O^2Mg)$　Magnesium aluminate.

General methods of formation and properties.—Hydrated alumina, which has not been too strongly heated, dissolves in strong acids forming salts which are mostly soluble in water. In the feebler acids and in all organic acids it is completely insoluble. The salts of these latter acids are formed best by decomposing solution of aluminium sulphate with the barium or lead salt of the acid in question. Alumina forms no carbonate. Most aluminium salts are soluble in water and rather difficult to crystallize: the few insoluble salts are white, gelatinous, and similar to the hydrate in appearance. In the neutral salts the acid is loosely held, for their solution strongly reddens litmus paper and their action is as if part of the acid were free in the salt. For instance, a solution of alum attacks iron giving off hydrogen, a soluble basic salt of aluminium being formed as well as sulphate of iron. The neutral salts of volatile acids give off acid simply by boiling their solutions, basic salts being formed. An aqueous solution of aluminium chloride loses its acid almost completely on evaporation. Gentle ignition is sufficient in most cases to completely decompose aluminium salts. Hydrated alumina dissolves easily in caustic alkali forming soluble aluminates; with baryta two aluminates are known, one soluble the other not; all other known aluminates are insoluble.

Neutral solutions of aluminium salts react as follows with the common reagents :—

Hydrogen sulphide produces no precipitate.

Ammonium sulphide precipitates aluminium hydrate with separation of free sulphur.

Caustic potash or soda precipitates aluminium hydrate, soluble in excess.

Aqua ammonia precipitates aluminium hydrate insoluble in excess, especially in presence of ammoniacal salts.

Alkaline carbonates precipitate aluminium hydrate insoluble in excess.

Sodium phosphate precipitates white gelatinous aluminium phosphate, easily soluble in acids or alkalies.

ALUMINIUM OXIDE.

Commonly called alumina. Composition Al^2O^3, and contains 52.95 per cent. of aluminium when perfectly pure. Colorless corundum is a natural pure alumina, in which state it is infusible at ordinary furnace heats, insoluble in acids, has a specific gravity of 4, and is almost as hard as the diamond. To get this into solution it must be first fused with potassium hydrate or bisulphate. The alumina made by igniting aluminium hydrate or sulphate is a white powder, easily soluble in acids if the ignition has been gentle, but becoming almost insoluble if the heat has been raised to whiteness. The specific gravity of this ignited alumina also varies with the temperature to which it has been raised; if simply to red heat, it is 3.75; if to bright-redness, 3.8; and if to whiteness, 3.9. In the last case it acquires almost the hardness of corundum. It can be melted to a clear, limpid liquid in the oxyhydrogen blowpipe; after cooling it forms a clear glass, often crystallized. Gaseous chlorine does not act on it even at redness, but if carbon is present at the same time aluminium chloride is formed. Similarly, although neither carbon nor sulphur, alone or mixed together, acts on aluminium, carbon bisulphide converts it into aluminium sulphide.

The preparation of alumina is described at length in the next chapter.

ALUMINIUM HYDRATES.

There are three natural hydrates of aluminium, which may be briefly described as follows :—

Diaspore, formula $Al^2O^3.H^2O$ or $Al^2O^2.(OH)^2$, containing 85 per cent. of alumina, occurs in crystalline masses as hard as quartz, with a specific gravity of 3.4. Bauxite, of the general formula $Al^2O^3.2H^2O$ or $Al^2O.(OH)^4$, with the aluminium replaced by variable quantities of iron. If perfectly pure, it would contain 74 per cent. of alumina. Hydrochloric acid removes from it only the iron, heated with moderately dilute sulphuric acid it gives up its alumina, a concentrated alkaline solution also dissolves the alumina. Calcined with sodium carbonate it forms sodium aluminate without melting. Gibbsite, formula Al^2O^3.-$3H^2O$ or $Al^2(OH)^6$, containing when pure 65 per cent. of alumina, is a mineral generally stalactitic, white, and with a specific gravity of 2.4. It loses two-thirds of its water at 300° and the rest at redness.

The artificial hydrates are of two kinds, the soluble and insoluble modifications. The latter is the common hydrate, such as is obtained by adding ammonia to a solution containing aluminium. The precipitate is pure white, very voluminous, and can be washed free from the salts with which it was precipitated only with great difficulty. Its composition is $Al^2(OH)^6$, corresponding to the mineral gibbsite. It is insoluble in water, but easily soluble in dilute acids or alkali solutions. It dissolves in small quantity in ammonia, but the presence of ammonia salts counteracts this action. When dissolved in caustic potash or soda the addition of ammoniacal salts reprecipitates it. It loses its water on heating, in the same manner as gibbsite. Many other properties of this hydrate, and its manufacture on a large scale, are given in the next chapter. The soluble modification can only be made by complicated processes, too long to be described here, and is principally of use in the dyeing industries; a full description can be found in any good chemical dictionary.

ALUMINATES.

Potassium aluminate.—Formula $K^2Al^2O^4$, crystallizes with 3 molecules of water, the crystals containing 40 per cent. alumina, 37.5 per cent. potassa and 21.5 per cent. of water. It is formed when precipitated alumina is dissolved in caustic potash, or by melting together alumina and caustic potash in a silver dish and dissolving in water. If the solution is evaporated in vacuo, brilliant hard crystals separate out. They are soluble in water but insoluble in alcohol.

Sodium aluminate has not been obtained crystallized. Obtained in solution by dissolving alumina in caustic soda or by fusing alumina with caustic soda or sodium carbonate and dissolving in water. If single equivalents of carbonate of soda and alumina are used, the aluminate seems to have the composition $Na^2Al^2O^4$; if an excess of soda is used, the solution appears to contain $Al^2(ONa)^6$, or $Al^2O.3Na^2O$. If a solution of sodium aluminate is concentrated to 20° or 30° B., alumina separates out; if carbonic acid gas is passed through it, aluminium hydrate is precipitated. For a description of its manufacture on a large scale, see next chapter.

Barium aluminate.—Formula $BaAl^2O^4$. Deville prepared it by calcining a mixture of nitrate or carbonate of barium with an excess of alumina, or by precipitating sulphate of aluminium in solution by baryta water in excess. The aluminate is soluble in about 10 times its weight of water and crystallizes out on addition of alcohol. The crystals contain 4 molecules of water. Gaudin obtained it by passing steam over a mixture of alumina and barium chloride, or of alumina, barium sulphate, and carbon, at a red heat. Tedesco claimed that by heating to redness a mixture of alumina, barium sulphate, and carbon, barium aluminate was extracted from the residue by washing with water. He utilized this reaction further by adding solution of alkaline sulphate, barium sulphate being precipitated (which was used over), while alkaline aluminate remained in solution.

Calcium aluminate.—Lime water precipitates completely a solution of potassium or sodium aluminate, insoluble gelatinous calcium aluminate being formed, of the formula $Al^2(O^6Ca^3)$ or $Al^2O^3.3CaO$. At a red heat it melts to a glass, which, treated after cooling with boiling solution of boric acid, affords a compound appearing to contain $2Al^2O^3.3CaO$. (Tissier.) Lime water is also completely precipitated by hydrated alumina, the compound formed having the composition $CaAl^2O^4$ or $Al^2O^3.CaO$. Also, by igniting at a high temperature an intimate mixture of equal parts of aluminia and chalk, Deville obtained a fused compound corresponding to the formula $CaAl^2O^4$.

Zinc aluminate occurs in nature as the mineral Gahnite, formula $ZnAl^2O^4$. Berzelius has remarked that when a solution of zinc oxide in ammonia and a saturated solution of alumina in caustic potash are mixed, a compound of the two oxides is precipitated, which is redissolved by an excess of either alkali.

Copper aluminate.—On precipitating a dilute solution of sodium aluminate with an ammoniacal solution of copper sulphate, the clear solution remaining contained neither copper nor aluminium. Whether the precipitate contained these combined as an aluminate I was not able to determine.

Magnesium aluminate occurs in nature as Spinell; *iron aluminate* as Hercynite; *beryllium aluminate* as Chrysoberyl. I can find no certain information of their artificial production except in grains at a very high heat.

ALUMINIUM CHLORIDE.

Formula Al^2Cl^6 contains 20.2 per cent. of aluminium. The commercial chloride is often yellow or even red from the presence of iron, but the pure salt is quite white. It absorbs water very rapidly from the air. It usually sublimes without melting, especially when in small quantity, but if a large mass is rapidly heated, it may melt and even boil, but its melting point is very close to its boiling point. Its vapor condenses at 180° to 200°. When sublimed it deposits in brilliant, hexagonal crystals. A current of steam rapidly decomposes it into alumina and hydrochloric acid. Oxygen disengages chlorine from it at redness, but decomposes it incompletely. Potassium or sodium decomposes it explosively, the action commencing below redness. Anhydrous sulphuric acid convérts it into aluminium sulphate. Aluminium chloride combines with many other chlorides, forming the double salts.

On dissolving this salt in water, or by dissolving alumina in hydrochloric acid, a solution is obtained which on evaporation deposits crystals having the formula $Al^2Cl^6.12H^2O$. If these crystals are heated, they decompose, losing both water and acid and leaving alumina. Thus, it is not possible to obtain anhydrous aluminium chloride by evaporating its solution, and the anhydrous salt must be made by other methods, detailed at length in the next chapter.

ALUMINIUM-SODIUM CHLORIDE.

Formula $Al^2Cl^6.2NaCl$ contains 14 per cent. of aluminium. The commercial salt is often yellow or brown from the presence of ferric chloride, but the pure salt is perfectly white. Its melting point has been generally stated to be 180°, but Mr. Baker, chemist for the Aluminium Company, of London, states that when the absolutely pure salt is warmed it melts at 125° to 130°. That chemists should for thirty years have made an error of this magnitude seems almost incredible, and it would be satisfactory if Mr. Baker would advance some further information than the bare statement above. This salt volatilizes at a red heat without decomposition. It is less deliquescent in the air than aluminium chloride, and for this reason is much easier to handle on a large scale. It is recently stated that the absolutely pure salt deteriorates less than the impure salt in the air, and the inference is drawn that perhaps the greater deliquescence of the impure salt is due to the iron chlorides present. Its solution in water behaves similarly to that of aluminium chloride; it cannot be evaporated to dryness without decomposition, the residue consisting of alumina and sodium chloride.

The manufacture of this double salt on a large scale is described in the next chapter. It may be prepared in the laboratory by melting a mixture of the two component salts in the proper proportions. A similar salt with potassium chloride may be prepared by exactly analogous reactions.

ALUMINIUM-PHOSPHORUS CHLORIDE.

Formula $Al^2Cl^6.PCl^5$, contains 9 per cent. of aluminium. It is a white salt, easily fusible, volatilizes only about 400° and sublimes slowly, fumes in the air and is decomposed by water. Produced by heating the two chlorides together or by passing vapor of phosphorus perchloride over alumina heated to redness.

ALUMINIUM-SULPHUR CHLORIDE.

Formula $Al^2Cl^6.SCl^4$, contains 12.2 per cent. of aluminium. It forms a yellow crystalline mass, fuses at 100°, may be distilled without change, and is decomposed by water. May be obtained by distilling a mixture of aluminium chloride and ordinary sulphur chloride, SCl^2.

ALUMINIUM-SELENIUM CHLORIDE.

Formula $Al^2Cl^6.SeCl^4$. Obtained by heating the separate chlorides together in a sealed tube, when on careful distillation the less volatile double chloride remains. It is a yellow mass, melting at 100° and decomposed by water.

ALUMINIUM-AMMONIUM CHLORIDE.

Formula $Al^2Cl^6.3NH^3$. Solid aluminium chloride absorbs ammonia in large quantity, the heat developed liquefying the resulting compound. It may be sublimed in a current of hydrogen, but loses ammonia thereby and becomes $Al^2Cl^6.NH^3$.

ALUMINIUM-CHLOR-SULPHYDRIDE.

Formed by subliming aluminium chloride in a current of hydrogen sulphide. A current of hydrogen removes the excess of the gas used, leaving on sublimation fine colorless crystals. In air it deliquesces rapidly and loses hydrogen sulphide.

ALUMINIUM-CHLOR-PHOSPHYDRIDE.

Apparantly of the formula $3Al^2Cl^6.PH^3$. If phosphuretted hydrogen is passed over cold aluminium chloride very little is absorbed, but at its subliming point it absorbs a large quantity, the combination subliming and depositing in crystals. It is decomposed by water or ammonium hydrate, disengaging hydrogen phosphide.

ALUMINIUM BROMIDE.

Formula Al^2Br^6, containing 10.1 per cent. of aluminium. It is colorless, crystalline, melts at 93° to a clear fluid which boils at 260°. It is still more deliquescent than aluminium chloride. At a red heat in contact with dry oxygen, it evolves bromine and forms alumina ; it is also decomposed slowly by the oxygen of the air. It dissolves easily in carbon bi-sulphide, the solution fuming strongly in the air. It reacts violently with water, the solution on evaporation depositing the compound $Al^2Br^6.12H^2O$. The same result is attained by dissolving alumina in hydrobromic acid and evaporating. This hydrated chloride is decomposed by heat leaving alumina. The specific gravity of solid aluminium bromide is 2.5.

This compound is obtained by heating aluminium and bromine together to redness, or by passing bromine vapor over a mixture of alumina and carbon at bright redness.

ALUMINIUM IODIDE.

Formula Al^2I^6, containing 6.6 per cent. of aluminium. This compound is a white solid, fusible at 125° and boils at 350°. It dissolves easily in carbon bisulphide, the warm saturated solution depositing it in crystals on cooling. It dissolves also in alcohol and ether. Its behavior towards water is exactly analogous to that of aluminium bromide. It is prepared by heating iodine and aluminium together, or by passing iodine vapor over an ignited mixture of alumina and carbon.

ALUMINIUM FLUORIDE.

Formula Al^2F^6, containing 32.7 per cent. of aluminium. It is sometimes obtained in crystals which are colorless and slightly phosphorescent. They are insoluble in acids even in boiling sulphuric, and boiling solution of potash scarcely attacks them; they can only be decomposed by fusion with sodium carbonate at a bright red heat. Melted with boric acid, aluminium fluoride forms crystals of aluminium borate. L. Grabau describes the aluminium fluoride which he obtains in his process, as being a white powder, unalterable in air, unaffected by keeping, insoluble in water, infusible at redness, but volatilizing at a higher temperature.

Deville first produced this compound by acting on aluminium with silicon fluoride at a red heat. He afterwards obtained it by moistening pure calcined aluminium with hydrofluoric acid, drying and introduced into a tube made of gas carbon, protected by a refractory envelope. The tube was heated to bright redness, a current of hydrogen passing through meanwhile to facilitate the volatilization of the fluoride. Brunner demonstrated that aluminium fluoride is formed and volatilized when hydrofluoric acid gas is passed over red hot alumina. Finally, if a mixture of fluorspar and alumina is placed in carbon boats, put into a carbon tube, suitably protected, heated to whiteness and gaseous hydrofluoric acid passed over it, aluminium fluoride will volatilize and condense in the cooler part of the tube in fine cubical crystals, while calcium chloride remains in the boats.

ALUMINIUM FLUORHYDRATE.

When calcined alumina or kaolin is treated with hydrofluoric acid, alumina being in excess, soluble fluorhydrate of aluminium is formed, which deposits on evaporating the solution. It has the formula $Al^2F^6.7H^2O$, and easily loses its water when heated.

ALUMINIUM-HYDROGEN FLUORIDE.

If to a strongly acid solution of alumina in hydrofluoric acid alcohol is added, an oily material separates out and crystallizes, having the formula $3Al^2F^3.4HF.10H^2O$. If the acid solution is simply evaporated, acid fumes escape and a crystalline mass remains which, washed with boiling water and dried, has the formula $2Al^2F^3.HF.10H^2O$. On heating these compounds to 400° or 500° in a current of hydrogen, pure amorphous aluminium fluoride remains. The acid solution of alumina first used seems to contain an acid of the composition $Al^2F^6.6HF$, which is capable of forming salts with other bases. Thus, if this solution is neutralized with a solution of soda, a precipitate of artificial cryolite, $Al^2F^6.6NaF$, falls. The similar potash compound is formed in the same way.

ALUMINIUM-SODIUM FLUORIDE.

Formula $Al^2F^6.6NaF$, containing 12.85 per cent. of aluminium, occurs native as cryolite, a white mineral with a waxy appearance, as hard as calcite, specific gravity 2.9, melting below redness and on cooling looking like opaque, milky glass. If kept melted in moist air, or in a current of steam, it loses hydrofluoric acid and sodium fluoride and leaves a residue of pure alumina. When melted it is decomposable by an electric current or by sodium or magnesium. It is insoluble in water, unattacked by hydrochloric but decomposed by hot sulphuric acid. The native mineral is contaminated with ferrous carbonate, silica, phosphoric, and vanadic acids. An extended description of its utilization, manufacture, etc. will be found in the next chapter.

ALUMINIUM SULPHIDE.

Formula Al^2S^3, containing 36 per cent. of aluminium. The pure salt is light yellow in color and melts at a high temperature. In damp air it swells up and disengages hydrogen sulphide, forming a grayish white powder; it decomposes water very actively, forming hydrogen sulphide and ordinary gelatinous aluminium hydrate. Steam decomposes it easily, at red heat forming amorphous alumina, which is translucent and very hard. Gaseous hydrochloric acid transforms it into aluminium chloride. Elements having a strong affinity for sulphur reduce it, setting free aluminium, but it is doubtful if hydrogen or carburetted hydrogen has this effect.

It may be formed by throwing sulphur into red-hot aluminium, or by passing sulphur vapor over red-hot aluminium. Traces only of aluminium sulphide are formed by passing hydrogen sulphide over ignited alumina, but carbon-bisulphide vapor readily produces this reaction. For details of its formation see next chapter.

ALUMINIUM SELENIDE.

When aluminium is heated in selenium vapor, the two elements combine with incandescence, producing a black powder. In the air this powder evolves the odor of hydrogen selenide; in contact with water it disengages that gas abundantly and furnishes a red deposit of selenium along with aluminium hydrate. When a solution of an aluminium salt is treated with an alkaline polyselenide, a flesh-colored precipitate falls, the composition of which is not known, which is decomposed at redness leaving aluminium.

ALUMINIUM BORIDES.

AlB^2, containing 55.1 per cent. of aluminium, was first obtained by Deville and Wöhler by heating boron in contact with aluminium, or on reducing boric acid with the latter metal, the action not being long continued. Also, if a current of boron trichloride with carbonic oxide is passed over aluminium in boats in a tube heated to redness, aluminium chloride volatilizes and there remains in the boats a crystalline mass, cleavable, and covered with large hexagonal plates of a high metallic lustre. To remove the aluminium present in excess the mass is treated with hydrochloric acid and then by caustic soda. The final residue is composed of hexagonal tablets, very thin but perfectly opaque, of about the color of copper. These crystals do not burn in the air, even if heated to redness, but their color changes to dark-gray. They burn in a current of chlorine, giving chlorides of the two elements contained in them. They dissolve slowly in concentrated hydrochloric acid or in solution of caustic soda; nitric acid, moderately concentrated, attacks them quickly.

AlB^3, containing 45 per cent. of aluminium, has been obtained by Hampe by heating aluminium with boric acid for three hours at a high temperature, carbon being carefully kept away. On cooling very slowly, the upper part of the fusion is composed of aluminium borate, the centre is of very hard alumina containing a few black crystals of aluminium boride, while at the bottom is a button of aluminium also containing these crystals. To free these crystals, the aluminium is dissolved by hydrochloric acid. These crystals are the compound sought for, and contain no other impurity than a little alumina, which can be removed by boiling sulphuric acid. These purified crystals are black, but are thin enough to show a dark-red by transmitted light. Their specific gravity is 2.5, they are harder than corundum, but are scratched by the diamond. Oxygen has no action on them at a high temperature, solution of caustic potash or hydrochloric acid does not attack them, boiling sulphuric acid has scarcely any action, but they dissolve completely in warm, concentrated nitric acid.

If the operation by which this product is made is conducted in the presence of carbon, the compound formed contains less aluminium and also some carbon. Its composition corresponds to the formula $Al^3C^2B^{48}$, containing about 12 per cent. of aluminium and 3.75 per cent. of carbon. The crystals of this compound are yellow and as brilliant as the diamond. Their specific gravity is 2.6, hardness between that of corundum and the diamond. They are not attacked by oxygen, even at a high temperature; hot hydrochloric or sulphuric acid attacks them only superficially,

7

concentrated nitric acid dissolves them slowly but completely. They resist boiling solution of caustic potash or fused nitre, but take fire in fused caustic potash or chromate of lead.

ALUMINIUM NITRIDE.

Formula AlN, containing 66 per cent. of aluminium, is formed when aluminium is heated in a carbon crucible to a high temperature. Mallet* obtained it in quantity by heating aluminium with dry sodium carbonate at a high heat, for several hours, in a carbon crucible. The aluminium is partially transformed into alumina, some sodium vaporizes and some carbon is deposited. After cooling, there are found on the surface of the button little yellow crystals and amorphous drops, to recover which the whole is treated with very dilute hydrochloric acid. This product has the composition AlN. Calcined in the air it slowly loses nitrogen and forms alumina. It decomposes in moist air, loses its transparency, becomes a lighter yellow, and finally only alumina remains, the nitrogen having formed ammonia. Melted with caustic potash it disengages ammonia and forms potassium aluminate.

ALUMINIUM SULPHATE.

Anhydrous.—The salt obtained by drying hydrated aluminium sulphate at a gentle heat has the formula $Al^2(SO^4)^3$, containing 15.8 per cent. of aluminium, and of a specific gravity of 2.67. By heating this salt several minutes over a Bunsen burner it loses almost all its acid, leaving alumina. Hydrogen likewise decomposes it at redness, forming water and sulphur dioxide and leaving alumina with hardly a trace of acid. Melted with sulphur, Violi states that it is transformed into aluminium sulphide, evolving sulphurous acid gas.† Hot hydrochloric acid in excess partly converts it into aluminium chloride.

Hydrated.—This is the ordinary aluminium sulphate; its formula is $Al^2(SO^4)^3.18H^2O$, and it contains 8.4 per cent. of alu-

---

* Ann. der Chemie u. Pharmacie, 186, p. 155.
† Berichte des Deutschen Gesellschaft, X, 293.

minium and 47 per cent. of water. It has a white, crystalline appearance and tastes like alum. It dissolves freely in water, from which it crystallizes out at ordinary temperatures with the above formula; crystallized out at a low temperature it retains $27H^2O$, or one-half as much again. Water dissolves one-half its weight of this salt, the solution reacting strongly acid; it is almost insoluble in alcohol. At a gentle heat it melts in its water of crystallization, then puffs up and leaves a porous mass of anhydrous sulphate which is soluble with difficulty in water. If heated to redness it leaves only alumina. The salt with $18H^2O$ has a specific gravity of 1.76; it is the salt found in fibrous masses in solfataras, its mineralogical name being Halotrichite. A hydrated sulphate with $10H^2O$ is formed and precipitated when alcohol is added to an aqueous solution of aluminium sulphate. On heating it acts similarly to the other hydrated sulphates.

Basic.—On precipitating a solution of aluminium sulphate with alkaline hydrate or carbonate a series of basic salts are formed. On precipitating with ammonia, the compound formed has the formula $Al^2O^3.SO^3.9H^2O$, corresponding to the mineral Aluminite. If the ammonia is in insufficient quantity to entirely precipitate the solution, a precipitate is very slowly formed having the formula $3Al^2O^3.2SO^3.20H^2O$. On precipitating a cold solution of alum by alkaline carbonate not in excess, a precipitate is very slowly formed having the formula $2Al^2O^3.SO^3.12H^2O$. If a very dilute solution of acetate of alumina is precipitated by adding potassium sulphate, a compound deposits very slowly having the formula $2Al^2O^3.SO^3.10H^2O$. Native minerals are met with of analogous composition to these precipitates: Felsobanyte, $2Al^2O^3.SO^3.10H^2O$; Paraluminite, $2Al^2O^3.SO^3.15H^2O$. By heating a concentrated solution of aluminium sulphate with aluminium hydrate, and filtering cold, the solution deposits on further cooling a gummy mass having the formula $Al^2O^3.2SO^3.xH^2O$. On washing with water it deposits a basic salt having the formula $Al^2O^3.SO^3$. By letting stand a very dilute solution of sulphuric acid completely saturated with aluminium hydrate, Rammelsberg obtained transparent crystals having the formula $3Al^2O^3.4SO^3.30H^2O$. On boiling a solution of aluminium sulphate with zinc, Debray obtained a granular precipitate having the formula

$5Al^2O^3.3SO^3.20H^2O$. By leaving zinc a long time in a cold solution of aluminium sulphate, a gelatinous precipitate was obtained having the formula $4Al^2O^3.3SO^3.36H^2O$; the same compound was formed if the zinc was replaced by calcium carbonate.

The manufacture of aluminium sulphate from clay, aluminous earths, cryolite, beauxite, etc. is carried on industrially on a very large scale; descriptions of the processes used may be found in any work on industrial chemistry—they are too foreign to metallurgical purposes to be treated of here.

ALUMS.

Under this name are included a number of double salts containing water, crystallizing in octahedra and having the general formula $R^2SO^4.R^2(SO^4)^3.24H^2O$, in which the first R may be potassium, sodium, rubidium, cäesum, ammonium, thallium or even organic radicals; the second R may be aluminium, iron, manganese or chromium; the acid may even be selenic, chromic or manganic, instead of sulphuric. We will briefly describe the most important alums' consisting of double sulphates of aluminium and another metal, remarking, as with aluminium sulphate, that their preparation may be found at length in any chemical treatise.

POTASH ALUM.

Formula $K^2SO^4.Al^2(SO^4)^3.24H^2O$, containing 10.7 per cent. of alumina or 5.7 per cent of aluminium. Dissolves in 25 parts of water at 0° and in two-sevenths part at 100°. The solution reacts acid. It forms colorless, transparent octahedrons, insoluble in alcohol. On exposure to air they become opaque, being covered with a white coating, which is said not to be efflorescence—a loss of water—but to be caused by absorption of ammonia from the air. The crystals melt in their water of crystallization, but lose it all above 100°. Heated to redness it swells up strongly, becomes porous and friable, giving the product called *calcined alum*; at whiteness it loses a large part of its sulphuric acid, leaving a residue of potassium sulphate and alumina. If it is mixed with one-third its weight of carbon and calcined, the residue inflames spontaneously in the air. If a mixture of alumina and bi-sulphate of potassium is fused and afterwards washed with warm water, a residue is obtained of anhydrous alum, or $K^2SO^4.Al^2(SO^4)^3$. The mineral Alunite is a basic potash alum, $K^2SO^4.3(Al^2O^3.SO^3).6H^2O$.

AMMONIA ALUM.

Formula $(NH^4)^2SO^4.Al^2(SO^4)^3.24H^2O$, containing 11.3 per cent. of alumina or 6.0 per cent. of aluminium. Dissolves in 20 parts of water at 0° and in one-fourth part at 100°. When heated, the crystals swell up strongly, forming a porous mass, losing at the same time water and sulphurous acid; if the temperature is high enough there remains a residue of pure alumina. The temperature necessary for complete decomposition is higher than that required for volatilising ammonium sulphate alone.

SODA ALUM.

Formula $Na^2SO^4.Al^2(SO^4)^3.24H^2O$, containing 11.1 per cent. of alumina or 5.9 per cent. of aluminium. Dissolves in an equal weight of water at ordinary temperatures. The crystals effloresce and fall to powder in the air. It is insoluble in absolute alcohol. On account of its great solubility in water it cannot be separated from ferrous sulphate by crystallization, and therefore it is either contaminated with much iron or else, to be obtained pure, special expensive methods must be adopted. These difficulties cause the manufacture of soda alum to be insignificant in amount when compared with potash alum.

ALUMINIUM-METALLIC SULPHATES.

Sulphate of aluminium forms double sulphates with iron, manganese, magnesium and zinc, but these compounds are not analogous to the alums. They are extremely soluble in water, do not crystallize in octahedrons or any isometric forms, and their composition is different from the alums in the amount of water of crystallization. It has been determined that the double sulphates with manganese and zinc contain 25 equivalents of water, which would permit their being considered as combinations of sulphate of aluminium, $Al^2(SO^4)^3.18H^2O$, with a sulphate of the magnesian series containing seven equivalents of water, as $ZnSO^4.7H^2O$.

ALUMINIUM SELENITES.

By adding a solution of selenite of soda to one of sulphate of aluminium maintained in excess, an amorphous, voluminous precipitate forms having the composition $4Al^2O^3.9SeO^2.3H^2O$. This substance decomposes on being heated, leaving alumina. If varying quantities of selenious acid are added to this first salt, other salts of the formulas $Al^2SO^3.3SeO^2.7H^2O, 2Al^2O^3.9SeO^2.12H^2O$, $Al^2O^3.6SeO^2.5H^2O$ are formed. These are mostly insoluble in water and decompose on being heated, like the first.

ALUMINIUM NITRATE.

Formula $Al^2(NO^3)^6.18H^2O$ is obtained on dissolving aluminium hydrate in nitric acid. If the solution is evaporated keeping it strongly acid, it deposits on cooling voluminous crystals having the formula $Al^2(NO^3)^6.15H^2O$. This salt is deliquescent, melts at 73° and gives a colorless liquid which, on cooling, becomes crystalline. It is soluble in water, nitric acid, and alcohol; on evaporating these solutions it is obtained as a sticky mass. It is easily decomposed by heat; if kept at 100° for a long time it loses half its weight, leaving as residue a soluble salt of the formula $2Al^2O^3.3N^2O^5.3H^2O$. Carried to 140°, this residue loses all its nitric acid, leaving alumina. On this property is based a separation of alumina from lime or magnesia, since the nitrates of these latter bases resist the action of heat much better than aluminium nitrate.

ALUMINIUM PHOSPHATES.

The normal phosphate, $Al^2(PO^4)^2$, is obtained as a white, gelatinous precipitate when a neutral aluminium solution is treated with sodium phosphate. It is soluble in alkalies or mineral acids but not in acetic acid. If a solution of this salt in acids is neutralized with ammonia, a basic phosphate is precipitated having the composition $3Al^2(OH)^3PO^4 + Al^2(OH)^6$. The mineral Wavellite has this composition, with nine molecules of water. The mineral Kalait contains $Al^2(PO^4)^2 + Al^2(OH)^6 + 2H^2O$, and when it is colored azure blue by a little copper it forms the Turquois.

ALUMINIUM CARBONATE.

If to a cold solution of alum a cold solution of sodium carbonate is added drop by drop, stirring constantly until the solution reacts feebly alkaline, a precipitate is obtained which, after being washed with cold water containing carbonic acid gas, contains when damp single equivalents of alumina and carbonic acid, $Al^2O^3.CO^2$. If the precautions indicated are not used, the precipitate contains a very small proportion of carbonic acid.

ALUMINIUM BORATE.

Formula $3Al^2O^3.BO^3$. Prepared by Ebelman by heating together alumina, oxide of cadmium, and boric acid. After three days' heating the platinum capsule containing the mixture was found covered with transparent crystals of the above composition, hard enough to scratch quartz and having a specific gravity of 3. Troost obtained the same substance by heating alumina in the vapor of boron trichloride. Fremy prepared it by heating fluoride of aluminium with boric acid. Ebelman also obtained it by heating a mixture of alumina and borax to whiteness; under these conditions crystals of corundum were formed at the same time.

By precipitating a cold solution of alum with sodium borate, double salts are obtained containing soda, but which leave on washing with warm water two compounds having the formulæ $2Al^2O^3.BO^3.5H^2O$ and $3Al^2O^3.2BO^3.8H^2O$. If the washing is prolonged too far the two salts are completely decomposed, leaving a residue of pure alumina.

ALUMINIUM SILICATES.

Compounds of alumina and silica, or aluminium, silicon, and oxygen are of wide occurrence in nature. In them these bases occur in many proportions; the following proportions of alumina to silica, Al^2O^3 to SiO^2, have been observed: 2–1, 3–2, 1–1, 2–3, 1–2, 1–3, 1–4, 1–8, thus varying from tri-basic silicates to pent-acid silicates. Some are anhydrous, others hydrous; in many of these silicates ferric oxide replaces varying quantities of alumina and an immense number of silicates are known containing other metallic bases besides aluminium. These by their various combinations form the basis of most rocks.

$Al^2O^3.SiO^2$, or Al^2SiO^5, occurs in nature as Disthene, Andalusite, and Fibrolite. They are not attacked by acids, are infusible before the blowpipe, specific gravity 3 to 3.5, and hardness about that of quartz.

$Al^2O^3.2SiO^2.2H^2O$ or $Al^2Si^2O^7 + 2H^2O$ forms kaolin or white china clay, and mixed with various impurities forms the basis of many common clays. It is produced mostly from orthoclase, a feldspar containing silica, alumina, and potash, by the decomposing influence of the atmosphere. The moisture and carbonic acid of the air produce the following reaction—

$$K^2Al^2Si^6O^{16} + 2H^2O + CO^2 = Al^2Si^2O^7.2H^2O + K^2CO^3 + 4SiO^2.$$

Kaolin acquires a certain plasticity when mixed with water. Hydrochloric or nitric acids have no action on it, but cold sulphuric acid dissolves its alumina setting the silica at liberty. It is infusible unless contaminated with particles of feldspar or calcium sulphate, carbonate, or phosphate. The specific gravity of kaolin is 2.3. If fused with six times its weight of caustic potash the resulting mass gives up potassium aluminate when washed with water. An analogous result is obtained with sodium carbonate. Pure kaolin with the formula $Al^2Si^2O^7.2H^2O$ contains 39.4 per cent. of alumina or 20.9 per cent. of aluminium; if it is calcined enough to drive off the water, the residue will contain 45.9 per cent. of alumina or 24.3 per cent. of aluminium.

Common clays contain from 50 to 70 per cent. of silica and 15 to 35 per cent. of alumina, and are not often amenable to a formula. Two of somewhat constant composition have been given the formulæ $Al^2O^3.3SiO^2.4H^2O$ and $Al^2O^3.5SiO^2.3H^2O$.

CHAPTER VI.

PREPARATION OF ALUMINIUM COMPOUNDS FOR REDUCTION.

WE will consider this division under four heads :—

I. Alumina.
II. Aluminium chloride and aluminium-sodium chloride.
III. Aluminium fluoride and aluminium-sodium fluoride.
IV. Aluminium sulphide.

I.

THE PREPARATION OF ALUMINA.

We will treat this subject in three divisions :—

1. From Aluminium Sulphate or Alums.
2. From Beauxite.
3. From Cryolite.

1. PREPARATION OF ALUMINA FROM ALUMS OR ALUMINIUM SULPHATE.

Hydrated alumina can be precipitated from a solution of any aluminium salt by ammonium hydrate, an excess of which re-dis-solves a portion. Its chemical formula is ordinarily written $Al^2O^3.3H^2O$ or $Al^2(OH)^6$. The aluminium hydrate thus precipitated is a pure white, very voluminous, almost pasty mass, very hard to wash. By boiling and washing with boiling water it becomes more dense, but always remains very voluminous. Washing on a fil-ter with a suction apparatus gives the best results. At a freezing temperature this hydrate changes into a dense powder which is more easily washed. On drying it shrinks very much in volume and forms dense, white pieces, transparent on the edges. When dried at ordinary temperatures it has the composition $Al^2O^3.H^2O$.

On ignition, the other molecule of water is driven off, leaving anhydrous alumina.   After gentle ignition it remains highly hygroscopic, and in a very short time will take up from the air 15 per
cent. of water.   In this condition it is easily soluble in hydrochloric or
sulphuric acid.   On stronger ignition it becomes harder and soluble
only with difficulty in concentrated acid ; after ignition at a high
temperature it is insoluble, and can only be brought into solution
again by powdering finely and fusing with potassium acid-sulphate or alkaline carbonate.   At ordinary furnace temperatures
alumina does not melt, but in the oxy-hydrogen blow-pipe or
the electric arc it fuses to a limpid liquid and appears crystalline
on cooling.

The precipitation in aqueous solution and subsequent ignition
is not economical enough to be practised on a large scale, and for
industrial purposes the aluminium sulphate or alum is ignited
directly.   About the easiest way to proceed is to take ammonia
alum crystals, put them into a clean iron pan and heat gently, when
the salt melts in its water of crystallization.   When the water
has evaporated, a brittle, shining, sticky mass remains, which on
further heating swells up and decomposes into a dry, white
powder.   This is let cool, powdered, put into a crucible and heated
to bright redness.   All the ammonia and almost all the sulphuric
acid are thus removed.   The rest of the acid can be removed by
moistening the mass with a solution of sodium carbonate, drying
and again igniting ; on washing with water the acid is removed
as sodium sulphate.   The residue, however, will contain some
caustic soda, which for its further use in making aluminium chloride is not harmful.   Potash alum can be treated in a similar way,
the potassium sulphate being washed away after the first ignition.   Still more easily and cheaply can alumina be made by igniting a mixture of 4 parts aluminium sulphate and 1 of sodium
carbonate.   On washing, sodium sulphate is removed from the alumina.*

Deville used the following method at Javel : Ammonia alum
or even the impure commercial aluminium sulphate was calcined,
the residue appearing to be pure, white alumina, but it still con-

* Kerl and Stohman, 4th Ed. p. 739.

tained sulphuric acid, potassium sulphate, and a notable proportion of iron. This alumina is very friable, and is passed through a fine sieve and put into an iron pot with twice its weight of solution of caustic soda of 45 degrees. It is then boiled and evaporated, and the alumina dissolves even though it has been strongly calcined. The aluminate of soda produced is taken up in a large quantity of water, and if it does not show clear immediately a little sulphuretted hydrogen is passed in, which hastens the precipitation of the iron. The liquor is let stand, the clear solution decanted off and subjected while still warm to the action of a stream of carbonic acid gas. This convers the soda into carbonate and precipitates the alumina in a particularly dense form which collects in a space not one-twentieth of the volume which would be taken up by gelatinous alumina. This precipitate is best washed by decantation, but a large number of washings are necessary to remove all the sodium carbonate from it; it is even well, before finishing the washing, to add a little sal-ammoniac to the wash-water in order to hasten the removal of the soda. The well-dried alumina is calcined at a red heat.

*Tilghman decomposes commercial sulphate of alumina, $Al^2(SO^4)^3.18H^2O$, by filling a red-hot fire-clay cylinder with it. This cylinder is lined inside with a magnesia fettling, is kept at a red heat, the sulphate put in in large lumps, and steam is passed through the retort, carrying with it vapor of sodium chloride. This last arrangement is effected by passing steam into a cast-iron retort in which the salt named is kept melted, and as the steam leaves this retort it carries vapor of the salt with it. It is preferable, however, to make a paste of the sulphate of alumina and the sodium chloride, forming it into small hollow cylinders, which are well dried, and then the fire-clay cylinder filled with these. Then, the cylinder being heated to whiteness, highly superheated steam is passed over it. The hydrochloric acid gas which is formed is caught in a condensing apparatus, and there remains a mass of aluminate of soda, which is moistened with water and treated with a current of carbon dioxide and steam. By washing

* Mierzinski.

the mass, the soda goes into solution and hydrated alumina remains, which is washed well and is ready for use.

Mr. Webster's process for making pure alumina at a low price is now incorporated as a part of the Aluminium Co. Ld.'s processes, but whether it is that there have been no advances made in this line during the past few years or that valuable advances have been made but are sedulously kept secret, I am unable to say. The late descriptions of the Deville-Castner processes all commence with the sentiment: In the beginning we have aluminium hydrate. Such being the case, the only description we can give of Webster's process is one dated 1883.

*Three parts of potash alum are mixed with one part of pitch, placed in a calcining furnace and heated to 200° or 250°. About 40 per cent. of water is thus driven off, leaving sulphate of potash and aluminium, with some ferric oxide. After heating about three hours, the pasty mass is taken out, spread on a stone floor and when cold broken to pieces. Hydrochloric acid (20 to 25 per cent.) is poured upon these pieces, placed in piles, which are turned over from time to time. When the evolution of sulphuretted hydrogen has stopped, about five per cent. of charcoal-powder or lampblack, with enough water to make a thick paste, is added. The mass is thoroughly broken up and mixed in a mill, and then worked into balls of about a pound each. These are bored through to facilitate drying, and heated in a drying chamber at first to 40°, then in a furnace from 95° up to 150°. The balls are then kept for three hours at a low red heat in retorts while a mixture of two parts steam and one part air is passed through, so that the sulphur and carbon are converted into sulphurous oxide and carbonic oxide, and thus escape. The current of gas carries over some potassium sulphate, ferrous sulphate, and alumina, and is therefore passed through clay condensers.

The residue in the retorts consists of alumina and potassium sulphate; it is removed, ground to fine powder in a mill, treated with about seven times its weight of water, boiled in a pan or boiler by means of steam for about one hour, then allowed to stand till cool. The solution containing the potassium sulphate

* Austrian Patent, Sept. 28, 1882 ; English patent, No. 2580, 1881. Dingler, 1883, vol. 259, p. 86.

is run off and evaporated to dryness, the alumina is washed and dried. The potassium sulphate, as a by-product, is said to pay one-half the cost of the process.

This deposit contains about 84 per cent. of alumina, while that obtained by the old process of precipitation has only 65 per cent. Thus a large saving is effected in cost and 19 per cent. more alumina is obtained. In addition to this, the whole of the by-products are recovered, consisting of potassium sulphate, sulphur (which is used in making sulphuric acid), and aluminate of iron.

2. PREPARATION OF ALUMINA FROM BEAUXITE.

At Salindres, the alumina used in the Deville process is obtained from beauxite by the following processes, which are in general use for extracting pure alumina from this mineral.* Beauxite is plentiful enough in the south of France, principally in the departments of Herault, Bouches-du-Rhone, and Var. It contains at least seventy-five per cent. alumina. To separate the alumina from ferric oxide, it is treated with carbonate of soda, under the influence of a sufficiently high temperature, the alumina displacing the carbonic acid and forming an aluminate of soda, $Al^2O^3.3Na^2O$, while the ferric oxide remains unattacked. A simple washing with water then permits the separation of the former from the insoluble ferric oxide. The beauxite is first finely pulverized by means of a vertical mill-stone, then intimately mixed with some sodium carbonate. The mixture is made, for one operation, of—

    480 kilos. beauxite.
    300   "   sodium carbonate of 90 alkali degrees.

This mixture is introduced into a reverberatory furnace, resembling in form a soda furnace, and which will bear heating strongly. The mass is stirred from time to time, and it is kept heated until all the carbonate has been attacked, which is recognized by a test being taken which does not effervesce with acids. The operation lasts from five to six hours.

The aluminate thus obtained is separated from ferric oxide by a washing with warm water. This washing is made at first with

* Fremy's Ency. Chimique.

a feeble solution which has served for the complete exhaustion of the preceding charge, which was last washed with pure water, forming thus this feeble solution. This gives, on the first leaching, solutions of aluminate concentrated enough to be called strong liquor, which are next treated by the current of carbonic acid gas to precipitate the hydrated alumina. The charge is next washed with pure water, which completely removes the aluminate; this solution is the weak liquor, which is put aside in a special tank, and used as the first leaching liquor on the next charge treated. This treatment takes place in the following apparatus (see Fig. 1): *B* is a sheet-iron vessel, in the middle of which is a metallic

Fig. 1.

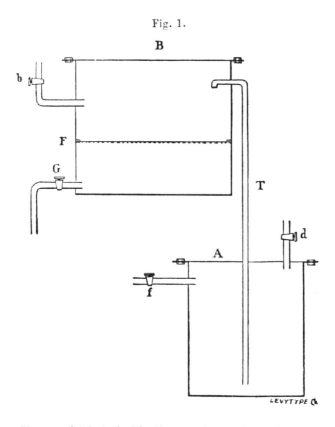

grating, *F*, on which is held all round its edges, by pins, a cloth, serving as a filter. The upper part of this vessel is called simply the filter. *A* ought to be closed by a metallic lid held on firmly by bolts. To work the apparatus, about 500 kilos of the charge to be washed is placed on the filter cloth, the lid is closed, then the steam-cock f of the reservoir A is opened.   In A is the weak solution from the last washing of the preceding charge. The pressure of the steam makes it rise by the tube T into the filter; another jet of steam, admitted by the cock b, rapidly warms the feeble liquor as it soaks into the charge.   After filtering through, the strong liquor is drawn off by turning the stopcock G.   The weak solution of the reservoir A is put into the filter in successive portions, and not all at once; and after each addition of solution has filtered through, its strength in B.° is taken, before any more solution is run in; then, when the solution marks 3° to 4°, it is placed in a special tank for weak liquor, with all that comes through afterwards.   Just about this time, the weak liquor of the reservoir A is generally all used up, and is replaced by pure water introduced by the tube d.   All the solutions which filtered through, marking over 3° to 4° B., are put together, and form the strong liquor which marks about 12° B. This extraction of the aluminate being completed by the pure water, the residue on the filter is taken out, and a new operation may be commenced.

The strong liquor is introduced into a vessel having an agitator, where a strong current of carbonic acid gas may precipitate the alumina from it.   The gas is produced by small streams of hydrochloric acid continuously falling on some limestone contained in a series of earthenware jars.   The precipitation vessel is called a baratte.   The carbonic acid after having passed through a washing flask, is directed to a battery of three barattes, where the precipitation is worked methodically, so as to precipitate completely the alumina of each baratte, and utilize at the same time all the carbon dioxide produced.   In order to do this, the gas always enters first into a baratte in which the precipitation is nearest completion, and arrives at last to that in which the solution is freshest.   When the gas is not all absorbed in the last baratte, the first is emptied, for the precipitation in it is then completed, and it is made the last of the series, the current being now directed first into the baratte which was previously second, while the newly charged one is made the last of the series.   The process is thus kept on continuously.   The apparatus used is shown in Fig. 2.

Each baratte holds about 1200 litres of solution, and the complete precipitation of all the alumina in it takes five to six hours. A mechanical agitator stirs the contents continually, and a current of steam is let into the double bottom so as to keep the tempera-

Fig. 2.

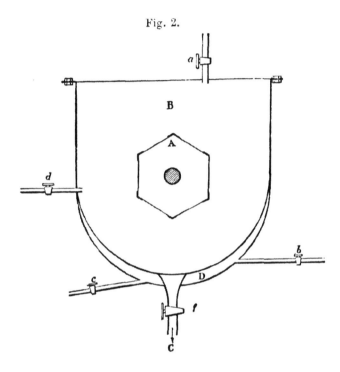

a. Charging pipe. *b.* Steam pipe. *c.* Steam drip. *d.* CO$_2$ enters. *f.* Discharge pipe. *A.* Agitator, made of iron rods. *C.* Tank in which the precipitate settles. *B.* Baratte body. *D.* Steam jacket.

ture of the solution about 70°. The precipitated alumina and the solution of sodium carbonate which remain are received in a vat placed beneath each baratte. The solution is decanted off clear, after standing, and then evaporated down to dryness, regenerating the sodium carbonate used in treating the beauxite to make the aluminate, less the inevitable losses inseparable from all industrial operations. The deposit of alumina is put into a conical strainer to drain, or else into a centrifugal drying machine, which rapidly drives out of the hydrated alumina the solution of sodium carbonate which impregnates it; a washing with pure water in the drier itself terminates the preparation of the alumina. At the works at Salindres, a part of this alumina is converted into sulphate of alumina, which is sold, the remainder being used for the aluminium manufacture. After washing in the dryer, the alumina presents this composition :—

Alumina	47.5
Water	50.0
Sodium carbonate	2.5

Behnke* produces alumina by igniting beauxite or a similar mineral with sodium sulphate, carbon, and ferric oxide, using for each equivalent of alumina present at least one equivalent of alkali and one-half an equivalent of ferric oxide. The mixture is heated in a muffle or reverberatory furnace. The fritted product is ground, exposed to the air, and washed with water. Sodium aluminate goes into solution along with some sodium sulphate, while ferrous sulphide and undecomposed material remains as a residue. By passing carbonic acid gas or gases from combustion through the solution, the alumina is precipitated. The residue spoken of is roasted, the sulphurous oxide given off utilized, and the residue used over in place of fresh ferric oxide.

R. Liebert† proposes to treat beauxite, aluminous iron ore, etc. in a somewhat similar way. These materials are to be ground fine, mixed with sodium chloride and magnesium sulphate (Kieserite), moistened with water, and pressed into bricks or balls. These are dried and put into a retort heated red-hot by generator gas. Hydrochloric acid gas is first given off, sodium sulphate and magnesium chloride being formed. In a further stage of the process sulphurous oxide is evolved, the alumina reacting on the sodium sulphate to form sodium aluminate. The latter is washed out of the residue, and its alumina precipitated by the ordinary methods.

H. Muller‡ proposes to extract the alumina from silicates containing it by mixing them with limestone, dolomite, or magnesite, also with alkali caustic, carbonate, or sulphate (in the last case also with carbon), and heating the mixture to bright-redness.

---

* German Patent (D. R. P.), No. 7256.
† German Patent (D. R. P.), No. 5670.
‡ German Patent (D. R. P.), No. 12,947.

Alkaline aluminate is washed out of the resulting mass, while the residue, consisting of lime, magnesia, iron oxide, etc., is mixed with water-glass and moulded into artificial stone.

Common salt is said not to react on beauxite if fused with it alone, but will decompose it if steam is used. Tilghman[*] first used this reaction in 1847. It is said that it was also used at Nanterre and Salindres previously to 1865. A mixture of sodium chloride and beauxite was treated in a closed retort and steam passed through, or, better, in a reverberatory furnace and steam passed over it, at a high temperature. Much sodium chloride must have been lost by the latter arrangement. The fused mass was treated with water, when sodium aluminate dissolved out.

R. Wagner[†] proposed to make a solution of sodium sulphide, by reducing sodium sulphate by carbon bisulphide, and to boil the beauxite in it. The sulphuretted hydrogen evolved was to be absorbed by ferric hydrate; while the sodium aluminate was converted into soda and alumina by any of the ordinary methods.

According to Löwig's experiments, solution of sodium aluminate can be precipitated by calcium, barium, strontium, or magnesium hydrates, forming caustic soda and hydrated alumina, the latter being precipitated, together with lime, baryta, strontia, or magnesia. The precipitate is washed by decantation and then divided into two portions, one of which is dissolved in hydrochloric acid, the other made into a mush with water and gradually added to the solution of the first half until the filtrate shows only a very little alumina in solution. Chloride of calcium, barium, strontium, or magnesium has been formed, and the alumina all precipitated.

Dr. K. J. Bayer has made an improvement in the process of extracting alumina from beauxite, which has received great commendation from those directly interested in the business, and who may be supposed to have proved its merits. Dr. Bayer thus describes it :[‡] Beauxite is fused with sodium carbonate or sulphate, and the solution obtained by washing, containing sodium

* Polytechnisches Journal, 106, p. 196.
† Wagner's Jahresb., 1865, p. 332.
‡ Stahl und Eisen, Feb. 1889, p. 112.

aluminate, is not decomposed by carbonic acid as formerly, but by the addition of aluminium hydrate with constant stirring. The decomposition of the solution goes on until the quantity of alumina remaining in solution is to the sodium protoxide as 1 to 6. This precipitation takes place in the cold, and the pulverulent aluminium hydrate separated out is easily soluble in acids. The alkaline solution remaining is concentrated by evaporation, taken up by ground beauxite, dried, calcined, and melted, and thus goes through the process again. The use of this caustic soda solution containing alumina is thus much more profitable than using soda, because by using the latter only 75 per cent. of the beauxite used is utilized, whereas by the former all the alumina dissolved by the solution is obtained again.

3. Preparation of Alumina from Cryolite.

By the dry way. — The following method was invented by Julius Thomson; the description is taken principally from Mierzinski's "Fabrikation des Aluminiums:" The cryolite is pulverized, an easy operation, and to every 100 parts, 130 to 150 parts of chalk are added, and a suitable quantity of fluorspar is also used, which remains in the residue on washing after ignition. More chalk is used than is theoretically necessary, in order to make the mass less fusible and keep it porous. But, to avoid using too much chalk merely for this purpose, a certain quantity of coke may be put into the mixture. It is of the first importance that the mixture be very intimate and finely pulverized. It is of greater importance that the mixture be subjected to just the proper well-regulated temperature while being calcined. The cryolite will melt very easily, but this is to be avoided. On this account, the calcination cannot take place in an ordinary smelting furnace, because, in spite of stirring, the mass will melt at one place or another, while at another part of the hearth it is not even decomposed, because the heat at the fire-bridge is so much higher than at the farther end of the hearth. Thomson constructed a furnace for this special purpose (see Figs. 3 and 4), in which the flame from the fire first went under the bed of the furnace, then over the charge spread out on the bed, and finally in

a flue over the roof of the hearth. The hearth has an area of
nearly 9 square metres, being 4 metres long and 2.5 metres wide.

Fig. 3.

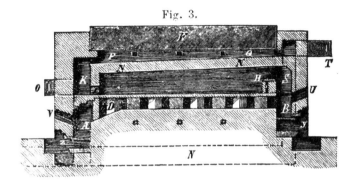

It is charged twelve times each day, each time with 500 kilos of
mixture, thus roasting 6000 kilos daily, with a consumption of
800 kilos of coal. The waste heat of the gases escaping from

Fig. 4.

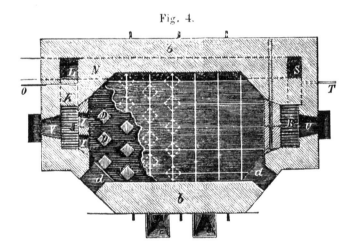

the furnace is utilized for drying the soda solution to its crystallizing point, and the gases finally pass under an iron plate on
which the chalk is dried. In this furnace the mass is ignited
thoroughly without a bit of it melting, so that the residue can be
fully washed with water.

The decomposition takes place according to the formula—

$$Al^2F^6.6NaF + 6CaCO^3 = Al^2O^3.3Na^2O + 6CaF^2 + 6CO^2,$$

the resultant product containing aluminate of soda, soluble in water, and insoluble calcium fluoride. The reaction commences at a gentle heat, but is not completed until a red heat is reached. Here is the critical point of the whole process, since a very little raising of the temperature above a red heat causes it to melt. However, it must not be understood that the forming of lumps is altogether to be avoided. These lumps would be very hard and unwork-able when cold, but they can be broken up easily while hot, so that they may be drawn out of the furnace a few minutes before the rest of the charge is removed, and broken up while still hot without any trouble. The whole charge, on being taken out, is cooled and sieved, the hard lumps which will not pass the sieve are ground in a mill and again feebly ignited, when they will become porous and may be easily ground up. However, the formation of these lumps can be avoided by industrious stirring of the charge in the furnace. A well-calcined mixture is porous, without dust and without lumps which are too hard to be crushed between the fingers. We would here remark that mechanical furnaces of similar construction to those used in the manufacture of soda, potash, sulphate of soda, etc., are more reliable and give the best results if used for this calcination. The mixture, or ashes, as the workmen call it, is drawn still hot, and washed while warm in conical wooden boxes with double bottoms, or the box may have but one bottom, with an iron plate about 76 milli-metres above it. A series of such boxes, or a large apparatus having several compartments, may be so arranged that the wash-ing is done methodically, *i. e.*, the fresh water comes first in con-tact with a residue which is already washed nearly clean, and the fresh charge is washed by the strong liquor. This is known as the "Lessiveur méthodique," and an apparatus constructed espe-cially for this purpose is described in Dingler 186, 376, by P. J. Havrez, but the subject is too general and the description too long to be given here. A very suitable washing apparatus is also that of Schank, used in the soda industry for washing crude soda, and described in "Lunge's Handbook of the Soda Industry," Book II. p. 410. Since the ashes are taken warm from the fur-nace the washing water need not be previously heated, but the final wash-water must be warmed, as the ashes have been cooled down by the previous washings. As soon as the strong liquor does not possess a certain strength, say 20° B., it is run over a fresh charge and so brought up. The solution contains sodium aluminate.

The carbon dioxide necessary for precipitating the hydrated alumina may be made in different ways. The gases coming from the furnace in calcining the cryolite might be used if they were not contaminated with dust; and there is also the difficulty that exhausting the gases from the furnace would interfere with the calcination. It has also been recommended to use the gases from the fires under the evaporating pans, by exhausting the air from the flues and purifying it by washing with water. This can only be done where the pans are fired with wood or gas. However, the lime-kiln is almost exclusively used to furnish this gas. The kiln used is shaped like a small blast furnace. Leading in at the boshes are two flues from five fire-places built in the brickwork of the furnace, and the heat from these calcines the limestone. The gases are taken off by a cast-iron down-take at the top. At the bottom of the furnace, corresponding with the tap hole in a blast furnace, is an opening, kept closed, from which lime is withdrawn at intervals. A strong blast is blown in just above the entrance of the side flues, and by keeping up a pressure in the furnace, leakings into it may be avoided. The gas is sucked away from the top by a pump, which forces it through a cleaning apparatus constructed like a wash-bottle, and it is then stored in a gasometer. Instead of the pump, a steam aspirator may be used, which is always cheaper and takes up less room.

The precipitation with carbonic acid gas is made by simply forcing it through a tube into the liquid. The apparatus used at Salindres is one of the most improved forms. (See p. 112.) The precipitate is granular, and settles easily. However, it is not pure hydrated alumina, but a compound of alumina, soda, carbonic acid, and water, containing usually about—

Alumina	45 per cent.
Sodium carbonate	20 "
Water	35 "

The sodium carbonate can be separated by long-continued boiling with water, but by this treatment the alumina becomes very gelatinous and very difficult of further treatment. The precipitate was formerly separated on linen filters, but centrifugal machines are now preferred. The evaporated solution gives a high grade of carbonate of soda free from iron. The heavy residue which is left after the ashes have been lixiviated consists of calcium fluoride with small quantities of ferric oxide, lime, undecomposed cryolite, and aluminate of soda, and has not been utilized for any purpose.

R. Biederman* states that if steam is passed over molten cryolite at a white heat, hydrofluoric acid gas and sodium fluoride are formed and driven over, while a white, pure crystalline mass of alumina remains.

Utilization of aluminous fluoride slags.—At Nanterre, Deville used the following process for utilizing in one operation the slags from the aluminium manufacture and the residues from the sodium manufacture.

"The slags from making aluminium contain 60 per cent. of sodium chloride and 40 per cent. of insoluble matter; the former can be removed by a single washing. The insoluble material is almost entirely aluminium fluoride, with a little alumina and undecomposed cryolite. When fluorspar is used as a flux, the sodium chloride in the slag is in part replaced by calcium chloride; but, in general, all the fluorine in the slag is found combined with the aluminium, which shows the great affinity between these two elements. The residues left in the sodium retorts deteriorate quickly when exposed to the air, and contain ordinarily, according to my analysis—

Carbon	20.0
Carbonate of soda	14.5
Caustic soda	8.3
Sulphate of soda	2.4
Carbonates of lime and iron	29.8
Water	25.0
	100.0

"To utilize these two materials, 5 to 6 parts of the sodium residues are mixed carefully with one part of the washed slag, and the whole calcined at a red heat. The fusion becomes pasty;

* Kerl and Stohman, 4th Ed., p. 819.

it is cooled and washed, when aluminate of soda goes into solution and on treatment with carbon dioxide gives sodium carbonate and alumina.　According to my laboratory experiments—

1000 grammes of sodium residues
160　　"　　" washed slags

have given

110　　"　　" calcined alumina
225　　"　　" dry sodium carbonate.

" The residue left on washing the fusion weighs about one-half the weight of the soda residues used, and contains—

Carbon	30.0
Calcium fluoride	32.0
Alumina	0.6
Various other materials	37.4

" The latter item is formed of ferric oxide, oxide of manganese, a little silica and some oxysulphide of calcium."

Decomposition of cryolite in the wet way.—Deville used the following method at Javel, which he thus describes :—

" In the Greenland cryolite there are to be found numerous pieces containing siderite (ferrous carbonate). It is necessary to extract all these pieces before using the mineral as a flux in producing the aluminium. The rejected fragments are then utilized by pulverizing them finely, mixing with about three-fourths of their weight of pure, burnt lime and the whole carefully slaked. After the slaking, water is added in large quantity, and the material is heated in a large cast-iron vessel by means of a steam-coil. A reaction takes place at once, and is complete if the process is well conducted. Some insoluble aluminate of lime may be formed, but it can be recovered from the residue by digesting it with some solution of carbonate of soda. The residue remaining is calcium fluoride, which settles easily, and the clear liquor decanted off contains aluminate of soda, from which alumina can be precipitated as before. The calcined alumina obtained may contain iron when the cryolite used contains a large amount of ferrous carbonate. It has appeared to me that the latter mineral may be decomposed by the lime, and some protoxide of iron be thus dissolved by the soda in small quantity.

"We make alumina by this method at Nanterre only because it utilizes the impure pieces of cryolite and works in conveniently with the previously-described processes for utilizing the slags."

An ingenious modification of the above process was devised by Sauerwein. The first reaction is the same, five parts of finely-powdered cryolite being boiled with four parts of burnt lime, as free as possible from iron, producing a solution of sodium aluminate and a residue of insoluble calcium fluoride. Tissier recommended using two parts of cryolite to one of lime, but with these proportions only about one-third of the aluminium in the cryolite is converted into soluble aluminate. Hahn claims that complete decomposition takes place by using 100 parts of cryolite to 88 parts of burnt lime. The solution is settled, washed by decantation, and these washings put with the strong solution first poured off; the next washings are reserved for the fresh wash-water of another operation. The solution of sodium aluminate is then boiled with a quantity of cryolite equal to the amount first used, when sodium fluoride is formed and alumina precipitated. This operation is in no way difficult, only requiring a little more attention than the first. The alumina thus made is very finely divided. The reactions involved are :—

$$Al^2F^6.6NaF + 6CaO = Al^2O^3.3Na^2O + 6CaF^2.$$
$$Al^2F^6.6NaF + Al^2O^3.3Na^2O + 6H^2O = 2(Al^2O^3.3H^2O) + 12NaF.$$

During this last operation it is best to add an excess of cryolite, and keep the liquid in motion to prevent that mineral from caking at the bottom. Lead is the best material to make these precipitating tanks of, since iron would contaminate the alumina. The precipitate is washed as in the previous operation. The solution of sodium fluoride is boiled with the requisite quantity of burnt lime, which converts it into caustic soda, NaOH, which is separated from the precipitated calcium fluoride by decantation and washing.

In the establishment of Weber, at Copenhagen, where at one time all the cryolite produced in Greenland was received, the mineral was decomposed by acid. Hydrochloric acid attacks the mineral slowly, but sulphuric acid immediately dissolves the sodium fluoride, with disengagement of hydrofluoric acid; gelatinous aluminium fluoride separates out and is attacked more slowly. The cryolite requires nearly $1\frac{1}{2}$ parts of sulphuric acid to dissolve it, the reaction being—

$$Al^2F^6.6NaF + 6H^2SO^4 = Al^2(SO^4)^3 + 3Na^2SO^4 + 12HF.$$

The solution is evaporated and crystallized, when the sodium sulphate crystallizes out and the mother liquor is treated for its alumina. This method is too costly when compared with more recent processes to be used at present.

According to Schuch* very finely-powdered cryolite is dissolved by a large excess of hot dilute soda solution, but is thrown down unaltered when carbonic acid gas is passed through the solution. An excess of concentrated soda liquor converts the mineral into sodium aluminate and sodium fluoride, the former being soluble but the latter almost insoluble in the soda solution.

II.

The Preparation of Aluminium Chloride and Aluminium-Sodium Chloride.

Anhydrous aluminium chloride cannot be made by evaporating the solution of alumina in hydrochloric acid, for, as we have seen, decomposition of the salt sets in, hydrochloric acid is evolved and alumina remains. The same phenomena occur on evaporating a solution of the double chloride. These anhydrous chlorides are prepared by a method discovered by Oerstedt, applicable to producing a number of similar metallic chlorides, which consists in passing a current of dry chlorine gas over an ignited mixture of alumina and carbon.

Wöhler† proceeded as follows in preparing the aluminium chloride which was used in his early experiments : "Alumina is mixed with charcoal powder and made plastic with oil. Cylinders of about 5 millimetres diameter are made of this paste, placed in a crucible with charcoal powder and heated until no more combustible gases distil. After cooling the crucible the

* Polytechnisches Journal, 165, p. 443.
† Pogg. Ann., 11, p. 146.

cylinders are taken out, and a porcelain or glass tube open at both ends filled with them. This is then placed in a combustion furnace, connected at one end with a chlorine generator, and at the other with a tubular extension from the further end of which the gases escape, either into the air or into a flask filled with milk of lime. When the whole apparatus is ready, and filled with well-dried chlorine gas, the tube is heated to glowing, when aluminium chloride is formed and condenses in the extension of the tube."

Deville paid great attention to the production and purification of aluminium chloride; the following is his account of the processes used at Javel :—

Manufacture on a small scale.—"I took 5 kilos of alumina and mixed it with 2 kilos of carbon and a little oil; the paste was made into balls and ignited at a bright-red heat. The compact, coke-like mass resulting was broken in pieces and put, with its powder, into a stoneware retort, *C* (Fig. 5), having a capacity of

Fig. 5.

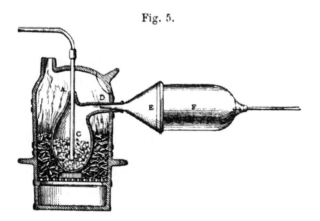

about 10 litres, and terminating in a neck, *D*. This retort was put in a furnace and heated to redness, while a current of dry chlorine gas passed in by the tube *A*. During the first few moments considerable quantities of water vapor escape from the neck. When aluminium chloride distils, as is shown by dense, white fumes, a porcelain or stoneware funnel, *E*, is adjusted to the neck *D*, and kept in place by filling the joint with fine asbestos and then luting it over with a little potter's clay mixed with hair.

Against this funnel fits a globular vessel, *F*, the joint being made tight in a similar way. This apparatus condenses and holds all the chloride distilled. However fast the chlorine may pass into the retort, it is so well absorbed during three-fourths of the operation that not a trace is mixed with the carbonic oxide escaping. However, the gas always fumes a little because of a small quantity of silicon chloride being formed by the chlorine and carbon attacking the sides of the retort, or from chloride of sulphur or a little chloroxycarbonic acid. When the globe *F* is filled it is taken away to extract the coherent, crystalline aluminium chloride it contains, and is replaced immediately by another. During one operation three jars were thus filled, and altogether a little over 10 kilos of chloride obtained. In the retort there remained almost a kilo of coke mixed with alumina in the proportion of two of carbon to one of alumina, making 330 grammes of the latter remaining unattacked out of 5 kilos. This coke contains also some double chloride of alumina and potassium and a little calcium chloride, which render it deliquescent. This residue was washed, mixed with a fresh quantity of alumina, and employed in a new operation."

Manufacture on a large scale.—" In applying this process on a large scale, the oil and carbon were replaced by tar, the alembic by a gas-retort and the glass receiver by a small brick chamber lined with glazed tiles. The alumina was obtained by calcining ammonia alum in iron pots; the residue obtained by one calcination at a bright-red heat was mixed with pitch, to which a little charcoal dust was added. The paste was well mixed, introduced into iron pots, covered carefully and heated until all vapors of tar ceased burning. The aluminous carbon is used while it is still hot, if possible, as it is quite hygroscopic. (This aluminous carbon conducts electricity wonderfully well; it is the best electrode to use in making aluminium by the battery, since its alumina regenerates the bath.) The residue is hard, porous, and cracked, and contains sulphur from the sulphuric acid of the alum, a little iron, phosphoric acid in small quantity, a perceptible proportion of lime, and finally potash, which is always present in alums made from clay. The chlorine gas used was conducted by lead pipes and passed over calcium chloride before being used. The

retort used was of about 300 litres capacity, and was placed vertically in a sort of chimney, C (Fig. 6), the flame circulating all
around it. In the bottom was a square opening, x, about 20
centimetres square, which could be closed by a tile kept in place

Fig. 6.

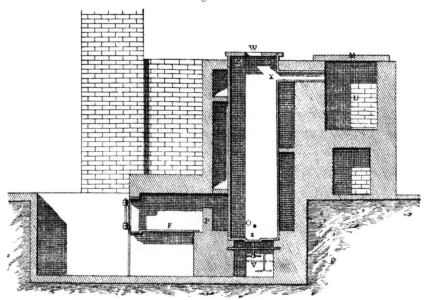

by a screw, V. A porcelain tube pierced the sides of the furnace
and entered the retort at O; it was protected from the flame by a
fire-clay cylinder inclosing it. At the top, the retort was closed
by a tile, Z, of refractory brick, in the centre of which was made
a square opening, W, of 10 to 12 centimetres side. Finally, an
opening, X, placed 30 centimetres below the plate Z, gave issue
to the vapors distilled, conducting them into the chamber L.
This condension chamber was about 1 metre cube; it had one
wall of bricks in common with the furnace, thus keeping it rather
hot. The other walls should be thin and set with close joints
and very little mortar. The cover, M, was movable; it and the
sides of the chamber were of glazed tiles. An opening 20–30
centimetres square in the lower part of the chamber communicated
with flues lined with lead, for a little chloride was drawn into
them. The uncondensed gas passed to a chimney.

"To work such an apparatus it is necessary, first of all, to dry

it with the greatest care in all its parts, especially the condensation chamber. The retort is slowly heated and is left open at Z
until it is judged quite dry, and is then filled with red-hot, freshly
calcined mixture of carbon and alumina. The top cover is then
replaced and the fire urged until the retort is at a dark-red heat
all over. Finally, chlorine is passed in, but the opening at W is
kept open; the gas is allowed to pass into the condensation
chamber only when fumes of aluminium chloride appear very
abundantly at W. When the operation proceeds right, almost
all the aluminium chloride is found attached in a dense, solid
mass to the cover M. I have taken out at one time a plate weighing almost 50 kilogrammes, which was less than 10 centimetres
thick; it was made up of a large number of sulphur-yellow crystals penetrating each other and looking like stalactites and long
soda crystals. When it is judged that the material in the retort
is almost exhausted, the hole x is opened, the residue scraped out,
and fresh mixture put in. During the operation there should be
no white vapors coming from the condensation chamber, but the
odor of the gas will always be sharp because of the silicon
chloride present, formed unavoidably by the chlorine attacking
the retort. A gas retort, handled well, should last continuously
two or three months, or even more. The furnace should be constructed so as to permit its easy replacement without much
expense. When in use, the retort is closely watched through spy-
holes in the wall, and any cracks which may appear promptly
plastered up, if not large, with a mixture of fine asbestos and
soda glass."

Purification of aluminium chloride.—" It often happens that
the chloride obtained is not pure, either from the nature of the
apparatus employed, or from neglect of the many precautions
which should be taken. In this case, to purify it, it is heated in
an earthen or cast-iron vessel with fine iron turnings. When the
hydrochloric acid, hydrogen and permanent gases are driven from
the apparatus, it is closed and heated hotter, which produces a
light pressure under which influence the aluminium chloride melts
and enters into direct contact with the iron. The ferric chloride,
which is as volatile as aluminium chloride, is transformed into
ferrous chloride, which is much less volatile, and the aluminium

chloride can be obtained pure by being volatilized away or distilled in an atmosphere of hydrogen."

When the processes just described were put in use at the chemical works at La Glaciére, great care had to be taken to avoid letting vapors and acid gases escape into the air, since the works were surrounded by dwellings. To avoid these inconveniences, the vapor of aluminium chloride was made to enter a heated space in which was sodium chloride, in order to produce the less volatile double chloride; but the apparatus choked up so persistently that the attempt was given up. It then occurred to Deville to put sodium chloride into the mixture itself in the retort. The same apparatus was used as before, except that the large gas-retort had to be replaced by a smaller earthen one which could be heated much hotter, the grate being carried half way up the retort.* The condensation chamber had to be replaced by a small earthen vessel. The double chloride produced is fusible at about 200°, and is quite colorless when pure, but colored yellow by iron. It is, moreover, very little altered in dry air when in compact masses and can be easily handled. When the double chloride is obtained quite pure, it gives up its aluminium completely when reduced by sodium.

The following description by M. Margottet† will show the form of apparatus used in 1882 by the French company carrying on the Deville process at Salindres:—

The double chloride may be obtained in the same manner as the simple chloride; it is sufficient to put some common salt, NaCl, into a mixture of alumina and carbon, and, on heating this mixture strongly, there is formed by the action of the chlorine, aluminium-sodium chloride, which distils at a red heat and condenses in a crystalline mass at about 200°. The hydrated alumina obtained in the preceding operation is mixed with salt and finely pulverized charcoal, in proper proportions, the whole is sifted, and a mixture produced as homogeneous as possible; then it is agglomerated with water and made into balls the size of the

* It was when first using this process that Deville borrowed some zinc retorts from the Vielle Montagne works, and since they contained a little zinc in their composition the aluminium made for a while was quite zinciferous.

† Fremy's Ency. Chimique.

fist. These balls are first dried in a drying stove, at about 150°, then calcined at redness in retorts, where the double chloride should commence to be produced just as the balls are completely dried. These retorts are vertical cylinders of refractory earth, each one is furnished with a tube in its lower part for the introduction of chlorine, and with another towards its upper end for the exit of the vapor of double chloride. (See Fig. 7.) A lid

Fig. 7.

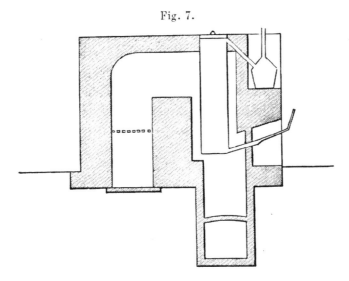

carefully luted during the operation with a mixture of fine clay and horse-dung serves for the charging and discharging of the retort. The double chloride is condensed in earthen pots like flower pots, made of ordinary clay, and closed by a well-luted cover, into which passes a pipe of clay to conduct the gas resulting from the operation into flues connected with the main chimney. Each retort is heated by a fire, the flame of which circulates all round it, and permits keeping it at a bright red heat. An operation is conducted as follows : The retort is filled with stove dried balls, the lid is carefully luted, and the retort is heated gently till all the moisture is driven off. This complete desiccation is of great importance, and requires much time. Then chlorine, furnished by a battery of three generating vessels, is passed in. During the first hours, the gas is totally absorbed by the balls; the double chloride distils regularly for about three hours, and runs into the earthen pots where it solidifies.   Toward the end, the distillation is more difficult and less regular, and the chlorine is then only incompletely absorbed.   After each operation there remains a little residue in the retort, which accumulates and is removed every two days, when two operations are made per day. One operation lasts at least twelve hours, and a retort lasts sometimes a month.   The double chloride is kept in the pots in which it was condensed until the time it is to be used in the next operation ; it is almost chemically pure, save traces of iron, and is easy to keep and handle.

The following estimate was made by Wurtz, in 1872, showing the cost of a kilo of aluminium-sodium chloride as made by the above process :—

Anhydrous alumina 0.59 kilos @ 86 fr. per 100 kilos = 0 fr. 50.7 cent.
Manganese dioxide 3.74 "   " 14 "   "   "   "   = 0 " 52.3 "
Hydrochloric acid 15.72 "   " 3 "   "   "   "   = 0 " 47.1 "
Coal   .   .   .   . 25.78 "   " 1.40 "   "   "   "   = 0 " 36.1 "
                                   Wages .   .   .   . 0 " 23.8 "
                                   Expenses .   .   . 0 " 38.0 "
                                                        ─────────────
                                   Total . . . . . 2 " 48.0 "

This is equal to about 22½ cents per pound.   An average of 10 kilos of this was used to produce one kilo of aluminium, which shows a yield of only 70 per cent. of the contained aluminium, and an increased cost of 67 cents on every pound of aluminium from the imperfection of reduction.   In this respect there certainly seems large room for improvement.

The largest plant ever erected for the manufacture of aluminium-sodium chloride is that of the Aluminium Co. Ltd. at Oldbury near Birmingham, England.   The plant was commenced in the latter part of 1887, and was in working order in July, 1888. The process is in principle identical with that used at Salindres, but the whole is on such a large commercial scale that the apparatus deserves description.

Twelve large regenerative gas furnaces are used, in each of which are placed five horizontal fire-clay retorts about 10 feet in length, into which the mixture is placed.   These furnaces are in two rows, of six each, along each side of a building about 250 feet long, leaving a clear passage down the centre 50 feet wide.

9

Above this central passage is a platform swung from the roof, which carries the large lead mains to supply chlorine to the retorts; opposite each retort is a branch pipe controlled by a valve. The valves are designed so that the chlorine must pass through a certain depth of (non-aqueous) liquid, thus regulating the flow and preventing any back pressure in the retort from forcing vapor into the main. The opposite or back ends of the retorts are fitted with pipes which convey the vapor of the double chloride into cast-iron condensers and thence into brick chests or boxes, the outsides or ends of which are closed by wooden doors fitting tightly. Convenient openings are arranged for clearing out the passages, which may become choked because of the quickness with which the double chloride condenses. On looking down the centre of the building it presents the appearance of a double bank of gas retorts for making ordinary illuminating gas, except that the retorts are only one high.

The chlorine plant is on a correspondingly large scale, the usual manganese-dioxide method being employed and the spent liquor regenerated by Weldon's process. The chlorine gas is stored in large gasometers from which it is supplied to the retorts at a certain pressure. The mixture for treatment is made by mixing hydrated alumina with common salt and carbon in the form of charcoal powder or lamp-black. This being well mixed is moistened with water, thrown into a pug-mill from which it is forced out as solid cylinders, and cut into about 3 inch lengths by a workman. The lumps are then piled on top of the large chloride furnaces to dry. In a few hours they are hard enough to allow handling, and are put into large wagons and wheeled to the front of the retorts.

When the retorts are at the proper temperature for charging, the balls are thrown in until the retort is quite full, the fronts are then put up and luted tightly with clay, and the charge left alone for about four hours, during which the water of the hydrated alumina is completely expelled, the rear end of the retort being disconnected from the condensing chamber, which must be kept perfectly dry, and connected directly with the chimney. At the end of this time the chlorine is turned on and the retort connected with the receiver. At first the chlorine passed in is all absorbed by the charge and only carbonic oxide escapes into the boxes, where it is ignited and burns, thus warming them up. After a certain time dense fumes are evolved, and then the condensers are shut tightly and the uncondensed gases pass into the chimney. The chlorine is passed in for 72 hours in varying quantity, the boxes at the rear being opened from time to time by the workmen to note the progress of the distillation. The greater part of the double chloride liquefies and trickles down to the floor of the chambers, but a portion sublimes and condenses on the walls as a yellow crystalline powder. These chambers are emptied from time to time and the contents packed away in air-tight wooden chests that it may keep without absorbing moisture from the air. At the end of the distillation the chlorine valves are closed and the condenser boxes cleaned out; the retorts are also opened at their front end and the residue raked out. This residue consists of a small quantity of alumina, charcoal and salt, and is remixed in certain proportions with fresh material and used over again. The retorts are then immediately re-charged and the operations repeated. Each set of five retorts produces about 1600 to 1800 lbs. in one operation, or say 3500 lbs. per week. The twelve furnaces are therefore capable of producing easily 1,500,000 lbs. of double chloride per annum. Since 10 lbs. of this salt are required to produce 1 lb. of aluminium, the capacity of the works is thus seen to be 150,000 lbs. or over of metal per year.

This last remark as to the proportion of chloride required to form the metal will show the absolute necessity there is to keep iron from contaminating the salt. This gets in, in varying proportions, from the iron in the materials used and in the fire-clay composing the retort, and exists as ferrous and ferric chlorides. Exercising the utmost care as to the purity of the alumina and charcoal used, and after having the retorts made of a special fire-clay containing a very small percentage of iron, it was found impossible to produce a chloride on a large scale containing less than 0.3 per cent. of iron. This crude chloride is highly deliquescent and varies in color from light yellow to dark red—the color depending not so much on the absolute amount of iron present as on the proportion of iron present as ferric salt, which has a high color. Since practically all the iron present in the salt passes into the aluminium, it is seen that the latter would contain 3 per cent. or more of iron. For some time the only way to obviate this difficulty was to resort to purifying the aluminium, by which the content of iron was finally reduced to 2 per cent. Mr. Castner has since perfected a process for purifying the double chloride by which only 0.01 per cent. of iron is left in it. The principle employed in doing this is described in the patent claims* to be the reduction of the iron salts to metallic iron by melting the chloride (single or double) with a quantity of metallic aluminium or sodium sufficient for this purpose. The purified chloride is quite white and far less deliquescent than the crude salt, which seems to indicate that the iron chlorides have a large share in rendering the crude salt so deliquescent. The purified chloride is preserved by melting and running into tight iron drums.

The success of the manufacture of the double chloride is said to depend on the proportions of the mixture, the temperature of the furnace, the quantity of chlorine introduced and the details of construction of the retorts; but very little information on these points has been made public. The following figures may give some idea of the quantities of materials used: The production of 100 lbs. of double chloride is said to require—

Common salt	357 lbs.
Hydrated alumina	491 "
Chlorine gas	674 "
Coal	1800 "

The salt and hydrated alumina are therefore mixed in about the same proportions as those indicated by the formula which represents the reaction

$$Al^2O^3 + 2NaCl + 3C + 6Cl = Al^2Cl^6.2NaCl + 3CO.$$

for if we assume the hydrated alumina used to contain 90 per cent. of that compound, the 491 lbs. of it used would correspond to very nearly the amount of salt said to be used. As to the cost of this double chloride, so many uncertain elements enter into it that it cannot be satisfactorily estimated from the data at

* U. S. Patent, No. 409668, Aug. 27, 1889.

hand. We are informed, however,* that the double chloride used represents 43 per cent. of the cost of aluminium to this company. If we place the total cost at 8 shillings per lb. this would indicate a trifle over 4 pence per lb. as the cost of the double chloride. I think it is probably not over 3 pence.

H. A. Gadsden,† of London, has patented a method of obtaining aluminium in which the aluminium chloride used is obtained by a method similar in all respects to the process as described by Deville except that the corundum or beauxite used is mixed with about 10 per cent. of sodium or potassium fluoride and a small quantity of fluorspar. After this has been mixed and calcined it is pulverized, 10 per cent. of charcoal dust added, made into balls and heated in a muffle until pasty. Taken out of the muffle they are then put into a retort, heated highly, and chlorine gas passed over them, when aluminium chloride distils.

Count R. de Montgelas‡ patents a process for producing aluminium chloride and the double chloride with sodium, in which the only difference from the preceding methods is that molasses is used instead of pitch for moulding the mixture into balls, the mixture otherwise containing alumina, charcoal, and sodium chloride, and it is claimed that by regulating the heat at which chlorine is passed over this mixture, previously calcined, aluminium chloride can be volatilized while aluminium-sodium chloride remains in the retort. The use of horizontal retorts is recommended, and these certainly possess advantages over the vertical ones, but I am unable to say if this process has the merit of being the pioneer in this direction.

Prof. Chas. F. Mabery,§ of the Case School of Applied Science, Cleveland, patented and assigned to the Cowles Bros. the process of making aluminium chloride, consisting in passing dry chlorine or hydrochloric acid gas over an alloy of aluminium and some other metal kept in a closed vessel at a temperature sufficient to volatilize the aluminium chloride formed, which is caught in a condenser. Or, hydrochloric acid gas is passed through an elec-

---

* Zeitschrift des Vereins Deutscher Ingenieure, 1889, p. 301.

† German Patent (D. R. P.) No. 27572 (1884).

‡ English Patent, Nos. 10011, 10012, 10013, Aug. 4, 1886.

§ U. S. Patent, Oct. 26, 1886.

trically heated furnace in which alumina is being decomposed by carbon, a condenser being attached to the opposite end of the furnace.

Mr. Paul Curie* states that aluminium chloride can be made by passing vapors of carbon disulphide and hydrochloric acid either simultaneously or successively over ignited alumina or clay. The first forms aluminium sulphide which the latter converts into the volatile chloride, which distils.

H. W. Warren† recommends the following process as of general application in producing anhydrous metallic chlorides: Petroleum is saturated with either chlorine or hydrochloric acid gas, both gases being soluble in it to a large extent, particularly the latter gas. This operation is performed at a low temperature, as more of the gases is then dissolved. The oxide of the metal, alumina for instance, is put into large earthenware retorts and raised to red heat. The saturated oil is then boiled and its vapor passed into the retort. On contact with the oxide a strong reaction commences, fumes of aluminium chloride are at once evolved and distil into a condenser, the operation being continued until no more white fumes appear. Then fresh alumina is supplied, and the reaction continues. The aluminium chloride may be purified from any oil by gentle application of heat. Mr. Warren also used naphthaline chloride with advantage, as also chloride of carbon, but their high price rendered them unable to compare with petroleum in economy. Aluminium bromide can be similarly prepared by substituting bromine for chlorine.

Camille A. Faure, of New York, the well-known inventor of the Faure storage battery, has patented‡ a process for producing aluminium chloride which is very similar to the above method. The manipulation is described as follows: An oxygenated ore of aluminium is brought to about a red heat by bringing it, in a furnace, into direct contact with the flame. When at proper heat the flame is cut off and a gas containing carbon and chlorine introduced. A mixture of petroleum vapor or a similar hydro-

* Chemical News, 1873, p. 307.
† Chemical News, April 29, 1887.
‡ U. S. Patent, No. 385345, July 3, 1888.

carbon with hydrochloric acid gas is preferred. Vaporized chloride of aluminium immediately passes off into a condenser.

In a paper written by Mr. Faure, and read before the French Academy of Sciences by M. Berthelot,* it was stated that the aim of this process was to suppress the prominent disadvantages of the older methods : viz., cost and wear and tear of retorts, great consumption of fuel, slowness of the operation, large amount of labor, and cost of the chlorine. For this purpose the chlorine is replaced by hydrochloric acid gas and the carbon by a hydrocarbon. Since all pure hydrocarbons are decomposed at a red heat with deposition of carbon the process would appear impracticable, but a proper mixture of hydrochloric acid gas and naphthaline vapor is said not to decompose by the highest temperature alone, a new compound being formed, a sort of naphthaline chloride, which is exceedingly corrosive and powerful enough to attack any oxide and convert it into chloride. To carry out the process a gas furnace with large bed is used. On this is spread a layer of small pieces of beauxite about two feet deep. The flame comes in over the ore, passes downward through it and through numerous holes arranged in the hearth, and thence to a chimney. In this way the heat of the gases is well utilized, while the layer of beauxite is heated to whiteness on top and to low-red at the bottom. The flames are then turned off and the mixture of naphthaline and hydrochloric acid vapors passed upward through the bed, and by their reaction producing aluminium chloride, which is diverted by suitable flues into a condenser. It is claimed that by careful fractional condensation the chlorides of silicon, iron, calcium, etc., formed from impurities in the beauxite, can be easily separated, that of silicon being more volatile and those of iron and calcium less volatile than aluminium chloride. As naphthaline is a bye product from gas-works, it is claimed that it can be bought for $1\frac{1}{2}$ cents per lb., and that only $\frac{1}{4}$ of a lb. is used per lb. of aluminium chloride produced. It is also claimed that one furnace, with two men to work it, will produce 4000 lbs. of chloride a day. The estimated cost of the chloride is about $1\frac{1}{2}$ cents per pound, of which 17 per cent. is for beauxite, 47 per

---

* July 30, 1888.

cent. for hydrochloric acid, 27 per cent. for naphthaline, and 9 per
cent. for labor.   Mr. Faure has been experimenting in the vicinity
of New York during the last few months, and is sanguine of
having the process at work commercially in 1890. (See further,
Chap. XI.)

In all the processes for producing aluminium chloride so far
considered the use of common clay was not recommended, since
silicon chloride is formed as well as aluminium chloride.   The
only method proposed for using clay for this purpose is that of
M. Dullo, nearly twenty years ago, and which cannot have been
very successful, since it has not been heard of in operation.   We
will repeat his remarks, however, for there is still a large field
open in the utilization of clay for the manufacture of aluminium,
and since the metal is becoming so cheap the manufacturers are
not above looking for and utilizing the cheapest raw material
available.

* " Aluminium chloride may be obtained easily by direct treatment of clay.   For this purpose a good clay, free from iron and
sand, is mixed with enough water to make a thick pulp, to which
are added sodium chloride and pulverized carbon.   For every 100
parts of dry clay there are taken 120 parts of salt and 30 of carbon.   The mixture is dried and broken up into small fragments,
which are then introduced into a red-hot retort traversed by a
current of chlorine.   Carbonic oxide is disengaged, while at the
same time aluminium chloride and a little silicon chloride are
formed.   It is not necessary that the chlorine should be absolutely
dry, it may be employed just as it comes from the generator.   The
gas is absorbed very rapidly, because between the aluminium and
silicon there are reciprocal actions under the influence of which
the chemical actions are more prompt and energetic.   The aluminium having for chlorine a greater affinity than silicon has, aluminium chloride is first formed, and it is only when all the aluminium is thus transformed that any silicon chloride is formed.
When the latter begins to form the operation is stopped, the incandescent mixture is taken out of the retort and treated with
water.   The solution is evaporated to dryness to separate out a

* Bull. de la Soc. Chem. 1860, vol. v. p. 472.

small quantity of silica which is in it, the residue is taken up with water, and aluminium-sodium double chloride remains when the filtered solution is evaporated to dryness."

We must say of M. Dullo's suggestions that it is the general experience that the more volatile silicon chloride is formed first; it is also very improbable that a solution of aluminium-sodium chloride can be evaporated without decomposition.

III.

The Preparation of Aluminium Fluoride and Aluminium-Sodium Fluoride. (Cryolite.)

Natural cryolite is too impure for use in many operations which aim to produce very pure aluminium. Schuh has proposed boiling the mineral in solution of soda. Under certain conditions sodium aluminate is formed (see p. 122), but if the solution of soda is dilute the liquor remains clear after taking up the cryolite, and on passing a current of carbonic acid gas through it aluminium-sodium fluoride is precipitated. In this way the pure double fluoride can be separated from impure cryolite.

Berzelius recommended preparing artificial cryolite by decomposing aluminium hydrate by a solution of sodium fluoride and hydrofluoric acid, the hydrate being added to the liquid until its acidity was just neutralized :—

$$Al^2O^3.3H^2O + 6NaF + 6HF = Al^2F^6.6NaF + 6H^2O.$$

If a solution of sodium fluoride alone is used, half the aluminium and half the sodium will remain in the solution as sodium aluminate :—

$$2Al^2O^3.3H^2O + 12NaF = Al^2F^6.6NaF + Al^2O^3.3Na^2O + 3H^2O.$$

Deville states that on adding sodium chloride to a solution obtained by dissolving alumina in an excess of hydrofluoric acid, a precipitate of cryolite is obtained. Since cryolite is hardly attacked at all by hydrochloric acid, it is probable that the reaction occurring is

$$Al^2F^6 + 6HF + 6NaCl = Al^2F^6.6NaF + 6HCl.$$

The process which Deville recommended as best, however, is the treatment of a mixture of calcined alumina and carbonate of soda, mixed in the proportions in which their bases exist in cryolite, by an excess of pure hydrofluoric acid :—

$$Al^2O^3 + 3Na^2CO^3 + 12HF = Al^2F^6.6NaF + 3CO^2 + 6H^2O.$$

100 parts of pure alumina requiring 310 parts of sodium carbonate and 245 of anhydrous hydrofluoric acid, there being 410 parts of cryolite formed. On drying the mass and melting it there results a limpid, homogeneous bath having all the characteristics of cryolite, being reduced by sodium or by an electric current, which would not result from a mere mixture of alumina and sodium fluoride melted together.

Deville also states that when anhydrous aluminium chloride is heated with sodium fluoride in excess, a molten bath results of great fluidity, and on cooling and dissolving away the excess of sodium fluoride by repeated washings the residue is similar to cryolite, while the solution contains no trace of any soluble aluminium salt :—

$$Al^2Cl^6 + 12NaF = Al^2F^6.6NaF + 6NaCl.$$

It is evident, however, that the above reaction would be the reverse of a profitable one, and is therefore not of economical utility.

Pieper* patents a very similar reaction but operates in the wet way. A solution of aluminium chloride is decomposed by adding to it a suitable quantity of sodium fluoride in solution. Sodium chloride is formed and cryolite precipitated, as in the last reaction given. By adding different proportions of sodium fluoride solution precipitates of double salts are obtained containing varying proportions of the two fluorides. The use of aluminium chloride in solution would dispense with the objection made to Deville's analogous method, and this process would very probably produce cryolite quite cheaply.

Bruner† produced aluminium fluoride by passing hydrofluoric acid gas in the required quantity over alumina heated red hot in a platinum crucible :—

$$Al^2O^3 + 6HF = Al^2F^6 + 3H^2O.$$

* German Patent (D. R. P.), No. 35212. † Pogg. Annalen, 98, p. 488.

Deville* states that it can be made by melting together equivalent quantities of cryolite and aluminium sulphate :—

$$Al^2F^6.6NaF + Al^2(SO^4)^3.xH^2O = 2Al^2F^6 + 3Na^2SO^4 + xH^2O.$$

On washing the fusion, sodium sulphate goes into solution. It is also stated that hydrochloric acid gas acting on a mixture of fluorspar and alumina at a high temperature will produce aluminium fluoride :—

$$Al^2O^3 + 3CaF^2 + 6HCl = Al^2F^6 + 3CaCl^2 + 3H^2O.$$

The calcium chloride would be partly volatilized and the remainder washed out of the fusion.

Hautefeuille† obtained crystallized aluminium fluoride by passing hydrofluoric acid gas and steam together over red-hot alumina.

Ludwig Grabau, of Hanover, bases his process of producing aluminium on the reduction of aluminium fluoride (see Chap. X.), which is prepared on a commercial scale by the following ingenious methods :—

‡The process is based on the conversion of aluminium sulphate into fluoride by reaction with cryolite, the fluoride being afterwards reduced by sodium in such manner that a double fluoride of sodium and aluminium results which is used over again, thus forming a continuous process. The purest obtainable cryolite is used to start the process, after which no more is needed, the material supplying the aluminium being its sulphate, which can be obtained cheaply in large quantities and almost perfectly pure. The process is outlined by the reactions—

$$Al^2F^6.6NaF + Al^2(SO^4)^3 = 2Al^2F^6 + 3Na^2SO^4$$
$$2Al^2F^6 + 6Na = 2Al + Al^2F^6.6NaF.$$

It is thus seen that theoretically the cryolite would be exactly reproduced, but the losses and incomplete reactions unavoidable in practice would cause less to be obtained and necessitate the continual addition of fresh cryolite; since, however, it is not desired to base the process on the continual use of cryolite, because of the

* Ann. de Chim. et de Phys. [3], 61, p. 333 ; [3], 49, p. 79
† Idem, [4], 4, p. 153.
‡ German Patent (D. R. P.) No. 48535, March 8, 1889.

impurities in that mineral, an indirect process is used consisting of two reactions, in place of the first given above, in which theoretically a larger quantity of cryolite is finally obtained than is used to begin with. This is operated by introducing fluorspar into the process, the base of which goes out as calcium sulphate or gypsum and so supplies the fluorine needed.

In practice, a solution of aluminium sulphate is heated with powdered fluorspar (obtained as pure as possible and further cleaned by treatment with dilute hydrochloric acid). The aluminium sulphate will not be entirely converted into fluoride, as has been previously observed by Friedel, but about two-thirds of the sulphuric acid is replaced by fluorine, forming a fluor-sulphate of aluminium. This latter compound remains in solution, while gypsum and undecomposed fluorspar remain as a residue and are filtered out.

$$Al^2(SO^4)^3 + 2CaF^2 = Al^2F^4(SO^4) + 2CaSO^4.$$

This solution is concentrated and mixed with cryolite in such proportion that the alkali in the latter is just equivalent to the sulphuric acid in the fluor-sulphate. The mass is dried and ignited, and the product washed and dried.

$$3Al^2F^4(SO^4) + Al^2F^6.6NaF = 4Al^2F^6 + 3Na^2SO^4.$$

On reduction with sodium, the 4 molecules of aluminium fluoride, treated with sodium as by the reaction given, produce 2 molecules of the double fluoride. It is thus seen that after allowing for reasonable losses in the process there is much more cryolite produced than is used, and the excess can be very profitably sold as pure cryolite, being absolutely free from iron or silica.

IV.

THE PREPARATION OF ALUMINIUM SULPHIDE.

Until the researches of M. Fremy, no other method of producing aluminium sulphide was known save by acting on the metal with sulphur at a very high heat. Fremy was the first to open up a different method, and it may be that his discoveries will yet be the basis of successful industrial processes. In order to understand just how much he discovered we here give all that his original paper contains concerning this sulphide.*

" We know that sulphur has no action on silica or boric oxide, magnesia, or alumina. I thought that it might be possible to replace the oxygen by sulphur if I introduced or intervened a second affinity, as that of carbon for oxygen These decompositions produced by two affinities are very frequent in chemistry; it is thus that carbon and chlorine, by acting simultaneously on silica or alumina, produce silicon or aluminium chloride, while either alone could not decompose it ; a similar case is the decomposition of chromic oxide by carbon bisulphide, producing chromium sesquisulphide. Reflecting on these relations, I thought that carbon bisulphide ought to act at a high heat on silica, magnesia, and alumina, producing easily their sulphides. Experiment has confirmed this view. I have been able to obtain in this way almost all the sulphides which until then had been produced only by the action of sulphur on the metals.

" To facilitate the reaction and to protect the sulphide from the decomposing action of the alkalies contained in the porcelain tube which was used, I found it sometimes useful to mix the oxides with carbon and to form the mixture into bullets resembling those employed in the preparation of aluminium chloride. I ordinarily placed the bullets in little carbon boats, and heated the tube to whiteness in the current of vaporized carbon bisulphide. The presence of divided carbon does not appear useful in the preparation of this sulphide.

" The aluminium sulphide formed is not volatile ; it remains in the carbon boats and presents the appearance of a melted vitreous mass. On contact with water it is immediately decomposed.

$$Al^2S^3 + 3H^2O = Al^2O^3 + 3H^2S.$$

" The alumina is precipitated, no part of it going into solution. This precipitated alumina is immediately soluble in weak acids. The clear solution, evaporated to dryness, gives no trace of alumina. It is on this phenomenon that I base the method of analysis.

* Ann. de Chem. et de Phys. [3] xxxviii. 312.

"Aluminium sulphide being non-volatile, it is always mixed with some undecomposed alumina. It is, in fact, impossible to entirely transform all the alumina into sulphide. I have heated less than a gramme of alumina to redness five or six hours in carbon bisulphide vapor, and the product was always a mixture of alumina and aluminium sulphide. The reason is that the sulphide being non-volatile and fusible coats over the alumina and prevents its further decomposition. The alumina thus mixed with the sulphide, and which has been exposed to a red heat for a long time, is very hard, scratches glass, and is in grains which are entirely insoluble in acids. By reason of this property I have been able to analyze the product exactly, for on treating the product with water and determining on the one hand the sulphuretted hydrogen evolved, and on the other the quantity of soluble alumina resulting, I have determined the two elements of the compound. One gramme of my product contained 0.365 grm. of aluminium sulphide, or 36.5 per cent., the remainder being undecomposed alumina." The composition of this sulphide was—

Aluminium . . . . . 0.137 grm. = 37.5 per cent.
Sulphur . . . . . 0.228 " = 62.5 "
_____ _____
0.365 " 100.0 "

The formula Al²S³ requires—

Aluminium . . . . 36.3 per cent.
Sulphur . . . . 63.7 "

The above is the substance of Fremy's investigations and results. Reichel* next published an account of further experiments in this line. He found that by melting alumina and sulphur together no reaction ensued. In the case of magnesia, the sulphide was formed if carbon was mixed with the magnesia and sulphur, but this change did not alter the alumina. Hydrogen gas passed over a mixture of alumina and sulphur likewise gave negative results. Sulphuretted hydrogen passed over ignited alumina did not succeed. By filling a tube with pure alumina, passing in hydrogen gas and the vapor of carbon bisulphide, the heating being continued until carbon bisulphide condensed in the

* Jahresb. der Chemie, 1867, p. 155.

outlet tube, and then hydrogen being passed through until the tube was cold, a product was obtained containing aluminium sulphide and undecomposed alumina.

In 1886, I made a series of experiments on the production and reduction of aluminium sulphide. Alumina, either alone or mixed with carbon or with carbon and sulphur, was put in porcelain or carbon boats into a hard glass or porcelain tube. This was then heated and vapor of carbon bisulphide passed over it. The product was analyzed according to Fremy's directions. The proportion of aluminium sulphide obtained in the product varied from 13 to 40 per cent. The best result was obtained at the highest heat—almost whiteness. The presence of sulphur or carbon, or both together, mixed with the alumina did not promote to any degree the formation of a richer product. The conditions for obtaining the best results seem to be high heat and fine division of the alumina to facilitate its contact with the carbon bisulphide vapor. The product was light-yellow when not mixed with carbon, easily pulverized, and evolved sulphuretted hydrogen gas energetically when dropped into water. Since carbon bisulphide can now be manufactured at a very low price, say 2 to 3 cents per lb., it is not impossible that it may be found profitable to produce aluminium from its sulphide. In such a case, large retorts would be used, a stirring apparatus would facilitate the formation of a richer product, and the unused carbon bisulphide could be condensed and saved.

M. Comenge,* of Paris, proposed to prepare aluminium sulphide by using a clay retort similar to those used in gas-works, filling it one-half its length with charcoal or coke and the other half with alumina. The retort being heated to redness, sulphur is introduced at the coke end, when in contact with the carbon it forms carbon bisulphide, which acts upon the alumina at the other end, producing the sulphide.

Messrs. Reillon, Montague, and Bourgerel† obtained a patent in England for producing aluminium, in which aluminium sulphide is obtained by mixing powdered alumina with 40 per cent.

---

* English Patent, 1858, No. 461.

† English Patent, No. 4756, March 28, 1887.

of its weight of charcoal or lampblack and formed into a paste with a sufficient quantity of oil or tar. This is then calcined in a closed vessel and an aluminous coke obtained. This is broken into pieces, put into a retort, and treated with carbon bisulphide vapor. The inventors state that the reaction takes place according to the formula $2Al^2O^3 + 3C + 3CS^2 = 2Al^2S^3 + 6CO$.

Petitjean* states that if alumina is mixed with tar or turpentine and ignited in a carbon-lined crucible, and the coke obtained mixed intimately with sulphur and carbonate of soda and ignited a long time at bright redness, there results a double sulphide of aluminium and sodium, from which aluminium can be easily extracted.

It has been stated† that if aluminium fluoride is heated with calcium sulphide, aluminium sulphide results. F. Lauterborn‡ also makes the same claim in a patent twenty years later, but the possibility of this reaction taking place is not yet beyond question.

CHAPTER VII.

THE MANUFACTURE OF SODIUM.

SOME years ago, in order to treat fully of the metallurgy of aluminium, it would have been as necessary to accompany it with all the details of the manufacture of sodium as to give the details of the reduction of the aluminium, because the manufacture of the former was carried on solely in connection with that of the latter. But now sodium has come out of the list of chemical curiosities and has become an article of commerce, used for many other purposes than the reduction of aluminium, though that is still its chief use. So we regard the manufacture of sodium as a separate metallurgical subject, still intimately con-

---

* Polytechnisches Central. Blatt., 1858, p. 888.
† Chemical News, 1860.
‡ German Patent (D. R. P.), No. 14495 (1880).

nected with that of aluminium, but yet so far distinct from it as to deserve a metallurgical treatise of its own.

Davy to Deville (1808–1855).

Sodium was first isolated by Davy by the use of electricity in the year 1808.* Later Gay Lussac and Thenard made it by decomposing at a very high temperature a mixture of sodium carbonate and iron filings.† In 1808, also, Curaudau announced that he had succeeded in producing potassium or sodium without using iron, simply by decomposing their carbonates by means of animal charcoal. Brünner, continuing this investigation, used instead of animal charcoal the so-called black flux, the product obtained by calcining crude tartar from wine barrels. He was the first to use the wrought-iron mercury bottles. The mixture was heated white hot in a furnace, the sodium volatilized, and was condensed in an iron tube screwed into the top of the flask, which projected from the furnace and was cooled with water. In Brünner's experiments he only obtained three per cent. of the weight of the mixture as metallic sodium, the rest of the metal being lost as vapor.

Donny and Mareska gave the condenser the form which with a few modifications it retains to-day. It was of iron, 4 millimetres thick, and was made in the shape of a book, having a length of about 100 centimetres, breadth 50, and depth 6 (see Fig. 8). This form is now so well known that a further description is unnecessary. With this condenser the greatest difficulty of the process was removed, and the operation could be carried on in safety. This apparatus was devised and used by Donny and Mareska in 1854, with the supervision of Deville.

Fig. 8.

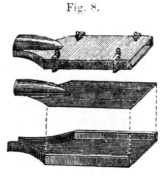

* Phil. Trans., 1808.
† Recherches Physico-chemiques, 1810.

Deville's Improvements at Javel (1855).

The following is Deville's own description of the attempts which he made to reduce the cost of producing sodium. As far as we can learn these experiments were commenced in 1854, but the processes about to be given are those which were carried out at Javel, March to June, 1855. As the description contains so many allusions to the difficulties met not only in producing but also in handling and preserving sodium, its perusal is yet of value to all concerned in this subject, although the actual methods here described have been superseded by much more economical ones.

Properties of sodium.—The small equivalent of sodium and the low price of sodium carbonate should long since have caused it to be preferred to potassium in chemical operations, but a false idea prevailed for a long time concerning the difficulties accompanying the reduction. When I commenced these researches the cost of sodium was at least double that of potassium. In this connection I can quote from my memoir published in the Ann. de Chim. et de Phys., Jan. 1, 1855: "I have studied with care the preparation of sodium and its properties with respect to oxygen and the air, in order to solve the difficulties which accompany its reduction and the dangers of handling it. In this latter respect, sodium is not to be compared to potassium. As an example of how dangerous the latter is, I will relate that being used to handle sodium and wishing once to replace it with potassium, the simple rubbing of the metal between two sheets of paper sufficed to ignite it with an explosion. Sodium may be beaten out between two sheets of paper, cut and handled in the air without accident if the fingers and tools used are not wet. It may be heated with impunity in the air, even to its fusing point, without taking fire, and, when melted, oxidation takes place slowly, and only at the expense of the moisture of the air. I have even concluded that the vapor alone of sodium is inflammable, but the vivid combustion of the metal can yet take place at a temperature which is far below its boiling point, but at which the tension of the metallic vapors has become sensible." I will add to these remarks that sodium possesses two considerable advantages: it is obtained pure at the first operation, and, thanks to a knack which I was a long time in finding out, the globules of the metal may be reunited and treated as an ordinary metal when melting and casting in the air. I have thus been able to dispense with the distillation of the raw products in the manufacture—an operation which had become to be believed necessary, and which occasioned a loss of 50 per cent. or so on the return without appreciable advantage to the purity of the metal. The manufacture of sodium is in no manner incumbered by the carburetted products, or perhaps nitrides, which are very explosive and render the preparation of potassium so dangerous. I ought to say, however, that by making potassium on a large scale by the processes I am about to describe for sodium, Rousseau Bros. have diminished the dangers of its preparation very much, and practise the process daily in their chemical works.

Method employed.—The method of manufacture is founded on the reaction of carbon on alkaline carbonate. This method has been very rarely applied to sodium, but is used every day for producing potassium. Brünner's process is, in fact, very difficult to apply, great trouble being met, especially, in the shape of condenser used. It is Donny and Mareska who have mastered the principles which should guide in constructing these condensers.

Composition of mixtures used.—The mixture which has given me excellent results in the laboratory is—

Sodium carbonate	717 parts.
Wood charcoal	175 "
Chalk	108 "
	1000

Dry carbonate of soda is used, the carbon and chalk pulverized, the whole made into a paste with oil and calcined in a crucible. The end of a mercury bottle, cut off, serves very well, and can be conveniently closed. Oil may be used altogether in place of charcoal, in which case the following proportions are used :—

Sodium carbonate	625 parts.
Oil	280 "
Chalk	95 "

That a mixture be considered good, it should not melt at the temperature at which sodium is evolved, becoming liquid at this point, and so obstructing the disengagement of the gas.* But it should become pasty, so as to mould itself evenly against the lower side of the iron vessel in which it is heated. The considerable latent heat required by carbonic oxide and sodium in assuming the gaseous state is one cause of cooling which retards the combustion of the iron. When soda salt is introduced in place of dried soda crystals, the mixture, whatever its composition, always melts, the gases making a sort of ebullition, the workmen saying that the apparatus "sputters." This behavior characterizes a bad mixture. It has been demonstrated to me that the economy made at the expense of a material such as carbonate of soda, the price of which varies with its strength in degrees, and which forms relatively a small portion of the cost of sodium, is annulled by a decrease of 20 to 25 per cent. in the return of sodium. The oil used ought to be dry and of long flame. It acts as a reducing agent, and also furnishes during the whole operation hydrogenous gases, and even, towards the close, pure hydrogen, which help to carry the sodium vapor rapidly away into the condenser, and to protect the condensed metal from the destructive action of the carbonic oxide. Oil renders a similar service in the manufacture of zinc. The rôle of the chalk is easy to understand. By its infusibility it decreases the liability of the mixture to melt. Further, it gives off carbonic acid, immediately reduced by the carbon present to carbonic oxide. Now, the sodium ought to be carried rapidly away out of the apparatus, because it has the property of decomposing carbonic oxide, which is simultaneously formed, within certain limits of temperature, especially if the sodium is disseminated in little globules and so presents a large surface to the destructive action of the gas. It is necessary, then, that the metallic vapors should be rapidly conducted into the condenser and brought into the *liquid* state—not into that state comparable to "flowers of sulphur," in which the metal is very oxidizable because of its fine division. A rapid current of gas, even of carbonic oxide, actively carries the vapors into the con-

---

* It seems plain, however, granting that vapors would be evolved most freely from a *perfectly* infusible charge, that a pasty condition, such as is recommended in the next sentence, would be the worst possible state of the charge for evolving gas, being manifestly inferior to a completely fluid bath.—J. W. R.

denser, which they keep warm and so facilitate the reunion of the globules of sodium. At La Glaciére and Nanterre a mixture was used in which the proportion of chalk, far from being diminished, was, on the contrary, increased. The proportions used were—

Sodium carbonate	. . . .	40 kilos	= 597 parts.
Oil	. . . . . .	18 "	= 269 "
Chalk	. . . . .	9 "	= 134 "
		67 "	1000

This quantity of mixture ought to give 9.4 kilos of sodium, melted and cast into ingots, without counting the metal divided and mixed with foreign materials, of which a good deal is formed. This return would be one-seventh of the weight of the mixture or one-quarter of the sodium carbonate used.

Use of these mixtures.—The carbonate of soda, charcoal, and chalk ought to be pulverized and sieved, well mixed and again sieved in order to make a very intimate mixture; the mixture ought to be used as soon as possible after preparing, that it may not take up moisture. The mixture may be put just as it is into the apparatus where it should furnish sodium, but it may very advantageously be previously calcined so as to reduce its volume considerably, and so permit a greater weight being put into the same vessel. I believe that whenever this calcination may be made with economy, as with the waste heat of a furnace, a gain is made by doing so, but the procedure is not indispensable. However, the utility of it may be judged when it is stated that a mercury bottle held 2 kilos of non-calcined mixture, but 3.6 kilos were put into one when previously calcined. These two bottles heated in the same fire for the same time gave quantities of sodium very nearly proportional to the weight of soda in them. In working under the direction of a good workman, who made the bottles serve for almost four operations, I have been able to obtain very fine sodium at as low a price as 9.25 francs per kilo ($0.84 per pound). In the manufacture of sodium by the continuous process, where the materials may be introduced red hot into the apparatus, this preliminary calcination is a very economical operation.

Apparatus for reducing, condensing, and heating.—M. Brünner had the happy idea of employing mercury bottles in manufacturing potassium, thus the apparatus for reduction was in the
hands of any chemist, and at such a low price that any one has
been able to make potassium without much trouble.   These bottles are equally suitable for preparing sodium, and the quantity
which may be obtained from such apparatus and the ease with
which they are heated, are such that they might have been used
a long time for the industrial manufacture, except for two reasons
which tend to increase the price of these bottles continually.
For some time a large number of bottles have been sent to the
gold workers of Australia and California, also large quantities
have been used in late years in preparing the alkaline metals;
these two facts have diminished the number to such a point that
from 0.5 to 1 franc the price has been raised to 2.5 or 3 francs.
It has thus become necessary to replace them, which has been
done by substituting large wrought iron tubes which have the
added advantage of being able to be worked continuously.   I
will first describe the manufacture in mercury bottles, which may
still be very advantageously used in the laboratory, and afterwards the continuous production in large iron cylinders as now
worked industrially.

Manufacture in mercury bottles.—The apparatus needed is composed of a furnace, a mercury bottle, and a condenser.

The form of furnace most suitable is a square shaft, *C* (Fig. 9),
the sides of which are refractory brick, while the grate *G* ought
to have movable grate bars, the furnace being connected above
with a chimney furnishing good draft.   The flue *F*, connecting
with the chimney, should have a damper, *R*, closing tightly and
should lead exactly from the centre of the top of the shaft, thus
dividing the draft equally all over the grate.   Coke is charged
through lateral openings at *O*.   A small opening closed by a
brick should be left a short distance above the grate bars, in
order to poke down the coke around the bottle should it not
fall freely.   The space between the grate and bottle should
always be full of fuel in order to keep the iron of the bottle
from being burnt.   In front of the furnace is a square opening,
P, closed with an iron plate, which has a hole in it by which
the tube *T* issues from the furnace.

The mercury bottle is supported on two refractory bricks, KK, cut on their top side to the curve of the bottle.   These should be

Fig. 9.

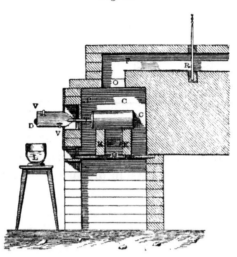

at least 20 centimetres high, to maintain between the grate and bottle a convenient distance.   The illustration gives the vertical dimensions correctly, but the horizontal dimensions are somewhat shortened.   There should be at least 12 centimetres between the bottle and the sides of the furnace.   However, all these dimensions should vary with the strength of the chimney draft and the kind of fuel used; the furnace should be narrower if the draft is very strong and the coke dense.   The iron tube T, which may conveniently be made of a gun barrel, is either screwed into the bottle or it may be simply fitted and forced into place, provided it holds tightly enough.   It should be about 5 to 6 centimetres long, and should project scarcely 1 centimetre from the furnace. The end projecting should be tapered off in order to fit closely into the neck of the condenser.

The condenser is constructed with very little deviation from that given by Donny and Mareska (see Fig. 8, p. 145).   I have tried my best to make this apparatus as perfect as possible, but have always reverted to the form described by those authors; yet even the very small differences I have made are indispensable and must be rigidly adhered to if it is wished to get the best results obtainable. Two plates of sheet iron, 2 to 3 millimetres thick, are taken and cut into the shape indicated by Fig. 10. One

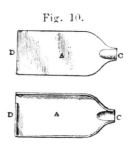

Fig. 10.

plate, *A*, remains flat except at the point *C*, where it is drawn by hammering into a semi-cylindrical neck of about 25 millimetres inside diameter. This corresponds with a similar neck in the other plate, so that on joining the two there is a short cylinder formed. The edges of the plate *A* are raised all around the sides about 5 to 6 millimetres, so that when the two plates are put together the longitudinal section from *D* to *C* is as in Fig. 11. As to the end, in one form the edge was not turned up, leav-

Fig. 11.

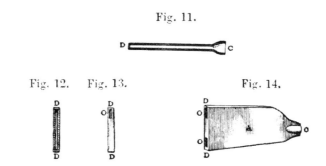

Fig. 12.   Fig. 13.   Fig. 14.

ing the end open as in Fig. 12. Another form, which I use when wishing to let the sodium accumulate in the condenser till it is quite full, is made by turning up the edge at the end all but a small space left free, thus giving the end the appearance of Fig. 13, this device also being shown in Fig. 10. The gas evolved during the reaction then escapes at the hole *O*. The most rational arrangement of the apparatus is that shown in Fig. 14. In this condenser the lower part, instead of being horizontal, is inclined, and the end having two openings, *O* and *O'*, the sodium trickles out at the lower one as it condenses, while the gas escapes by the slightly larger upper opening. In placing the plates together, the raised edges are washed with lime so as to form a good joint with the flat plate, and the plates are kept together by strong pressure grips.

To conduct the operation, the bottles are filled entirely with mixture, the tube T adjusted, and then placed on the two supports, there being already a good bed of fire on the grate. The front is put up, the shaft filled with coke, and the damper opened. The gases disengaged from the bottle are abundant, of a yellow color; at the end of half an hour white fumes of carbonate of soda appear. The condenser should not yet be attached, but it should be noted if any sodium condenses on a cold iron rod pushed into the tube, which would be indicated by it fuming in the air. As soon as this test shows that sodium is being produced the condenser is attached and the fire kept quite hot. The condenser soon becomes warm from the gases passing through it, while the sodium condenses and flows out at the end D (Fig. 9). It is received in a cast-iron basin L, in which some non-volatile petroleum is put. When at the end of a certain time the condenser becomes choked, it is replaced by another which has been previously warmed up to 200° or 300° by placing it on top of the furnace. If the closed condenser is used, care must be taken to watch when it becomes full, on the point of running from the upper opening, and the condenser then replaced and plunged into a cast-iron pot full of petroleum at a temperature of 150°. The sodium here melts at the bottom of this pot and is ladled out at the end of the day. The oil is generally kept up to 150° by the hot condensers being plunged in constantly. The pot ought to have a close cover, to close it in case the oil takes fire; the extinction of the fire can thus be assured and no danger results. If the oil fires just as a condenser is being introduced, the sodium is run out in the air without igniting, the only drawback being that the condenser must be cleaned before using again. This method occasions a large loss of oil, however, and has been completely abandoned for the other form of condensers. When the operation proceeds well only pure sodium is obtained, the carbonized products which accompany in so provoking a manner the preparation of potassium not occurring in quantity sufficient to cause any trouble. Before using a condenser a second time it is put on a grating over a basin of petroleum and rubbed with a chisel-pointed tool in order to remove any such carbonized products. From time to time this material is collected, put into a mercury bottle, and heated gently. The oil first distils, and is condensed in another cold bottle. The fire is then urged, a condenser attached, and the operation proceeds as with a fresh charge, much sodium being thus recovered.

The raw sodium is obtained from the bottles in quantities of over half a kilo; it is perfectly pure, dissolving in absolute alcohol without residue. It is melted and moulded into ingots just as lead or zinc. The operation I have described is executed daily, and only once has the sodium ignited. To prevent such accidents it is simply necessary to keep water away from the apparatus. The reduction of carbonate of soda and the production of sodium are such easy operations that when tried by those conversant with the manufacture of potassium or who have read about the difficulties of the production of sodium, success is only gained after several attempts—the failure being due solely to excess of precautions. The reduction should be carried on rapidly, so that a bottle charged with two kilos of mixture may be heated and emptied in, at most, two hours. It is unnecessary to prolong the operation after the yellow flame stops issuing from the condenser, for no more sodium is obtained and the bottle may frequently be destroyed. The temperature necessary for the reduction is not so high as it has been so far imagined. M. Rivot, who has assisted in these experiments, thinks that the bottles are not heated higher than the retorts in the middle of the zinc furnaces at Vieille Montagne. I have been even induced to try cast-iron bottles, but they did not resist the first heating, without doubt because they were not protected from the fire by any luting or covering. But I was immediately successful in using cast-iron bottles decarburised by the process used for making malleable castings. The mercury bottles heated without an envelope ought to serve three or four operations when entrusted to a careful workman. Besides all these precautions, success in this work depends particularly on the ability and experience of the workman, who can at any time double the cost of the sodium by carelessness in managing the fire.

Continuous manufacture in cylinders.—It might be thought that by increasing proportionately in all their parts the dimensions of the apparatus just described it would be easy to produce much larger quantities of sodium. This idea, which naturally presented itself to me at once, has been the cause of many unfruitful attempts, into the details of which I will not enter. I must, however, explain some details which may appear insignificant at first sight, but which were necessitated during the development of the process. For instance, it will perhaps look irrational for me to keep the same sized outlet tubes and condensers that were used with the mercury bottles, for tubes five times as large; but I was forced to adopt this arrangement after trying the use of tubes and condensers of all sizes; indeed, it is fortunate for the success of the operation that this was so, for it became very injurious to the workmen to handle the large and weighty apparatus in the face of a large sodium flame.

The mixture of sodium carbonate and carbon is made in the manner already described. I would say again that a previous strong calcination of the materials presents a great advantage, not only because it permits putting a much larger weight into the retorts at once but also that, being more compact, the mixture will not rise as powder and be violently thrown out of the strongly-

Fig. 15.

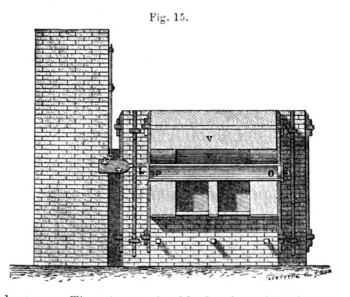

heated retorts. The mixture should also be calcined as needed and used to fill the tubes while still red hot. When cold, uncalcined mixture is used, it is put into large cartridges of thick paper or canvas, 8 centimetres diameter and 35 centimetres long.

The furnace and tubes are shown in section in Fig. 15. The tubes T are 120 centimetres long, 14 centimetres inside diameter, and 10 to 12 millimetres in thickness. They are formed from one piece of boiler iron, bent and welded along one side. The iron plate P which closes one end is about 2 centimetres thick, and pierced on one of its edges quite close to the side of the cylinder by a hole in which is screwed or fitted an iron tube, L, 5 to 6 centimetres long and 15 to 20 millimetres inside diameter, and tapering off at the end to fit into the condenser neck. The other end of the tube is closed by an iron plug, O, terminated by a knob. The welded side of the tube is kept uppermost. These iron tubes should not, like the mercury bottles, be heated in the bare fire; it is necessary to coat them with a resistant luting which is itself enveloped by a refractory jacket 1 centimetre thick, 22 centimetres interior diameter, and the same length as the retorts. This protection is commenced by plastering the retorts over with a mixture of equal parts of raw clay and stove ashes, which have been made into a paste with water and as much sand worked into the mixture as it will take without losing its plasticity, also adding some horse manure. This luting should be dried slowly, and the tube thus prepared is introduced into the refractory jacket, the open space between the two being filled with a powdered refractory brick. Finally, luting is put on the iron plate P, so that no part of it is exposed to the flame.

The furnace I have used is a reverberatory, but I do not recommend its use without important modifications because it does not realize all the conditions of easy and economic heating. The grate is divided into two parts by a little wall of refractory brick, on which the middle of the reduction cylinders rests. The tubes are thus seen to be immediately over the bed of fuel. The top of the bridge is a little higher than the upper edge of the cylinders, this and the very low arch making the flame circulate better all around the tubes. A third cylinder might easily be placed above these two, and be heated satisfactorily, without any more fuel being burnt. This reverberatory receives on its bed the mixtures to be calcined, placed in cast-iron or earthen pots according to their composition. When the furnace is kept going night and day producing sodium, the temperature rises on the bed to clear cherry-

·red, and experience has shown that other reducing cylinders might be placed there, under such conditions, and be heated sufficiently for the reduction.

All that I have said of the manufacture of sodium in mercury bottles applies equally to its manufacture in cylinders. The only difference consists in the charging and discharging, and I have only to add several precautions to be taken. On introducing the cartridges containing uncalcined mixture, only 8 to 9 kilos can be heated at once; double as much can be used of previously calcined mixture. The plug O is put in place, not so tightly that it cannot easily be taken out again; a little luting stops all leakages which show themselves. The reduction lasts about four hours. When it is finished, a little water is thrown on the plug O, and it is easily loosened and removed. On looking into the cylinder, the cartridges are seen to have kept their shape, but have shrunken so much that their diameter is only about 2 to 3 centimetres; they are very spongy. This shows that the mixture has not melted; the remainder is principally lime and carbon, and free from sodium carbonate. While opening the cylinder, a bright-red-hot iron is thrust into the outlet tube L, to keep dirt from getting into it, and it is kept in until the charging is finished. The cartridges are put in by means of semi-cylindrical shovels. The sudden heating of the mixture disengages soda dust from uncalcined mixtures,· which is very disagreeable to the workmen. The cylinders are closed, and when the sodium flame appears at the outlet tube the condenser is attached, and the operation proceeds as already described.

The envelopes of the cylinders are thick enough to prevent the distillation of the sodium being in any way affected by the accidental causes of cooling of the fire. So when fresh fuel is charged or the door of the reverberatory is opened causing the draft to cease almost entirely in the fire-place, the operation should not suffer by these intermittences provided that they are not too prolonged. In short, when operating in cylinders, the production of sodium is easier, less injurious to the workmen, and less costly in regard to labor and fuel than when working with mercury bottles. At times, after working a fortnight with many interruptions dangerous for the apparatus, my experiment has been suddenly ended. The furnace was intact; the envelopes of the tubes were split open, and the luting on the tubes found to be compact and coherent but without traces of fusion, showing perfect resistance. The iron tubes meanwhile had not suffered inside or out, and seemed as though they would last indefinitely. I attribute this success to the particular care given to the composition of the jackets, and to the perfection with which the tubes had been welded. Only on one of the tubes was a very slight crack found on a part not the most highly heated, and not sufficient to cause the tube to be discarded.

Tissier Bros.' method of procedure (1856). As related in the historical treatment of the subject (p. 22), Deville charged the Tissier Bros. with appropriating from him the process for the continuous production of sodium in cylinders, which, as just given, was devised during the experiments at Javel. On the other hand, the Tissier Bros. asserted their right to the process, patenting it, and using it in the works started at Rouen in the latter part of 1855. The following details are taken from Tissier's " Recherche de l'Aluminium," only such being selected as supplement Deville's description which has just been given.

The sodium carbonate is first well dried at a high temperature, then mixed with well-dried pulverized charcoal and chalk, ground to the finest powder, the success of the operation depending on the fineness of this mixture. The proportions of these to use are various. One simple mixture is of—

Sodium carbonate	566
Coal	244
Chalk	95
Coke	95
	1000

Another contains—

Sodium carbonate	615
Coal	277
Chalk	108
	1000

The addition of chalk has the object of making the mixture less fusible and more porous, but has the disadvantage that the residue remaining in the retort after the operation is very impure, and it is impossible to add any of it to the succeeding charge ; and also, some of it being reduced to caustic lime forms caustic alkali with some sodium carbonate, which is then lost. When the mixture is well made it is subjected to a preliminary calcination. This is done in cast-iron cylinders, two of which are placed side by side in a furnace and heated to redness (see Fig. 16). This is continued till all the moisture, carbonic acid, and

Fig. 16.

any carburetted hydrogen from the coal cease coming off. The mass contracts, becomes white and somewhat dense, so that a larger amount of the mixture can now be treated in the retorts where the sodium is evolved. As soon as the outcoming gases burn with a yellow flame, showing sodium coming off, the calcination is stopped. The mixture is then immediately drawn out on to the stone floor of the shop, where it cools quickly and is then ready for the next operation. This calcination yields a mixture which without any previous reactions is just ready to evolve sodium when brought to the necessary temperature. This material is made into a sort of cylinder or cartridge and put into the decomposition retorts (see Fig. 15). The charging should be done quickly. The final retorts are of wrought-iron, since cast-iron would not stand the heat. At each end this retort is closed with wrought-iron stoppers and made tight with fire-clay. Through one stopper leads the pipe to the condenser, the other stopper is the one removed when the retort is to be recharged. These retorts are placed horizontally in rows in a furnace. Usually four are placed in a furnace, preferably heated by gas, such as the Siemens regenerative furnace or Bicheroux, these being much more economical. In spite of all these precautions the retorts will be strongly attacked, and in order to protect them from the destructive action of a white heat for seven or eight hours they are coated with some kind of fire-proof material. The best for this purpose is graphite, which is made into cylinders inclosing the retorts, and which can remain in place till the furnace is worn out. These graphite cylinders not only protect the iron retorts, but prevent the diffusion of the gaseous products of the reaction into the hearth, and so support the retorts that their removal from the furnace is easily accomplished. Instead of these graphite cylinders the retorts may be painted with a mixture that melts at white heat and so enamels the outside. A mixture of alumina, sand, yellow earth, borax, and water-glass will serve very well in many cases. We would remark that the waste gases from this furnace can be used for the calcining of the mixture, or even for the reduction of the aluminium by sodium where the manufacture of the former is connected with the making of the sodium.

As for the reduction of the sodium, the retort is first heated to redness, during which the stopper at the condenser end of the retort is left off. The charge is then rapidly put in, and the stopper at once put in place. The reaction begins almost at once and the operation is soon under full headway, the gases evolved burning from the upper slit of the condenser tube with a flame a foot long. The gases increase in volume as the operation continues, the flame becoming yellower from sodium and so intensely bright as to be insupportable to look at. Now has come the moment when the workman must quickly adapt the condenser to the end of the tube projecting from the retort, the joint being greased with tallow or paraffin. The sodium collects in this in a melted state and trickles out. The length of the operation varies, depending on the intensity of the heat and the quantity of the mixture; a charge may sometimes be driven over in two hours, and sometimes it takes eight. We can say, in general, that if the reaction goes on quickly a somewhat larger amount of sodium is obtained. The higher the heat used, however, the quicker the retorts are destroyed. The operation requires continual attention. From time to time, a workman with a prod opens up the neck of the condenser. But if care is not taken the metal overflows; if this happens, the metal overflowing is thrown into some petroleum, while another man replaces the condenser with an empty one. The operation is ended when the evolution of gas ceases and the flame becomes short and feeble, while the connecting tube between the retort and condenser keeps clean and does not stop up. As soon as this occurs, the stopper at the charging end is removed, the charge raked out into an iron car, and a new charge being put in, the operation continues. After several operations the retorts must be well cleaned and scraped out. The sodium thus obtained is in melted bits or drops, mixed with carbon and sodium carbonate. It must, therefore, be cleaned, which is done by melting it in a wrought-iron kettle under paraffin with a gentle heat, and then casting it into the desired shapes. The sodium is kept under a layer of oil or any hydrocarbon of high boiling point containing no oxygen.

Tissier gives the reaction as—

$$Na^2CO^3 + 2C = 3CO + 2Na.$$

The sodium is condensed, while the carbonic oxide, carrying over some sodium, burns at the end of the apparatus. This would all be very simple if the reaction of carbonic oxide on sodium near the condensing point did not complicate matters, producing a black, infusible deposit of sodium monoxide (Na^2O) and carbon, which on being melted always gives rise to a loss of sodium.

Deville's Improvements at La Glacière (1857).

At this works Deville tried the continuous process of manufacturing sodium in cylinders on a still larger scale, with the following results, as described by Deville himself:—

We made no change in the composition of the mixtures used from those already described, or in the form or size of the iron tubes or the method of condensation; but we worked with six cylinders at a time in a furnace similar to the puddling furnaces of M. Guadillot, the tubes being protected by refractory envelopes. The cylinders were so arranged on the hearth that the flame bathed all parts of their surface. A low brick wall extends down the

11

centre of the hearth, supporting the middle of the cylinders, which extend across it. The hearth is well rammed with refractory sand, and the space between it and the bottom of the cylinders serves as a passage way for most of the flame.

Our six cylinders worked satisfactorily for five days. We were able to observe that they were all heated with remarkable uniformity, and that the heat was sufficient all round them. It also appeared that the rear end of the cylinders required only a hermetic seal. Indeed, as soon as the operation was well under way and sodium distilling off, some of it condensed and oxidized in the cool parts of the apparatus, forming a sort of plug of carbonate and carbides of sodium which the vapor and gases could no longer penetrate. We were thus able for a long time to distil sodium away from one of our tubes which was entirely opened at the rear.

This new furnace worked so well that we were hopeful of complete success when an accident happened which compelled the stopping of the experiment. The iron tubes had been ordered 1.20 metres long, the size of the hearth calculated accordingly, but they were delivered to us only 1.05 metres long. We made use of these, with the result that the rear ends became red-hot during the operation and allowed sodium vapors to leak through. These leaked through the luting, and escaping into the furnace melted the envelopes very rapidly.

In another attempt, in which this fault was avoided, we were unsuccessful because the envelopes gave way at the first heating up, both they and the iron tubes being of inferior quality. We were considerably inconvenienced by the failure of these experiments, which caused considerable expense and gave no very definite results. Just then a new sort of apparatus was devised, a description of which is given later on. It will be seen that we were compelled to employ tubes of very small value, so that their destruction in case of accident involved no great loss, and to heat each one by an independent fire, so that the stoppage or destruction of one cylinder would not necessitate the stoppage or endanger the safety of the neighboring ones.

Cast-iron vessels.—Deville tried at La Glacière, as well as at Javel, to utilize cast-iron vessels for producing sodium. Deville states the difficulties which caused their use to be unsuccessful to be as follows :—

The result was always unfavorable. Sodium is obtained, but as soon as its production becomes rapid the vessel melts and the operation is quickly ended. This follows because the temperature necessary for the production of the metal is far from being sufficient for producing it in large quantities at once ; and we know that this is the one condition for condensing the sodium well and obtaining it economically. This observation led me to think that by diminishing very much the temperature of the furnace, large apparatus of cast-iron with large working surface could be used, thus making at a time a large amount of metallic vapor which could be condensed in recipients of ordinary size. The whole large apparatus would thus have the output of a smaller one worked at a higher temperature. My experience has shown me that in large sized tubes heated to a low temperature there is formed in a given time about as much sodium as from a single mercury bottle at a much higher heat. This is the reason why larger condensers are not necessary with the larger tubes. Before knowing this fact, I tried a large number of useless experiments to determine the size of condensers suitable for large apparatus. It is on this principle that I have long been endeavoring to make sodium without working at high temperatures and using less costly and more easily protected apparatus.

Improvements used at Nanterre (1859).

The method used here was exactly that already described, the improvements being solely in details of the apparatus. These are described by Deville as follows :—

The experiments made at Javel and the continuous process used at Glacière have shown us in the clearest manner the absolute necessity of efficient protection for the iron cylinders, for without this protection the method cannot be practised with economy. Further, experiments in this direction are very costly, for the failure of a tube stops the working of a large number of cylinders and often compromises the brick work of the furnace itself. We therefore came to the conclusion that for making the small quantity of sodium we required, 300 to 500 kilos a month, it would be better to employ smaller apparatus independent of each other and easy to replace.

The iron tubes are made of thinner iron and at very little expense, by taking a sheet of iron, curving it into a cylinder and rivetting the seam. This tube resembles very closely those used at Javel, shown in Fig. 15, but of smaller dimensions. It is closed at each end by cast-iron plugs, one of which has a hole for the outlet tube. These cylinders are filled with sodium mixture and placed in furnaces of the form of Fig. 9, except it is necessary to have openings in the back and front of the furnace so that the cast-iron plugs closing the cylinders may be outside, to prevent their melting. We used coke at first for fuel, fed around the cylinders, but M. Morin has since placed the tubes out of direct contact with the fuel, uses soft coal, and heats the tubes by contact with the flame and by radiation. In the latest form used, two cylinders are placed in each furnace, and, in general, they serve for two or three operations. All that has been said in connection with the manufacture in mercury bottles is immediately applicable to the manufacture in cylinders of this kind, the capacities of which may vary from two to six or eight litres, without any change in the manner of using them. We have, however, adopted altogether condensers of cast iron. The neck is cylindrical and belongs only to one-half of the apparatus, the neck end of the other plate being bevelled and fitting closely against a recess in the other plate.

The foregoing shows the sodium industry as it was perfected by Deville, in 1859, and as it remained for twenty-five years without sensible change. The cost of sodium by this process is stated to have been, in 1872, as follows:—

Manufacture of one kilo of sodium.

Soda	. .	. 9.35 kilos @ 32 fr. per 100 kilos	= 3 fr. 9 cent.
Coal	. .	. 74.32 "	" 1.40 "	"	"	= 1 " 4 "
		Wages	. . .	. 1 " 73 "
		Expenses	. . .	. 3 " 46 "
		Total	. . .	. 11 " 32 "

which is equal to \$1 per lb. The larger part of the expense account is the cost of retorts or tubes in which the operation takes place, and which are so quickly destroyed that the replacing of them forms nearly one-quarter of the cost of the metal.

Minor Improvements (1859–1888).

R. Wagner* uses paraffin in preference to paraffin oil in which to keep the sodium after making it. Only pure paraffin, which has been melted a long time on a water bath, and all its water driven off, can be used. The sodium to be preserved is dipped in the paraffin melted on a water bath and kept at no higher heat than 55°, and the metal is thereby covered with a thick coat of paraffin which protects it from oxidation, and may then be put up in wooden or paper boxes. When the metal is to be used, it is easily freed from paraffin by simply warming it, since sodium melts at 95° to 96° C., and the paraffin at 50° to 60°.

The reduction of potassium carbonate by carbon requires a much less degree of heat than that of sodium carbonate, and, therefore, many attempts have been made to reduce potassium and sodium together, under circumstances where sodium alone would not be reduced. Dumas† added some potassium carbonate to the regular sodium mixture; and separated the sodium and potassium from each other by a slow, tedious oxidation. R. Wagner‡ made a similar attempt. He says that not only does the reduction of both metals from a mixture of their carbonates with carbon work easier than sodium carbonate alone with carbon, but even caustic soda may be used with potassium carbonate and carbon. Also, the melting point of potassium and sodium alloyed is much lower than that of either one alone, in consequence of which their boiling point and the temperature required for reduction are lower.

J. B. Thompson and W. White§ specify mixing dry sodium carbonate with a liquid carbonaceous material, preferably tar, driving off all volatile matter in iron pots at a low heat, and then distilling in a tubular fire-clay retort connected with a tightly-closed receiver containing a little paraffin oil to insure a non-oxidizing atmosphere, and also provided with a small escape pipe

* Dingler, 1883, p. 252.
† Handbuch der Angewandten Chemie, 1830, ii. 345.
‡ Dingler, 143, 343.
§ English Patent 8426, June 11, 1887.

for carbonic oxide. This process gave great prospects of success when tried in the laboratory, but on a manufacturing scale it failed for the reason (assigned by Mr. Thompson) that the sheet-iron tray, designed to keep the material from attacking the retort, absorbed carbon at about 1000° and fused, after which no sodium was produced, since the material took up silica from the retort, absorbing so much that the carbon no longer decomposed it.

H. S. Blackmore,* of Mount Vernon, U. S. A., patents the following process of obtaining sodium :—

$27\frac{1}{2}$ parts calcium hydrate,
31   "   ferric oxide,
31   "   dry sodium carbonate,
$10\frac{1}{2}$ "   charcoal

are intimately mixed and subjected to a red heat for 20 minutes, afterwards to a white heat. Caustic soda is first produced, the carbon reduces the ferric oxide, producing iron, which in its turn reduces the caustic soda and sodium vapors distil. The residue consists of ferric oxide and lime, and is slaked and used over.

O. M. Thowless† of Newark, N. J., claims to place a retort in a furnace, providing it on one side with an arm through which carboniferous material can be supplied, on the other side with a similar arm (surrounded by flues), into which caustic soda or sodium carbonate is charged—a valve controlling their flow into the retort. Outside the furnace and on top of it is a flat condenser into which the sodium vapor passes.

G. A. Jarvis‡ patents the replacement of the iron tubes or crucibles used in the manufacture of sodium, by fire-clay apparatus lined with basic material, such as strongly burnt magnesia with 10 per cent. of fluorspar.

Castner's Process (1886).

The first public announcement of this process was through one of the New York daily journals,§ and as the tone of the article is above that of the usual newspaper reports, and the expectations contained in it have been subsequently more than realized, we

---

* English Patent 15156, Oct. 22, 1888.
† English Patent 12486 (1887).
‡ English Patent 4842, March 31, 1888.
§ New York World, May 16, 1886.

cannot better introduce a description of this process than by quoting the paragraph referred to :—

"When sodium was reduced in price to $1.50 per lb. it was thought to have touched a bottom figure, and all hope of making it any cheaper seemed fruitless. This cheapening was not brought about by any improved or new process of reduction, but was owing simply to the fact that the aluminium industry required sodium, and by making it in large quantities its cost does not exceed the above-mentioned price. The retail price is now $4.00 per lb. The process now used was invented by Brünner, in 1808, and up to the present time nothing new or original has been patented except three or four modifications of his process which have been adopted to meet the requirements of using it on a large scale. Mr. H. Y. Castner, whose laboratory is at 218 West Twentieth Street, New York, has the first patent ever granted on this subject in the United States, and the only one taken out in the world since 1808. Owing to negotiations being carried on, Mr. Castner having filed applications for patents in various foreign countries, but not having the patents granted there yet, we are not at liberty to state his process fully. The metal is reduced and distilled in large iron crucibles, which are raised automatically through apertures in the bottom of the furnace, where they remain until the reduction is completed and the sodium distilled. Then the crucible is lowered, a new one containing a fresh charge is substituted and raised into the furnace, while the one just used is cleaned and made ready for use again. The temperature required is very moderate, the sodium distilling as easy as zinc does when being reduced. Whereas by previous processes only one-third of the sodium in the charge is obtained, Mr. Castner gets nearly all, for the pots are nearly entirely empty when withdrawn from the furnace. Thus the great items of saving are two or three times as much metal extracted from a given amount of salt, and cheap cast-iron crucibles used instead of expensive wrought-iron retorts. Mr. Castner expects to produce sodium at 25 cents per lb., thus solving the problem of cheap aluminium, and with it magnesium, silicon, and boron, all of which depend on sodium for their manufacture. Thus the production of cheap sodium means much more than cheap aluminium.

Mr. Castner is well known in New York as a chemist of good standing, and has associated with him Mr. J. H. Booth and Mr. Henry Booth, both well known as gentlemen of means and integrity."

The following are the claims which Mr. Castner makes in his patent :—*

1. In a process for manufacturing potassium or sodium, performing the reduction by diffusing carbon in a body of alkali in a state of fusion at moderate temperatures.

2. Performing the reduction by means of the carbide of a metal or its equivalent.

3. Mechanically combining a metal and carbon to increase the weight of the reducing material, and then mixing this product with the alkali and fusing the latter whereby the reducing material is held in suspension throughout the mass of fused alkali.

4. Performing the deoxidation by the carbide of a metal or its equivalent.

For an explanation of the principles made use of in the above outlined process we will quote from a lecture delivered by Mr. Castner at the Franklin Institute, Philadelphia, October 12th, 1886. That Institution has since bestowed on Mr. Castner one of its gold medals as a recognition of the benefit to science accruing from his invention.

" In the ordinary sodium process, lime is added to the reducing mixture to make the mass refractory, otherwise the alkali would fuse when the charge is highly heated, and separate from the light, infusible carbon. The carbon must be in the proportion to the sodium carbonate as four is to nine, as is found needful in practice, so as to assure each particle of soda in the refractory charge having an excess of carbon directly adjacent or in actual contact. Notwithstanding the well-known fact that sodium is reduced from its oxide at a degree of heat but slightly exceeding the reducing point of zinc oxide, the heat necessary to accomplish reduction by this process and to obtain even one-third of the metal in the charge, closely approaches the melting point of wrought iron.

" In my process, the reducing substance, owing to its composition and gravity, remains below the surface of the molten salt,

* U. S. Pat. No. 342897, June 1, 1886.   Hamilton Y. Castner, New York.

and is, therefore, in direct contact with the fused alkali. The metallic coke of iron and carbon contains about 30 per cent. carbon and 70 per cent. iron, equivalent to the formula FeC^2. I prefer to use caustic soda, on account of its fusibility, and mix with it such quantity of so-called 'carbide' that the carbon contained in the mixture shall not be in excess of the amount theoretically required by the following reaction :—

$$3NaOH + FeC^2 = 3Na + Fe + CO + CO^2 + 3H ;$$

or, to every 100 lbs. of pure caustic soda, 75 lbs. of 'carbide,' containing about 22 lbs. of carbon.

"The necessary cover for the crucible is fixed stationary in each chamber, and from this cover a tube projects into the condenser outside the furnace. The edges of the cover are convex, those of the crucible concave, so that when the crucible is raised into position and held there the tight joint thus made prevents all leaking of gas or vapor. Gas is used as fuel, and the reduction begins towards 1000° C. As the charge is fused, the alkali and reducing material are in direct contact, and this fact, together with the aid rendered the carbon by the fine iron, in withdrawing oxygen from the soda, explains why the reduction is accomplished at a moderate temperature. Furthermore, by reducing from a fused mass, in which the reducing agent remains in suspension, the operation can be carried on in crucibles of large diameter, the reduction taking place at the edges of the mass, where the heat is greatest, the charge flowing thereto from the centre to take the place of that reduced.

"I am enabled to obtain fully 90 per cent. of the metal in the charge, instead of 30 per cent. as formerly. The crucibles, after treatment, contain a little carbonate of soda, and all the iron of the 'carbide' still in a fine state of division, together with a small percentage of carbon. These residues are treated with warm water, the solution evaporated to recover the carbonate of soda, while the fine iron is dried, and used over again for 'carbide.'"

Mr. Castner having demonstrated in his New York laboratory the success of his process, went to England, and for several months during the winter of 1886–7 was engaged in building and working a large sodium furnace. This was successfully car-

ried out near London, the inventor being assisted by Mr. J. Mac-
Tear, F.C.S., who, in March, 1887, read a description of this
furnace and the results obtained before the Society of Chemical
Industry. During the working of this furnace it was inspected
by many chemical and metallurgical authorities, who were completely satisfied as to its success. As the furnace now used differs
in a few details from the one just referred to, it may be well to
extract the essential particulars from Mr. MacTear's paper—on
the ground that the importance of this invention justifies a complete discussion of its development :—

"Since Mr. Castner's paper upon his process, which was read
before the Franklin Institute of Philadelphia, October 12th,
1886, several slight changes in the mode of carrying on this
process have been made. These have been brought about by
the experience gained from the actual working of the process
upon a commercially large scale.

"The reactions by which the sodium is produced are somewhat difficult to describe, as they vary somewhat according to
the mixture of materials and temperature employed in the reduction. The mixture and temperature which it is now preferred to
use is represented by the reaction :—

$$6NaHO + FeC_2 = 2Na_2CO_3 + 6H + Fe + 2Na.$$

"In place of using an actual chemical compound of iron and
carbon, as expressed by the above reaction, a substitute or equivalent is prepared as follows : To a given quantity of melted pitch
is added a definite proportion of iron in a fine state of division.
The mixture is cooled, broken up into lumps, and cooked in large
crucibles, giving a metallic coke consisting of carbon and iron,
the proportions of each depending upon the relative quantities of
pitch and iron used. This metallic coke, after being finely ground,
provides a substance having the iron and carbon in a like proportion to an iron carbide, and from which neither the iron nor carbon can be separated by mechanical means. The fine iron is
conveniently prepared by passing carbonic oxide and hydrogen
in a heated state, as obtained from an ordinary gas producer, over
a mass of oxide of iron commercially known as 'purple ores,'
heated to a temperature of about 500° C.

"In producing sodium, caustic soda of the highest obtainable strength is used, and there is mixed with it a weighed quantity of the so-called 'carbide,' sufficient to furnish the proper amount of carbon to carry out the reaction indicated above. The crucibles in which this mixture is treated are made of cast-steel, and are capable of containing a charge of 15 lbs. of caustic soda, together with the proper proportion of the 'carbide.'

"After charging a crucible with the above mixture, it is placed in a small furnace where it is kept at a low heat for about thirty minutes, during which time the mass fuses, boils violently, and a large part of the hydrogen is expelled by the combined action of the iron and carbon, the 'carbide,' owing to its gravity, remaining in suspension throughout the fused soda. At the end of the time stated, the contents of the crucible have subsided to a quiet fusion. The crucible is then lifted by a pair of tongs on wheels and placed upon the platform of the elevating gear, as shown in the drawing, and raised to its position in the heating chamber of the main distilling furnace. The cover which remains stationary in the furnace has a convex edge, while the crucible has a groove round the edge into which the edge of the cover fits. A little powdered lime is placed in the crucible groove just before it is raised, so that when the edges of the cover and crucible come together they form a tight joint, and at the same time will allow the crucible to be lowered easily from the chamber when the operation is finished, to give place to another containing a fresh charge. From the cover projects a slanting tube (see Fig. 17), connected with the condenser. The condenser is provided with a small opening at the further end to allow the escape of hydrogen, and has also a rod fixed (as shown), by means of which any obstruction which may form in the tube during distillation, may be removed. After raising a crucible in its place in the furnace, the hydrogen escaping from the condenser is lighted, and serves to show by the size of the flame how the operation is progressing in the crucible, the sodium actually distilling soon after the crucible is in its place. The temperature of the reduction and distillation has been found to be about 823° C. The gas coming off during the first part of the distillation has been analyzed and found to consist of pure hydrogen. An analysis of the gas disengaged when the operation was almost completed, gave as a result, hydrogen 95 per cent., carbonic oxide 5 per cent. It

Fig. 17.

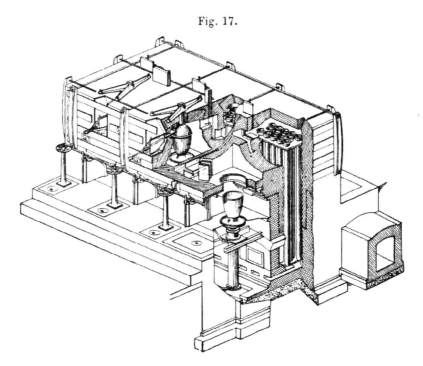

has been found advisable to use a little more 'carbide' than the reaction absolutely requires, and this accounts for the presence of the small quantity of carbonic oxide in the expelled gas, the free carbon acting upon the carbonate formed by the reaction, thus giving off carbonic oxide and leaving a very small percentage of the residue in the form of peroxide of sodium. This small amount of carbonic oxide rarely combines with any of the sodium in the tube, and so the metal obtained in the condensers is pure, and the tubes never become choked with the black compound. In the preparation of potassium a little less 'carbide' is used than the reaction requires, thus no carbonic oxide is given off, and all danger attached to the making of potassium is removed. After the reduction and distillation the crucible is lowered from the furnace and the contents poured out, leaving the crucible ready to be recharged. The average analyses of the residues show their composition to be as follows:—

Carbonate of soda . . . . . . 77 per cent.
Peroxide of sodium . . . . . . 2 "
Carbon . . . . 2 "
Iron . . . . 19 "

"The average weight of these residues from operating upon charges of 15 lbs. caustic soda and 5¼ lbs. of carbide is 16 lbs. These residues are treated either to produce pure crystallized carbonate of soda or caustic soda, and the iron is recovered and used again with pitch in the formation of the 'carbide.' From this residue weighing 16 lbs., is obtained 13 lbs. of anhydrous carbonate of soda, equivalent to 9.4 lbs. caustic soda of 76 per cent.

"Operating upon charges as above mentioned the yield has been—

Sodium, actual . . . 2.50 lbs. Theory 2.85 lbs.
Soda carbonate, actual . 13.00 lbs.   "   13.25 lbs.

"The average time of distillation in the large furnace has been 1 hour 30 minutes, and as the furnace is arranged for three crucibles, 45 lbs. of caustic soda are treated every 90 minutes, producing 7½ lbs. of sodium and 39 lbs. of carbonate of soda. The furnace is capable of treating 720 lbs. of caustic soda daily, giving a yield in 24 hours of 120 lbs. of sodium and 624 lbs. of anhydrous carbonate of soda. The furnace is heated by gas which is supplied by a Wilson Gas Producer, consuming 1 cwt. of fuel per hour. The small furnace in which the crucibles are first heated requires about ½ cwt. per hour. The following estimate of cost, etc., is given from the actual running of the furnace working with the above charges for 24 hours :—

	£	s.	d.
720 lbs. of caustic soda @ £11 per ton . . .	3	10	10
150 lbs. of "carbide" @ ½d. per lb. . . .	0	6	4
Labor	1	0	0
Fuel	0	17	0
Re-converting 624 lbs. of carbonate into caustic, at a cost of about £5 per ton on the caustic produced, say	1	0	0
Total . . .	6	14	2
Deducting value of 475 lbs. of caustic recovered	2	6	8
Cost of 120 lbs. of sodium . . .	£4	7	6

Cost per pound 8¼d.

"Regarding the item of cost relating to the damage caused to the crucibles by the heat, this question has been very carefully gone into, some of the crucibles have been used upwards of fifty times, and from present indications of their condition there is no doubt that they can continue to be used at least 150 times more before they become unfit for further use. In considering 200 operations to be the life of a crucible, the item of damage or wear and tear amounts to less than 1*d.* per lb. on the sodium produced, and if we take the furnace tear and wear at the same rate of 1*d.* per lb., we will see that the tear and wear of plant is only one-twelfth of that incurred in the ordinary process. It is upon these facts that Mr. Castner bases his claim to be able to produce sodium by his process upon the large scale, at a cost of less than 1*s.* per lb. The advantages of this process will be apparent to any one at all familiar with the manufacture of these metals as conducted heretofore. The first and most important end gained is their cheap production, and this is owing chiefly to the low heat at which the metals are produced, the quickness of the operation, non-clogging of the conveying tubes, and a very small waste of materials. The process furthermore admits of being carried on upon a very large scale, in fact it is intended ultimately to increase the size of the crucible so as to make the charges consist of 50 lbs. of caustic soda. Crucibles of cast iron have been found quite suitable, and it is intended in future to use crucibles made of this material in place of the more expensive steel."

Immediately on the demonstration of this success, a company was formed to unite Mr. Castner's sodium process with Mr. Webster's improvements in the production of aluminium chloride. The Aluminium Co., Ltd., first appeared before the public in June, 1887, and at the first meeting in the following September it was decided to build works at once. These were begun at Oldbury, near Birmingham, and were in working operation by the end of July, 1888. The furnaces here erected are larger than the one just described, and altogether have a producing capacity of nearly a ton of sodium a day. The following details respecting this latest plant and its working are taken mostly from an address delivered before the Society of Arts, March 13, 1889, by Mr.

William Anderson, and from a discourse at the Royal Institution, May 3, 1889, by Sir Henry Roscoe, president of the company.

There are four large sodium furnaces, each holding five pots or crucibles and heated by gas, applied on the regenerative principle. A platform about five feet above the floor allows the workmen to attend to the condensers, while the lifts on which the pots are placed sink level with the floor. The crucibles used are egg-shaped, about 18 inches diameter at their widest part and 24 inches high; when joined to the cover the whole apparatus is about 3 feet in height. The covers have vertical pipes passing through the top of the furnace, forming a passage for the introduction of part of the charge, and also a lateral pipe connecting with the condenser. The whole cover is fixed immovably to the roof of the furnace and is protected by brickwork from extreme heat; but it can easily be removed when necessary. The natural expansion of the vessels is accommodated by the water pressure in the hydraulic lifts on which the pots stand. When the lift is lowered and sinks with the lower part of the crucible to the floor level, a large pair of tongs mounted on wheels is run up, and catching hold of the crucible by two projections on its sides it is carried away by two men to the dumping pits, on the edge of which it is turned on its side, the liquid carbonate of soda and finely divided iron which form the residue are turned out, and the inside is scraped clean from the opposite side of the pit, under the protection of iron shields. When clean inside and out, it is lifted again by the truck and carried back to the furnace, receiving a fresh charge on its way. It is then put on the platform and lifted into place, having still retained a good red heat. It takes only $1\frac{1}{2}$ to 2 minutes to remove and empty a crucible, and only 6 to 8 minutes to draw, empty, recharge, and replace the five crucibles in each furnace. The time occupied in reducing a charge is one hour and ten minutes. It is thus seen that one bank of crucibles yields 500 pounds of sodium in twenty-four hours, the battery of four furnaces produces about a ton in that time.

The shape of the condenser has been altogether changed. Instead of the flat form used on the furnace at London (see Fig. 17), which resembled the condenser used in the Deville process, a peculiar pattern is used which is quite different. It consists in a tube-shaped cast-iron vessel 5 inches in diameter, nearly 3 feet long over all, and having a slight bend upwards at a point about 20 inches from the end. At this bend is a small opening in the bottom, which can be kept closed by a rod dropping into it; this rod, passing through a tight-fitting hole above, can be raised or lowered from outside. Thus the sodium can either run out continually into small pots placed beneath the opening or can be allowed to collect in the condenser until several pounds are present, then a small potful run out at once, by simply lifting the iron rod. The outer end of the condenser is provided with a lid, hinged above, which can be thrown back out of the way when required. This lid also contains a small peep-hole covered with mica. In the top of the condenser just before the end is a small hole through which the hydrogen and carbonic oxide gases escape when the end is closed, burning with the yellow sodium flame. The bend in the condenser is not acute enough to prevent a bar being thrust through the end right into the outlet tube projecting from the furnace, thus allowing the whole passage to be cleaned out should it become choked up. Previous to drawing the crucibles from the furnace for the purpose of emptying them and recharging, the small pots containing the metal distilled from one charge are removed and empty ones put in their place. Those removed each contain on an average about 6 lbs. of sodium, or 30 lbs. from the whole furnace. When sufficiently cool, petroleum is poured on top of the metal in the pots, and they are wheeled on a truck to the sodium casting shop, where the sodium is melted in large pots heated by oil baths and cast either into large bars ready to be used for making aluminium or into smaller sticks to be sold. The sodium is preserved under an oil such as petroleum, which does not contain oxygen in its composition, and the greatest care is taken to protect it from water.

Special care is taken to keep the temperature of the furnace at about 1000° C., and the gas and air-valves are carefully regulated so as to maintain as even a temperature as possible. The covers remain in the furnace from Sunday night to Saturday afternoon, and the crucibles are kept in use till worn out, when new ones, previously heated red-hot, are substituted without interrupting the general running of the furnace. These bottom halves of the crucibles are the only part of the plant liable to exceptional wear and tear, and their durability is found to depend very much on the soundness of the casting, because any pores or defects are rapidly eaten into and the pot destroyed. The average duration of each crucible is now 750 lbs. of sodium, or 125 charges.

Apropos of the reaction involved in the reduction, it has probably been observed that Mr. MacTear proposes a different formula from that suggested by Mr. Castner. Mr. Weldon remarked that when a mixture of sodium carbonate and carbon was heated the carbon did not directly reduce the soda, but at a high temperature the mixture gives off vapors of oxide of sodium (Na^2O) part of which dissociates into free oxygen and sodium vapor; as soon as this dissociation takes place the carbon takes up the oxygen, forming carbonic oxide, and thus, by preventing the recombination of the sodium and oxygen, leaves free sodium vapors.

Dr. Kosman, speaking in "Stahl und Eisen," January, 1889, on Castner's process, gives the following explanation of the reactions taking place :—

Ten kilos of caustic soda and 5 kilos of carbide (containing 1.5 kilos of carbon) give the following reaction :—

$$4NaOH + FeC^2 = Na^2CO^3 + Fe + 4H + CO + 2Na,$$

and half the sodium in the mixture is obtained.

Ten kilos of caustic soda and 10 kilos of carbide (containing 3 kilos of carbon) give this reaction—

$$2NaOH + FeC^2 = NaCO + Fe + 2H + CO + Na,$$

and half the sodium in the mixture is again obtained.

If 20 kilos of caustic soda and 15 kilos of carbide are mixed, both the above reactions take place, but if the ignition is continued, the sodium carboxyd (NaCO) reacts on the sodium carbonate according to the reaction—

$$Na^2CO^3 + NaCO = 3Na + 2CO^2,$$

and the entire reaction may be represented by

$$3NaOH + FeC^2 = 3Na + Fe + 3H + CO + CO^2,$$

and all the sodium in the mixture is obtained.

This is the reaction first proposed by Mr. Castner (see p. 169),

12

and the proportions indicated by it gave him the largest return of sodium. Mr. MacTear, however, states that the reaction which takes place is conditioned largely by the temperature, and that at 1000° C. it is probably to be represented by

$$6NaOH + FeC^2 = 2Na^2CO^3 + 6H + Fe + 2Na,$$

which is essentially the same as that given by Sir Henry Roscoe in his discourse, viz :—

$$3NaOH + C = Na^2CO^3 + 3H + Na.$$

This reaction would require $18\frac{3}{4}$ lbs. of carbide to 50 lbs. of caustic soda, and since the sodium carbonate is easily converted back into caustic by treatment with lime, the production of so much carbonate is offset by the ease with which the reaction takes place, and the added advantage that the gas evolved with the sodium is solely hydrogen, thus allowing the reduction to proceed in an atmosphere of that gas, and reducing the production of the usual deleterious sodium carbides to a minimum.

A further discussion of this subject will come up in considering Netto's process.

Netto's Process (1887).

Dr. Curt Netto, of Dresden, has taken out patents in several European countries,* which have been transferred to and are presumably being operated by the Alliance Aluminium Company, of London (see p. 38). The process is continuous, and is based on the partial reduction of caustic soda by carbon. Dr. Netto observes that carbon will reduce caustic soda at first at a red heat, but a white heat is necessary to finish the reduction, the explanation being that the reaction is at first—

$$4NaOH + C = Na^2CO^3 + 2H^2 + CO + Na^2,$$

and that the carbonate is only reduced at a white heat. To avoid any high temperature, the first reaction only is made use of, the carbonate being removed and fresh caustic supplied continuously, and without interrupting the operation or admitting air into the retort in which the reduction takes place.

A vertical cast-iron retort, protected by fire-clay coating, is

---

* German patent (D. R. P.) 45105 ; English patent, October 26, 1887, No. 14602.

surrounded by flues. The flame after heating the retort passes under an iron pot in which the caustic soda is kept melted, and situated just above the top of the retort. This pot has an outlet tube controlled by a stop-cock, by which the caustic may be discharged into a funnel with syphon-shaped stem fastened into the top of the retort. There is also a syphon-shaped outlet at the bottom of the retort, through which the molten sodium carbonate and bits of carbon pass. A hole with tight lid in the upper cover is provided for charging charcoal. A tube passes out just beneath the upper cover, connecting with a large condenser of the shape used by Deville (see Fig. 18). In operating, the retort is heated

Fig. 18.

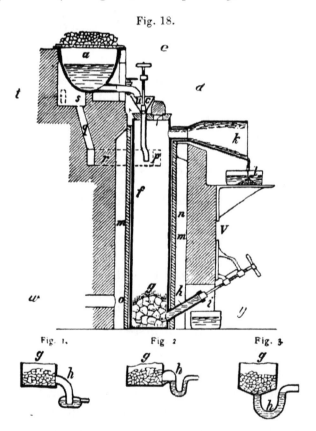

to bright redness, filled one-third with best wood charcoal, and then molten caustic soda tapped from the melting pot into the funnel, the feed being so regulated that the funnel is kept full and the retort closed. The lower opening is kept closed until enough sodium carbonate has accumulated to lock the syphon passage air tight. When after several hours' working the charcoal is almost all used up, the supply of caustic soda is shut off for a time and the retort recharged through the opening in the upper lid, when the operation goes on as before. The sodium carbonate produced is easily purified from carbon by solution. Since sodium vapor at a high temperature is very corrosive, all rivets and screw joints must be avoided in making the retort. On this account, the outlet tubes should be cast in one piece with the retort.

The process of O. M. Thowless, Newark, N. J.,* is essentially identical with Netto's process.

REDUCTION OF SODIUM COMPOUNDS BY ELECTRICITY.

The decomposition of fused sodium chloride by the electric current seems to promise the economic production of sodium, for not only is this metal formed but chlorine is obtained as a by-product, its value reducing very much the cost of the operation.

P. Jablochoff has devised the following apparatus for decomposing sodium or potassium chlorides.† (Fig. 19.)

The arrangement is easily understood. The salt to be decomposed is fed in by the funnel into the kettle heated by a fire beneath. The positive pole evolves chlorine gas, and the negative pole evolves vapor of the metal, for, as the salt is melted, the heat is sufficient to vaporize the metal liberated. The gas escapes through one tube and the metallic vapor by the other. The vapor is led into a condenser and solidified.

Prof. A. J. Rogers, of Milwaukee, Wis., has made a number of attempts to reduce sodium compounds electrolytically, using as a cathode a bath of molten lead and producing an alloy of lead and sodium which he makes use of for the reduction of aluminium compounds. Although these attempts are hardly past the experimental stage, yet the record of the results obtained may very probably be interesting and valuable to other investigators in this line.

Prof. Rogers reasons that from the known heat of combination

---

* U. S. Patent, Nos. 380775, 380776, April 4, 1888.   † Mierzinski.

of sodium and chlorine (4247 calories per kilo of sodium) there is enough potential energy in a pound of coal to separate nearly two pounds of sodium, if any mode of applying the combustion of the coal to this end without loss could be devised. If, however,

Fig. 19.

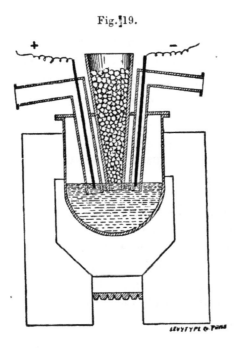

this energy is converted into mechanical work, this again into electrical energy, and this latter used to decompose sodium chloride, we can easily compute the amount of coal to be used in a steam boiler to produce a given amount of sodium by electrolysis. Now, if the electric current could be applied without loss in decomposing sodium chloride, 1 electric horse-power (746 Watts) would produce about 8 lbs. of sodium in 24 hours. But as in practice one mechanical horse-power applied to a dynamo yields only 80 or 90 per cent. of an electric horse-power, and as about 4 lbs. of coal are used per indicated horse-power per hour, from 105 to 120 lbs. of coal would be required per day to produce this result, or about 15 lbs. per lb. of sodium. Since, however, there is a transfer resistance in the passage of the electric current through the molten electrolyte, more than this will be required, in proportion to the amount of current thus absorbed.

The temperature of fusion of sodium chloride is given by Carnelly as 776° C., but Prof. Rogers remarks that the fusing point may be lowered considerably by the presence of other salts; for instance, it melts about 200° lower if a small amount of calcium chloride or potassium chloride is present. We will quote the results of some experiments as given by Prof. Rogers.*

"The following results were obtained among many others by using a Grove battery, a Battersea crucible to hold the sodium chloride, a carbon anode and an iron cathode terminating in a tube of lime placed in the melted salt. As soon as metallic sodium escaped and burnt at the surface of the liquid the current was stopped. A little sodium was oxidized but a considerable amount was found in the tube in metallic state. In six experiments the amount of sodium obtained was from 50 to 85 per cent. of the theoretical amount, averaging 65 per cent. It thus seemed that, with suitable apparatus, from 5 to 6 lbs. of sodium could be obtained in 24 hours per electric horse-power. Thus, if there were no practical difficulties in the construction of the crucibles and other apparatus involved, nor in working continuously on a large scale, the metal could be obtained at small cost. Various forms of crucibles were used and attempts made to distil the metal when formed at the negative electrode (sodium volatilizing at about 900°C.), but the sodium vapor carries with it a large amount of sodium chloride as vapor, and the distillation is attended with difficulty.

"During the last three years I have experimented on the reduction of sodium chloride using molten negative electrodes and especially lead. Lead, tin, zinc, cadmium and antimony all readily alloy with sodium, a large part of which can be recovered from the alloys by distillation in an iron crucible. They can be heated to a higher temperature than pure sodium in acid crucibles without the sodium attacking the crucible. In the following experiments a dynamo machine was used to supply the current.

"Experiment 1. A current averaging 72 amperes and 33 volts was passed through molten sodium chloride contained in two crucibles arranged in series, for two hours. Each contained 30

* Proceedings of the Wisconsin Natural History Society, April, 1889.

lbs. of salt; in the first was put 104 grammes of tin, in the second 470 grammes of lead, each serving as cathode and connection being made through the bottom of the crucible. A carbon anode passed through the cover and extended to within three inches of the molten cathode. The crucible containing the tin was nearer the fire and consequently hotter, and had an average potential across the electrodes of 12 volts, while that containing the lead cathode was 21 volts. When at the end of two hours the carbons were removed and the crucibles cooled and broken open, the lead alloy was found to contain 96 grammes of sodium, or 17 per cent. There was about 90 grammes of sodium found in the tin alloy, or between 45 and 50 per cent. Both these alloys rapidly oxidized in the air, and when thrown into water the action was very energetic, in the case of the tin alloy the liberated hydrogen being ignited, and after the reaction the metals were found at the bottom of the vessel in a finely divided state. Both these alloys reduce cryolite or aluminium chloride."

In Prof. Rogers' further experiments cryolite was added to the bath, so that sodium was produced and aluminium formed in one operation. (See under "Electrolytic Processes," Chap. XI.)

CHAPTER VIII.

THE REDUCTION OF ALUMINIUM COMPOUNDS FROM THE STANDPOINT OF THERMAL CHEMISTRY.

THE branch of chemical science called thermal chemistry may be said to be yet in its infancy. Although an immense mass of thermal data has been accumulated, yet the era of great generalizations in this subject has not yet been reached; and although we know with a fair degree of accuracy the heat of combination of thousands of chemical compounds, including nearly all the common ones, yet the proper way to use these data in predicting the possibility of any proposed reaction remains almost unknown. The principal barriers in the way are two: 1st, the unknown quantities entering into almost every chemical reaction thermally considered, *i. e.*, the heat of combination of elementary atoms to form molecules of the elements; 2d, the uncertainty as regards the critical temperature at which a given exchange of atoms and consequent reaction will take place. We will explain what is meant by these statements.

To illustrate, let us consider the case of hydrogen uniting with oxygen to form water according to the formula—

$$2(H—H) + (O = O) = 2 H^2O$$

where (H—H) and (O = O) represent respectively molecules of hydrogen and oxygen. Now, as 1 kilo of hydrogen unites with 8 of oxygen to form 9 of water, setting free 34462 units of heat (calories), if we take the atomic weights in the above reaction as representing kilos, we shall have the thermal value of the reaction $4 \times 34462 = +137848$ calories. But this quantity is evidently the algebraic sum of the heat evolved in the union of 4 kilos of hydrogen atoms with 32 kilos of oxygen atoms, and the heat absorbed in decomposing 4 kilos of hydrogen gas into atoms, and 32 kilos of oxygen gas into atoms. These two latter quantities are unknown, though a few chemists have concluded from studies on this question that they are probably very large. It has been calculated that the reaction—

H + H = (H—H) sets free 240,000 calories,

and　　O + O = (O = O) sets free 147,200 calories;

but no assurance can be placed on these numbers. If they were approximately true, then

4H + 2O = 2H²O would set free about 773,000 calories.

If these quantities are really anything like so large, and if they are at sometime determined with precision, thermo - chemical principles and conclusions will be greatly modified. Meanwhile, predictions based on the data we have lose all possibility of certainty, and so we need to keep in mind in our further discussion that our deductions at the best can be no more than probabilities. Further, suppose that we mix 1 kilo of hydrogen gas and 8 kilos of oxygen gas, put them in a tight vessel and keep them at the ordinary temperature. No reaction will take place in any length of time, even though 34,462 calories would be set free thereby. The explanation of this is probably that the atoms of hydrogen and oxygen are so firmly bound to each other in the molecules, that the dissimilar atoms have not strength of affinity sufficient to break away in order to combine. However this may be, it is well known that a spark only is necessary to cause an explosive combination of the gases under the above conditions, the temperature of the spark expanding the gases coming in contact with it, causing the atoms to swing with more freedom in the molecules, and as soon as two atoms of hydrogen come within the sphere of attraction of an atom of oxygen and form a molecule of water, the heat liberated is immediately communicated to the adjacent atoms, and almost instantaneously the entire gases have combined. The same principle undoubtedly holds true in cases of reduction. Carbon may be mixed with litharge and the mixture left in the cold forever without reacting, but at a certain temperature the carbon will abstract the oxygen. The temperatures at which reactions of this nature will take place are often determined experimentally, but I know of no theoretical grounds on which they can rationally be calculated.

There are other points which are somewhat indeterminate in these discussions, such as the influence of the relative masses of the reacting bodies, their physical states, i. e., solid, liquid or gaseous, also the influence of the physical conditions favoring the formation of a certain compound, but the nature of the subject and the meagreness of data in the particular phenomenon of reduction, render it inexpedient if not impracticable to take these points into consideration.

Starting with the above remarks in view, we will consider the heat generated by the combination of aluminium with certain other elements, as has been determined experimentally, and study from a comparison with the corresponding thermal data for other elements, what possibilities are shown for reducing these aluminium compounds.

The heat generated by the combination of aluminium with the different elements is given as follows ; the first column giving the heat developed by 54 kilos of aluminium (representing Al^2), and the second the heat per atomic weight of the other element, e. g., per 16 kilos of oxygen.

Element.			Compound.	Calories.	Calories.	Authority.
Oxygen	.	.	. *Al^2O^3	391,600	130,500	Bertholet.
				392,600	130,900	Baille & Féry.
Chlorine	.	.	Al^2Cl^6	321,960	53,660	Thomsen.
Bromine	.	.	Al^2Br^6	239,440	39,900	
Iodine	.	.	Al^2I^6	140,780	23,460	
Sulphur	.	.	Al^2S^3	124,400	41,467	

Let us consider the theoretical aspect of the reduction of Alumina. The heat given out by other elements or compounds which unite energetically with oxygen is as follows, the quantity given being that developed by combination with 16 kilos (representing one atomic weight) of oxygen.

Element.					Compound.	Calories.
Aluminium	.	.	.	.	**Al^2O^3**	**130,500**
Sodium	.	.	.	.	Na^2O	99,760
Potassium	.	.	.	.	K^2O	100,000 (?)
Barium	.	.	.	.	BaO	124,240
Strontium	.	.	.	.	SrO	128,440
Calcium	.	.	.	.	CaO	130,930
Magnesium	.	.	.	.	MgO	145,860
Manganese	.	.	.	.	MnO	95,000 (?)
Silicon	.	.	.	.	SiO^2	110,000
Zinc	.	.	.	.	ZnO	85,430
Iron	.	.	.	.	Fe^2O^3	63,700
Lead	.	.	.	.	PbO	50,300
Copper	.	.	.	.	CuO	37,160
"	.	.	.	.	Cu^2O	40,810
Sulphur	.	.	.	.	SO^2	35,540
Hydrogen	.	.	.	.	H^2O	68,360
Carbon	.	.	.	.	CO	29,000
"	.	.	.	.	CO^2	48,480
Carbonic anhydride	.	.	.	.	CO^2	67,960
Potassium cyanide	.	.	.	.	KCyO	72,000

On inspecting this list we find magnesium to be the only metal surpassing aluminium, while calcium is about the same. This would indicate that the reaction

$$Al^2O^3 + 3Mg = Al^2 + 3MgO$$

* Bertholet's number represented the formation of the hydrated oxide, $Al^2O^3.3H^2O$, and, for want of knowing the heat of hydration, has been generally used as the heat of formation of Al^2O^3. Recently, J. B. Baille and C. Féry (Ann. de Chim. et de Phys., June, 1889, p. 250) have, by oxidizing aluminium amalgam, obtained the above figure for the heat of formation of Al^2O^3, and determined that the heat of hydration is 3000 calories, which would make the heat of formation of the hydrated oxide 395,600.

would, if it were possible to bring the alumina and magnesium in the proper conditions for reacting, develop about

$$(145,860 — 130,500) \times 3 = 46,080 \text{ calories,}$$

and points to the possibility of reducing alumina by nascent, molten, or vaporized magnesium, under certain unknown conditions. It may be that molten alumina would be reduced by vapor of magnesium, but experiment only could establish or deny the possibility of the reaction. Even if this took place, it would probably not be of practical importance.

We notice further the fact that sodium or potassium could not reduce alumina without heat being absorbed in large quantity, and it is interesting to remember that some of the first attempts at isolating aluminium by using potassium were made on alumina, and were unsuccessful, so that it is practically acknowledged that while these metals easily reduce other aluminium compounds (according to reactions which are thermally possible, as we shall see later on) yet they cannot reduce alumina, under any conditions so far tried.

When we consider the case of reduction by the ordinary reducing agents, hydrogen, carbon, or potassium cyanide, we are confronted in every case with large negative quantities of heat, *i. e.,* deficits of heat. So large do these quantities appear that it is very small wonder that the impossibility of these reductions occurring under any conditions has been strongly affirmed. For instance

$$Al^2O^3 + 6H = Al^2 + 3H^2O$$

would require

$(130500 — 68360) \times 3 =$	186,420 calories.
$Al^2O^3 + 3C = Al^2 + 3CO$	304,500 calories.
$Al^2O^3 + 1\frac{1}{2}C = Al^2 + 1\frac{1}{2}CO^2$	246,060 calories.
$Al^2O^3 + 3KCy = Al^2 + 3KCyO$	175,500 calories.
$Al^2O^3 + 3CO = Al^2 + 3CO^2$	187,620 calories.

From these figures, however, we beg leave to disclaim predicting the absolute impossibility of the reactions taking place; the figures simply point to the probable impossibility of the reaction, or to its possibility only under very exceptional conditions. This position can be strengthened by considering that the reaction

$$ZnO + C = Zn + CO \quad \text{requires} \quad 56{,}430 \text{ cal.}$$
$$Fe^2O^3 + 3C = Fe^2 + 3CO \quad \text{``} \quad 104{,}100 \text{ cal.}$$
$$PbO + C = Pb + CO \quad \text{``} \quad 21{,}300 \text{ cal.}$$

yet these reactions are a matter of every day experience. If it be claimed that ferric oxide and litharge are really reduced by carbonic oxide, according to the reactions

$$Fe^2O^3 + 3CO = Fe^2 + 3CO^2 \text{ developing } 12{,}780 \text{ cal.}$$
$$PbO + CO = Pb \quad CO^2 \quad \text{``} \quad 17{,}660 \text{ ``}$$

and therefore that the reduction is possible because thermally positive, yet

$$ZnO + CO = Zn + CO^2 \text{ requires } 17{,}470 \text{ cal.}$$

and we still have before us a thermally negative reaction, which is practically carried out.

It is thus apparent that the reduction of alumina by the common reducing agents is, thermally considered, not an absolutely impossible question but one which presents, possibly, as much greater difficulty over the reduction of zinc oxide or iron ore as the heat deficit is greater in one case than in the other.

The question may be asked, "On what grounds has it been calculated that carbon will reduce alumina at a temperature of 10000° C.?" I have seen this statement in print, and was for some time at a loss to understand how this result was obtained, but came finally to the conclusion that it must have been deduced from the following premises:—

The reaction $Al^2O^3 + 3C = Al^2 + 3CO$ shows a deficit of 304500 calories. If, therefore, 304500 heat units can in some way be added to the alumina and carbon, then they might probably be induced to react. Evidently then, if we heat these substances they absorb a certain number of heat units for every degree rise of temperature, and at some certain temperature will have absorbed the required number of heat units to induce the reaction. The calculation seems to have been made thus—

Weight of alumina × specific heat.
$$102 \quad \times \quad 0.2 \quad = 20.4$$
Weight of carbon × specific heat.
$$36 \quad \times \quad 0.25 \quad = 9.$$

Caloric capacity per degree   29.4 calories.

The temperature to which the alumina and carbon must be heated in order to absorb 304,500 calories must be—

$$\frac{304500}{29.4} = 10350° \text{ C.}$$

There are several sources of error which will be immediately pointed out in this calculation. For instance, specific heats are known to increase with the temperature, while in the case of carbon its specific heat at a red heat is known to be about 0.46. Making this correction alone would reduce the temperature needed to about 8200° C.

That this method of figuring is not entirely unreasonable seems probable when we apply it to the reduction of oxide of zinc by carbonic oxide; for, in the case of the reaction, $ZnO + CO = Zn + CO_2$, the temperature calculated would be

$$\frac{17470 \text{ (Heat deficit)}}{(81 \times 0.125) + (28 \times 0.25)} = \frac{17470}{17.1} = 1020°, \text{ which is very}$$

close to the observed temperature. Yet we are constrained to regard this coincidence as fortuitous, since the same calculations for other oxides do not agree with the observed values. The method, if applied to the reduction of alumina by carbonic oxide, would give about 4500° C.; but it is certain, from what we know of the dissociation of carbonic acid by heat, that far below this temperature carbonic oxide loses almost all its affinity for oxygen, and this result must be rejected as mythical. The result obtained for reduction by carbon, forming CO, is open to a similar objection, owing to the fact that at very high temperatures carbonic oxide also is dissociated, but the dissociation takes place so slowly and to such a small degree within observable temperatures, that it is not impossible that alumina may be reduced by carbon at temperatures within the above-named limits.

The reduction of alumina by hydrogen, calculated by this method, would take place at 4500°, but since water is dissociated at high temperatures, this figure is open to the same criticism as that obtained for carbonic oxide. The lowest calculated value for the temperature of reduction of alumina is given by potassium cyanide, for the reaction

$$Al_2O_3 + 3KCy = Al_2 + 3KCyO$$

would require an addition of 175,600 cal., which would necessitate heating the substances to about 3000° (using the most probable value for the specific heat of potassium cyanide —0.2). Whether potassium cyanate (KCyO) dissociates sensibly at this temperature I cannot say, but the fact remains that if the temperatures calculated by this method are worthy of any credibility at all, they point to potassium cyanide as likely to reduce alumina at a lower temperature than either hydrogen, carbonic oxide, or carbon.

As the basis of our discussion of the reduction of aluminium chloride, bromide, or iodide, we give a table of the heat developed by the combination of some of the elements with one atomic weight (in kilos) of each of these haloids.

Element.	Compound.	Calories.	Compound.	Calories.	Compound.	Calories.
Aluminium . . .	Al^2Cl^6	**53,660**	Al^2Br^6	**40,000**	Al^2I^6	**23,460**
Potassium	KCl	105,600	KBr	95,300	KI	80,100
Sodium	NaCl	97,690	NaBr	85,700	NaI	69,000
Lithium	LiCl	93,810				
Barium	$BaCl^2$	97,370	$BaBr^2$	85,000		
Strontium	$SrCl^2$	92,270	$SrBr^2$	78,900		
Calcium	$CaCl^2$	84,910	$CaBr^2$	70,400		
Magnesium . . .	$MgCl^2$	75,500				
Manganese . . .	$MnCl^2$	56,000				
Zinc	$ZnCl^2$	$\begin{cases} 50,600† \\ 48,600* \end{cases}$	$ZnBr^2$	$\begin{cases} 43,100‡ \\ 40,640† \\ 37,500* \end{cases}$	ZnI^2	$\begin{cases} 30,000‡ \\ 26,600 \\ 24,500 \end{cases}$
Lead	$PbCl^2$	41,380	$PbBr^2$	32,200	PbI^2	20,000
Mercury	Hg^2Cl^2	41,275	Hg^2Br^2	34,150	Hg^2I^2	24,200
Tin	$SnCl^2$	40,400				
Iron	$FeCl^2$	41,000	$FeBr^2$	24,000	FeI^2	8,000†
Copper	Cu^2Cl^2	32,875	Cu^2Br^2	25,000	Cu^2I^2	16,000
Hydrogen	HCl	22,000	HBr	8,400	HI	—6,000

* Thomsen. † Andrews.
‡ Jahresbericht der Chemie, 1878, p. 102.

On inspecting this table we notice that, in general, all the metals down to zinc develop more heat in forming chlorides and very probably also in forming bromides and iodides. A reaction, then, between aluminium chloride and any of these metals, forming aluminium and a chloride of the metal, would be exothermic, which means, generally speaking, that if aluminium chloride and any one of these metals were heated together to the critical point at which the reaction could begin, the reaction would then proceed of itself, being continued by the heat given out by the first portions which reacted. Zinc seems to lie on the border line, and the evidence as to whether zinc will practically reduce these aluminium compounds is still contradictory, as may be seen by examining the paragraphs under "Reduction by Zinc." (Chap. XII.)

Of the first six metals mentioned in the table after aluminium, only potassium and sodium are practically available. The reaction

$$Al^2Cl^6 + 6K = Al^2 + 6KCl \text{ develops } 311,640 \text{ cal.}$$
$$Al^2Cl^6 + 6Na = Al^2 + 6NaCl \text{ develops } 264,180 \text{ cal.}$$

and the result of this strong disengagement of heat is seen when, on warming these ingredients together, the reaction once commenced at a single spot all external heat can be cut off, and the resulting fusion will become almost white hot with the heat developed. In fact, the heat developed in the second reaction would theoretically be sufficient to heat the aluminium and sodium chloride produced to a temperature between 3000° and 4000° C.

Magnesium should act in a similar manner, though not so violently, since

$$Al^2Cl^6 + 3Mg = Al^2 + 3MgCl^2 \text{ develops } 131,000 \text{ cal.}$$

And manganese possibly also, since

$$Al^2Cl^6 + 3Mn = Al^2 + 3MnCl \text{ develops } 14,040 \text{ cal.}$$

The reduction of aluminium chloride, bromide, or iodide by hydrogen is thermally strongly negative, which would indicate a very small possibility of the conditions ever being arranged so as to render the reaction possible. For instance, taking the most probable case,

$$Al^2Cl^6 + 6H = Al^2 + 6HCl \text{ requires } 189,960 \text{ calories.}$$ Moreover, a calculation similar to those made on the reduction of alumina by carbon would show a theoretical temperature of 2500° C. necessary to cause the reaction, if the energy required were added in the shape of heat.

The only probable substitutes for sodium in reducing aluminium chloride are thus seen to be magnesium (whose cost will probably be always greater than that of aluminium), manganese (which may sometime be used in the form of ferro-manganese for producing ferro-aluminium), and zinc (whose successful application to this purpose would be a most promising advance in the metallurgy of aluminium).

The heat of combination of fluorides is unknown, and so what would be an inquiry of interest with regard to these salts is out of our reach. We know, however, from experiment, that sodium will displace aluminium in its fluoride, developing a great deal of heat in the reaction, so that it is probable that the thermal relations of elements towards fluorine are similar to those towards chlorine. We can venture nothing further than this general observation.

In order to discuss the thermal relations of aluminium sulphide, we will make use of the following data, the heat developed being per atomic weight (32 kilos) of sulphur combining:—

Element.					Compound.	Calories.
Aluminium	.	.	.	.	. Al^2S^3	**41,467**
Potassium	.	.	.	.	. K^2S	103,700
Sodium	.	.	.	.	. Na^2S	88,200
Calcium	.	.	.	.	. CaS	92,000
Strontium	.	.	.	.	. SrS	99,200
Magnesium	.	.	.	.	. MgS	79,600
Manganese	.	.	.	.	. MnS	46,400
Zinc	.	.	.	.	. ZnS	41,326
Iron	.	.	.	.	. FeS	23,576
Copper	.	.	.	.	. Cu^2S	20,270
Lead	.	.	.	.	. PbS	20,430
Hydrogen	.	.	.	.	. H^2S	4,740
Carbon	.	.	.	.	. CS^2	—26,010

These figures point to the easy reduction of aluminium sulphide by potassium, sodium, or magnesium, and possibly by manganese and zinc. The other metals would require exceptional conditions, perhaps of temperature, for their action. It is interesting to note, as illustrating the many difficult points to be mastered by a consistent theory of the thermo-chemistry of reduction, that two observers at least have determined (probably from the deposition of the metals from solution by hydrogen sulphide), that the order of the affinity of the metals for sulphur is first the alkaline metals, then the others in the following order: copper, lead, zinc, iron, manganese, and then aluminium and magnesium—with the remark that the affinities of the latter two for sulphur appear quite insignificant. We are unable to suggest the meaning of the discrepancy here seen, it may be that when the metals are in solution some other circumstances beside the heat of combination may have the controlling influence in deciding which one of the metals would be first precipitated, such as the degree of acidity of the solution, etc. It is altogether probable that in reactions in the dry way, by heat, the order of affinity of the metals for sulphur would more nearly correspond to the order seen in the heats of combination.

The reduction of aluminium sulphide by hydrogen is seen to appear highly improbable.

Before closing this study of the thermal aspect of the reduction of aluminium compounds, it may be interesting to notice some of the reactions which are of use in the aluminium industry. It is well known that while chlorine gas can be passed over ignited alumina without forming aluminium chloride, and while carbon can be in contact with alumina at a white heat without reducing it, yet the concurrent action of chlorine and carbon will change the alumina into its chloride, a compound with a lower heat of formation. Thus—

$$Al^2O^3 + 3C = Al^2 + 3CO \text{ requires } 304,500 \text{ cal.}$$

But $Al^2O^3 + 3C + 6Cl = Al^2Cl^6 + 3CO$ requires a quantity of heat equal to the 304500 cal. minus 321960, the heat of formation of aluminium chloride, or in other words 17360 cal. is evolved, showing that the reaction is one of easy practicability. If it be inquired whether there is not some chloride which would act on alumina to convert it into chloride, we would remark that if we can find a chloride whose heat of formation is as much greater than the heat of formation of the corresponding oxide as the heat of formation of aluminium chloride is greater than that of alumina, then such a chloride might react. To be more particular, to convert alumina into aluminium chloride, a deficit of 391600 —

13

321960 or 69640 calories must be made up. If we know of an element which in uniting with 3 atom weights (48 kilos) of oxygen gives out 69640 calories more heat, or a still greater excess, than in uniting with 6 atom weights (213 kilos) of chlorine, then the chloride of that element might perform the reaction. Now—

$$6Na + 3O = 3Na^2O \text{ evolves } 299,280 \text{ cal.}$$
$$\text{and} \quad 6Na + 6Cl = 6NaCl \text{ evolves } 586,140 \text{ cal.}$$

leaving evidently a balance of 286,860 calories in the opposite direction to what we are looking for. And so for every metal except aluminium, I find the heat of formation of its chloride greater than that of an equivalent quantity of its oxide. The only element which I know of which possesses the opposite property is hydrogen, for—

$$6H + 3O = 3H^2O \text{ evolves } 205,080 \text{ cal.}$$
$$\text{and} \quad 6H + 6Cl = 6HCl \text{ evolves } 132,000 \text{ cal.}$$

and therefore the reaction—

$$Al^2O^3 + 6HCl = Al^2Cl^6 + 3H^2O$$

would evolve according to our calculations (205,080 — 132,000) — 69640 or 3440 calories, and would be thermally considered a possible reaction. Moreover, as a secondary effect, the water formed is immediately seized by the aluminium chloride, for the reaction

$$Al^2Cl^6 + 3H^2O \quad Al^2Cl^6.3H^2O \text{ evolves } 153,690 \text{ cal.}$$

and thus increases the total heat developed in the decomposition of the alumina to 158,130 calories. The result of this reaction is therefore the hydrated chloride, which is of no value for reduction by sodium, since when heated it decomposes into alumina and hydrochloric acid again, that is, it will decompose before giving up its water, and the water if undecomposed, or the acid if it decomposes, simply unites with the sodium without affecting the alumina. The immense heat of hydration, 153,690 calories, is so much greater than that of any other known substance, that it is in vain that we seek for any material which might abstract the water and leave anhydrous aluminium chloride.

Analogous to the reaction by which aluminium chloride is

formed from alumina is the reaction made use of for obtaining aluminium sulphide, yet with some thermal considerations of a different and highly interesting kind. If a mixture of alumina and carbon is ignited and, instead of chlorine, sulphur vapor is passed over it, no aluminium sulphide will be formed. An explanation of this fact is seen on discussing the proposed reaction thermally.

$$Al^2O^3 + 3C + 3S = Al^2S^3 + 3CO \text{ requires } 180,200 \text{ cal.}$$

It will be remembered that the similar reaction with chlorine evolved 17,360 calories; the quantity causing this difference is the heat of combination of aluminium sulphide, which is $321,960 - 124,400 = 197,560$ calories less than that of aluminium chloride, changing the excess of 17,360 calories into a deficit of 180,200 calories. This large negative quantity shows *à priori* that the reaction could be made to occur only under exceptional conditions, and its non-occurrence under all conditions so far tried gives evidence of the utility of the study of thermochemistry, at least as a guide to experiment. However, while carbon and sulphur cannot convert alumina into aluminium sulphide, carbon bisulphide can, for a current of the latter led over ignited alumina converts it into aluminium sulphide. The reaction taking place is

$$Al^2O^3 + 3CS^2 = Al^2S^3 + 3COS.$$

Now, since carbon and sulphur by themselves could not perform the reaction, we should be very apt to reason that a compound of carbon and sulphur would be still less able to do so, since the heat absorbed in dissociating the carbon-sulphur compound would cause a still greater deficit of heat. But here is precisely the explanation of the paradox. Carbon bisulphide is one of those compounds, not frequent, which has a negative heat of formation (— 26,010 calories), *i. e.*, heat is absorbed in large quantity in its formation, and therefore, per contra, heat is given out in the same quantity in its decomposition. The heat of formation of carbon oxysulphide being 37,030 calories, we can easily compute the thermal value of the reaction just given.

Heat absorbed.

Decomposition of alumina . . . . . 391,600 cal.

Heat developed.

Decomposition of carbon bisulphide . 78,030
Formation of carbon oxysulphide . . 111,090
"   of aluminium sulphide . . 124,400
                                      ─────── 313,520 "

Deficit of heat . . . . 78,080 "

It is thus seen that the reaction with carbon bisulphide is less than one-half as strongly negative as the reaction with carbon and sulphur alone, and in accordance with this we have the fact that aluminium sulphide is produced when carbon bisulphide vapor is passed over alumina heated white hot, while it is still further interesting to note that the presence of carbon mixed with the alumina is of no aid at all to the reaction.

CHAPTER IX.

REDUCTION OF ALUMINIUM COMPOUNDS BY MEANS OF POTASSIUM OR SODIUM.

THE methods comprised under this heading may be conveniently divided into three classes :—

I. Methods based on the reduction of aluminium chloride or aluminium-sodium chloride.

II. Methods based on the reduction of cryolite.

III. Methods based on the reduction of aluminium fluoride.

I.

The methods here included can be most logically presented by taking them in chronological order.

Oerstedt's Experiments (1824).

After Davy's unsuccessful attempts to isolate aluminium by the battery, in 1807, the next chemist to publish an account of attempts in this direction was Oerstedt, who published a paper in 1824 in a Swedish periodical.* Oerstedt's original paper is thus translated into Berzelius's "Jahresbericht:"†

"Oerstedt mixes calcined and pure alumina, quite freshly prepared, with powdered charcoal, puts it in a porcelain retort, ignites and leads chlorine gas through. The coal then reduces the alumina, and there results aluminium chloride and carbonic oxide, and perhaps also some phosgene, $COCl^2$; the aluminium chloride is caught in the condenser and the gases escape. The sublimate is white, crystalline, melts about the temperature of boiling water, easily attracts moisture, and evolves heat when in contact with water. If it is mixed with a concentrated potassium amalgam and heated quickly, it is transformed; there results potassium chloride, and the aluminium unites with the mercury. The new amalgam oxidizes in the air very quickly, and gives as residue when distilled in a vacuum a lump of metal resembling tin in color and lustre. In addition, Oerstedt found many remarkable properties of the metal and of the amalgam, but he holds them for a future communication after further investigation."

Oerstedt did not publish any other paper, and the next advance in the science is credited to Wöhler, whom all agree in naming as the true discoverer of the metal.

Wöhler's Experiments (1827).

In the following article from Poggendorff's Annalen,‡ Wöhler reviews the article of Oerstedt's given above, and continues as follows:—

"I have repeated this experiment of Oerstedt, but achieved no very satisfactory result. By heating potassium amalgam with aluminium chloride and distilling the product, there remained behind a gray, melted mass of metal, but which, by raising the heat to redness, went off as green vapor and distilled as pure

---

* Oversigt over det K. Danske Videnskabemes Selkabs Forhandlingar og dets Medlemmers Arbeider. May, 1824, to May, 1825, p. 15.

† Berz. Jahresb. der Chemie, 1827, vi. 118.

‡ Pogg. Ann., 1827, ii. 147.

potassium. I have therefore looked around for another method or way of conducting the operation, but, unpleasant as it is to say it, the reduction of the aluminium fails each time. Since, however, Herr Oerstedt remarks at the end of his paper that he did not regard his investigations in aluminium as yet ended, and already several years have passed since then, it looks as if I had taken up one of those researches begun auspiciously by another (but not finished by him), because it promised new and splendid results. I must remark, however, that Herr Oerstedt has indirectly by his silence encouraged me to try to attain to further results myself. Before I give the art how one can quite easily reduce the metal, I will say a few words about aluminium chloride and its production (see p. 122).

"I based the method of reducing aluminium on the reaction of aluminium chloride on potassium, and on the property of the metal not to oxidize in water. I warmed in a glass retort a small piece of the aluminium salt with some potassium, and the retort was shattered with a strong explosion. I tried then to do it in a small platinum crucible, in which it succeeded very well. The reaction is always so violent that the cover must be weighted down, or it will be blown off; and at the moment of reduction, although the crucible be only feebly heated from outside, it suddenly glows inside, and the platinum is almost torn by the sudden shocks. In order to avoid any mixture of platinum with the reduced aluminium, I next made the reduction in a porcelain crucible and succeeded then in the following manner: Put in the bottom of the crucible a piece of potassium free from carbon and oil, and cover this with an equal volume of pieces of aluminium chloride. Cover, and heat over a spirit lamp, at first gently, that the crucible be not broken by the production of heat inside, and then heat stronger, at last to redness. Cool, and when fully cold put it into a glass of cold water. A gray powder separates out, which on nearer observation, especially in sunlight, is seen to consist of little flakes of metal. After it has separated, pour off the solution, filter, wash with cold water, and dry; this is the aluminium."

In reality, this powder possessed no metallic properties, and moreover, it contained potassium and aluminium chloride, which

gave to it the property of decomposing water at 100°.   To avoid
the loss of aluminium chloride by volatilization at the high heat
developed during the reaction, Liebig afterwards made its vapor
pass slowly over some potassium placed in a long glass tube.
This device of Liebig is nearly the arrangement which Wöhler
adopted later, in 1845, and which gave him much better results.

Wöhler's Experiments (1845).

The following is Wöhler's second paper, published in 1845 :—*
"On account of the violent incandescence with which the reduction of aluminium chloride by potassium is accompanied, this
operation requires great precautions, and can be carried out only
on a small scale.   I took for the operation a platinum tube, in
which I placed aluminium chloride, and near it some potassium
in a platinum boat.   I heated the tube gently at first, then to
redness.   But the reduction may also be done by putting potassium in a small crucible which is placed inside a larger one, and
the space between the two filled with aluminium chloride.   A close
cover is put over the whole and it is heated.   Equal volumes of
potassium and the aluminium salt are the best proportions to
employ.   After cooling, the tube or crucible is put in a vessel of
water.   The metal is obtained as a gray metallic powder, but on
closer observation one can see even with the naked eye small tin-
white globules, some as large as pins' heads.   Under the microscope magnifying two hundred diameters the whole powder
resolves itself into small globules, several of which may sometimes
be seen sticking together, showing that the metal was melted at the
moment of reduction.   A beaten out globule may be again melted
to a sphere in a bead of borax or salt of phosphorus, but rapidly
oxidizes during the operation, and if the heat is continued disappears entirely, seeming either to reduce boric acid in the borax bead
or phosphoric acid in the salt of phosphorus bead.   I did not
succeed in melting together the pulverulent aluminium in a
crucible with borax, at a temperature which would have melted
cast-iron, for the metal disappeared entirely and the borax became

* Liebig's Annalen, 53, 422.

a black slag. It seems probable that aluminium, being lighter than molten borax, swims on it and burns. The white metallic globules had the color and lustre of tin. It is perfectly malleable and can be hammered out to the thinnest leaves. Its specific gravity, determined with two globules weighing 32 milligrammes, was 2.50, and with three hammered-out globules weighing 34 milligrammes, 2.67. On account of their lightness these figures can only be approximate. It is not magnetic, remains white in the air, decomposes water at 100°, not at usual temperatures, and dissolves completely in caustic potash (KOH). When heated in oxygen almost to melting, it is only superficially oxidized, but it burns like zinc in a blast-lamp flame."

These results of Wöhler's, especially the determination of specific gravity, were singularly accurate when we consider that he established them working with microscopic bits of the metal. It was just such work that established Wöhler's fame as an investigator. However, we notice that his metal differed from aluminium as we know it in several important respects, in speaking of which Deville says : " All this time the metal obtained by Wöhler was far from being pure ; it was very difficultly fusible, owing without doubt to the fact that it contained platinum taken from the vessel in which it had been prepared. It is well known that these two metals combine very easily at a gentle heat. Moreover it decomposed water at 100°, which must be attributed either to the presence of potassium or to aluminium chloride, with which the metal might have been impregnated : for aluminium in presence of aluminium chloride in effect decomposes water with evolution of hydrogen."

After Wöhler's paper in 1845, the next improvement is that introduced by Deville in 1854–55, and this is really the date at which aluminium, the metal, became known and its true properties established.

Deville's Experiments (1854).

The results of this chemist's investigations and success in obtaining pure aluminium were first made public at the séance of the French Academy, August 14, 1854, and included mention of an electrolytic method of reduction (see Chap. XI.), as well as of following on reduction by sodium.*

"The following is the best method for obtaining aluminium chemically pure in the laboratory. Take a large glass tube about four centimetres in diameter, and put into it 200 to 300 grammes of pure aluminium chloride free from iron, and isolate it between two stoppers of amianthus (fine, silky asbestos). Hydrogen, well dried and free from air, is brought in at one end of the tube. The aluminium chloride is heated in this current of gas by some lumps of charcoal, in order to drive off hydrochloric acid or sulphides of chlorine or of silicon, with which it is always impregnated. Then there are introduced into the tube porcelain boats, as large as possible, each containing several grammes of sodium, which was previously rubbed quite dry between leaves of filter paper. The tube being full of hydrogen, the sodium is melted, the aluminium chloride is heated and distils, and decomposes in contact with the sodium with incandescence, the intensity of which can be moderated at pleasure. The operation is ended when all the sodium has disappeared, and when the sodium chloride formed has absorbed so much aluminium chloride as to be saturated with it. The aluminium which has been formed is held in the double chloride of sodium and aluminium, $Al^2Cl^6.2NaCl$, a compound very fusible and very volatile. The boats are then taken from the glass tube, and their entire contents put in boats made of retort carbon, which have been previously heated in dry chlorine in order to remove all silicious and ferruginous matter. These are then introduced into a large porcelain tube, furnished with a prolongation and traversed by a current of hydrogen, dry and free from air. This tube being then heated to bright redness, the aluminium-sodium chloride distils without decomposition and condenses in the prolongation. There is found in the boats, after the operation, all the aluminium which had been reduced, collected in at most one or two small buttons. The boats when taken from the tube should be nearly free from aluminium-sodium chloride and

* Ann. de Phys. et de Chem., xliii. 24.

also from sodium chloride. The buttons of aluminium are united in a small earthen crucible which is heated as gently as possible, just sufficient to melt the metal. The latter is pressed together and skimmed clean by a small rod or tube of clay. The metal thus collected may be very suitably cast in an ingot mould."

The later precautions added to the above given process were principally directed towards avoiding the attacking of the crucible, which always takes place when the metal is melted with a flux, and the aluminium thereby made more or less siliceous. The year following the publication of these results, this laboratory method was carried out on a large scale at the chemical works at Javel.

Deville's Methods (1855).

The methods about to be given are those which were devised in Deville's laboratory at the Ecole Normale, during the winter of 1854–55, and applied on a large scale at Javel, during the spring of 1855 (March–July). The Emperor Napoleon III. defrayed the expenses of this installation. Descriptions of the methods used for producing alumina, aluminium chloride and sodium at Javel can be found under their respective headings (pp. 106, 120, 123, 146). We here confine our description to the mode of reducing the aluminium chloride by sodium, and the remarks incident thereto. The process has at present only an historic interest, as it was soon modified in its details so as to be almost entirely changed. The following is Deville's description:—

" *Perfectly pure aluminium.*—To obtain aluminium perfectly pure it is necessary to employ materials of absolute purity, to reduce the metal in presence of a completely volatile flux, and finally never to heat, especially with a flux, in a siliceous vessel to a high temperature."

" *Pure materials.*—The necessity of using absolutely pure materials is easy to understand; all the metallic impurities are concentrated in the aluminium, and unfortunately I know no absolute method of purifying the metal. Thus, suppose we take an alum containing 0.1 per cent. of iron and 11 per cent. of alumina; the alumina derived from it will contain 1 per cent.

of iron, and supposing the alumina to give up all the aluminium in it, the metal will be contaminated with 2 per cent. of iron."

"*Influence of flux or slag.*—The flux, or the product of the reaction of the sodium on the aluminous material, ought to be volatile, that one may separate the aluminium by heat from the material with which it has been in contact, and with which it remains obstinately impregnated because of its small specific gravity."

"*Influence of the vessel.*—The siliceous vessels in which aluminium is received or melted give it necessarily a large quantity of silicon, a very injurious impurity. Silicon cannot be separated from aluminium by any means, and the siliceous aluminium seems to have a greater tendency to take up more silicon than pure aluminium, so that after a small number of remeltings in siliceous vessels the metal becomes so impure as to be almost infusible."

In order to avoid the dangers pointed out above, Deville recommended following scrupulously the following details in order to get pure aluminium.

"*Reduction by solid sodium.*—The crude aluminium chloride placed in the cylinder *A* (Fig. 20), is vaporized by the fire and

Fig. 20.

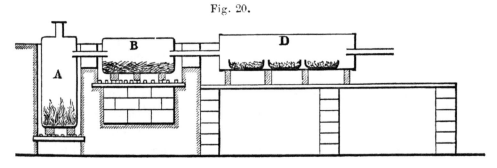

passes through the tube to the cylinder *B*, containing 60 to 80 kilos of iron-nails heated to a dull-red heat. The iron retains as relatively fixed ferrous chloride, the ferric chloride and hydrochloric acid which contaminate the aluminium chloride, and likewise transforms any sulphur dichloride (SCl^2) in it into ferrous chloride and sulphide of iron. The vapors on passing out of *B* through the tube, which is kept at about 300°, deposit spangles of ferrous chloride, which is without sensible tension at that temperature. The vapors then enter D, a cast-iron cylinder in which are three cast-iron boats each containing 500 grms. of sodium. It is sufficient to heat this cylinder barely to a dull-red heat in its lower part, for the reaction once commenced disengages enough heat to complete itself, and it is often necessary to take away all the fire from it. There is at first produced in the first boat some aluminium and some sodium chloride, which latter combines with the excess of aluminium chloride to form the volatile aluminium-sodium chloride, $Al^2Cl^6.2NaCl$. These vapors of double chloride condense on the second boat and are decomposed by the sodium into aluminium and sodium chloride. A similar reaction takes place in the third boat when all the sodium of the second has disappeared. When on raising the cover it is seen that the sodium of the last boat is entirely transformed into a lumpy black material, and that the reactions are over, the boats are taken out, immediately replaced by others, and are allowed to cool covered by empty boats. In this first operation the reaction is rarely complete, for the sodium is protected by the layer of sodium chloride formed at its expense. To make this disappear, the contents of the boats are put into cast-iron pots or earthen crucibles, which are heated until the aluminium chloride begins to volatilize, when the sodium will be entirely absorbed and the aluminium finally remains in contact with a large excess of its chloride, which is indispensable for the success of the operation. Then the pots or crucibles are cooled, and there is taken from the upper part of their contents a layer of sodium chloride almost pure, while underneath are found globules of aluminium which are separated from the residue by washing with water. Unfortunately, the water in dissolving the aluminium chloride of the flux exercises on the metal a very rapid destructive action, and only the globules larger than the head of a pin are saved from this washing. These are gathered together, dried, melted in an earthen crucible, and pressed together with a clay rod. The button is then cast in an ingot mould. It is important in this operation to employ only well purified sodium, and not to melt the aluminium if it still contains any sodium, for in this case the metal takes fire and burns as long as any of the alkaline metal remains in it. In such a case it is necessary to remelt in presence of a little aluminium-sodium chloride.

" Such is the detestable process by means of which were made the ingots of aluminium sent to the Exhibition (1855). To complete my dissatisfaction at the process, pressed by time and ignorant of the action of copper on aluminium, I employed in almost all my experiments reaction cylinders and boats of copper, so that the aluminium I took from them contained such quantities of this metal as to form a veritable alloy. Moreover, it had lost almost all ductility and malleability, had a disagreeable gray tint, and finally at the end of two months it tarnished by becoming covered with a black layer of oxide or sulphide of copper, which could only be removed by dipping in nitric acid. But, singular to relate, an ingot of virgin silver which had been put alongside the aluminium that the public might note easily the difference in color and weight of the two metals, was blackened still worse than the impure aluminium. Only one of the bars exhibited, which contained no copper, remained unaltered from the day of its manufacture till now (1859). It was some of this cupreous aluminium that I sent to Mr. Regnault, who had asked me for some in order to determine its specific heat. I had cautioned him at the time, of the number and nature of the impurities which it might contain, and the analysis of M. Salvêtat, which is cited in the memoir of Regnault, accords with the mean composition of the specimens that I had produced and analyzed at that time (see p. 54, Analysis 1). It is to be regretted that I gave such impure material to serve for determinations of such splendid precision; I was persuaded to do so only by the entreaties of M. Regnault who could not wait until I prepared him better. It is also this cupreous aluminium which M. Hulot has called 'hard aluminium,' in a note on the physical properties of this metal which he addressed to the Academy. Hulot has remarked that this impure metal, which is crystalline in structure, after having been compressed between the dies of the coining press may lose its crystalline structure, to which it owes its brittleness, and become very malleable. It possesses then such strength that it works well in the rolls of a steel-rolling mill. Further, this 'hard aluminium' becomes quite unalterable when it has thus lost its texture."

"*Reduction by sodium vapor*.—This process, which I have not perfected, is very easy to operate, and gave me very pure metal at the first attempt. I operate as follows : I fill a mercury bottle with a mixture of chalk, carbon, and carbonate of soda, in the proportions best for generating sodium. An iron tube about ten centimetres long is screwed to the bottle, and the whole placed in a wind furnace, so that the bottle is heated to red-white and the tube is red to its end. The end of the tube is then introduced into a hole made in a large earthen crucible about one-fourth way from the bottom, so that the end of the tube just reaches the inside surface of the crucible. The carbonic oxide (CO) disengaged burns in the bottom of the crucible, heating and drying it ; afterwards the sodium flame appears, and then pieces of aluminium chloride are thrown into the crucible from time to time. The salt volatilizes and decomposes before this sort of tuyere from which issues the reducing vapor. More aluminium salt is added when the vapors coming from the crucible cease to be acid, and when the flame of sodium burning in the atmosphere of aluminium chloride loses its brightness. When the operation is finished, the crucible is broken and there is taken from the walls below the entrance of the tube a saline mass composed of sodium chloride, a considerable quantity of globules of aluminium, and some sodium carbonate, which latter is in larger quantity the slower the operation was performed. The globules are detached by plunging the saline mass into water, when it becomes necessary to notice the reaction of the water on litmus. If the water becomes acid, it is renewed often ; if alkaline, the mass impregnated with metal must be digested in nitric acid diluted with three or four volumes of water, and so the metal is left intact. The globules are reunited by melting with the precautions before given."

Deville's Process (1859).

The process then in use at Nanterre was based on the use of aluminium-sodium chloride, which was reduced by sodium, with cryolite or fluorspar as a flux. The methods of preparing each of these materials were carefully studied out at the chemical works of La Glacière, where from April 1856, to April 1857, the manufacture of aluminium was carried on by a company formed by Deville and a few friends, and from thence proceeded the actual system which was established at Nanterre under the direction of M. Paul Morin when the works at La Glacière were closed. The methods of preparation of alumina, aluminium-sodium chloride and sodium used at Nanterre are placed under their respective headings (pp. 120, 127, 161, 163).

In the first year that the works at Nanterre were in operation there were made

Aluminium-sodium chloride	10,000 kilos.
Sodium	2,000 "
Aluminium	600 "

The metal prepared improved constantly in quality, M. Morin profiting continually by his experience and improving the practical details constantly, so that the aluminium averaged, in 1859, 97 per cent. pure. (See Analysis 9, p. 54.)

As to the rationale of the process used, aluminium chloride was replaced by aluminium-sodium chloride because the latter is less deliquescent and less difficult of preservation; but the small amount of moisture absorbed by the double chloride is held very energetically, at a high temperature giving rise to some alumina, which incloses the globules of metal with a thin coating, and so hinders their easy reunion to a button. Deville remarked that the presence of fluorides facilitated the reunion of these globules, which fact he attributed to their dissolving the thin coat of alumina; so that the employment of a fluoride as a flux became necessary to overcome the effect produced primarily by the aluminium-sodium chloride holding moisture so energetically. Deville gives the following account of the development of these improvements: "The facility with which aluminium unites in fluorides is due without doubt to the property which these possess of dissolving the alumina on the surface of the globules at the moment of their formation, and which the sodium is unable to reduce. I had experienced great difficulty by obtaining small quantities of metal poorly united, when I reduced the aluminium-

sodium chloride by sodium ; M. Rammelsberg, who often made
the same attempts, tells me he has had a like experience. But I
am assured by a scrupulous analysis that the quantity of metal
reduced by the sodium is exactly that which theory indicates,
although after many operations there is found only a gray powder, resolving itself under the microscope into a multitude of
small globules. The fact is simply that aluminium-sodium chloride is a very poor flux for aluminium. MM. Morin, Debray,
and myself have undertaken to correct this bad effect by the introduction of a solvent for the alumina into the saline slag which
accompanies the aluminium at the moment of its formation. At
first, we found it an improvement to condense the vapors of aluminium chloride, previously purified by iron, directly in sodium
chloride, placed for this purpose in a crucible and kept at a red
heat. We produced in this way, from highly colored material, a
double chloride very white and free from moisture, and furnishing on reduction a metal of fine appearance. We then introduced fluorspar (CaF^2) into the composition of the mixture to be
reduced, and we obtained good results with the following proportions :—

Aluminium-sodium chloride . . .	400 grammes.
Sodium chloride	200 "
Calcium fluoride	200 "
Sodium	75 to 80 "

"The double chloride ought to be melted and heated almost to
low red heat at the moment it is employed, the sodium chloride
calcined and at a red heat or melted, and the fluorspar pulverized and well dried. The double chloride, sodium chloride and
calcium fluoride are mixed and alternated in layers in the crucible
with sodium. The top layer is of the mixture, and the cover is
sodium chloride. Heat gently, at first, until the reaction ends,
and then to a heat about sufficient to melt silver. The crucible,
or at least that part of it which contains the mixture, ought to
be of a uniform red tint, and the material perfectly liquid. It is
stirred a long time and cast on a well dried, chalked plate.
There flows out first a very limpid liquid, colorless and very fluid,
then a gray material, a little more pasty, which contains alu-

minium in little grains, and is set aside, and finally a button with small, metallic masses which of themselves ought to weigh 20 grms. if the operation has succeeded well. On pulverizing and sieving the gray slag, 5 or 6 grms. of small globules are obtained, which may be pressed together by an earthen rod in an ordinary crucible heated to redness. The globules are thus reunited, and when a sufficient quantity is collected the metal is cast into ingots. In a well-conducted operation, 75 grms. of sodium ought to give a button of 20 grms. and 5 grms. in grains, making a return of one part aluminium from three of sodium. Theory indicates one to two and a half, or 30 grms. of aluminium from 75 of sodium. But all the efforts which have been made to recover from the insoluble slag the 4 or 5 grms. of metal not united but easily visible with a glass, have been so far unsuccessful. There is, without doubt, a knack, a particular manipulation on which depends the success of an operation which would render the theoretical amount of metal, but we lack it yet. These operations take place, in general, with more facility on a large scale, so that we may consider fluorspar as being suitable for serving in the manufacture of aluminium in crucibles. We employed very pure fluorspar, and our metal was quite exempt from silicon. It is true that we took a precaution which is necessary to adopt in operations of this kind; we plastered our crucibles inside with a layer of aluminous paste, the composition of which has been given in 'Ann. de Chim. et de Phys.' xlvi. 195. This paste is made of calcined alumina and an aluminate of lime, the latter obtained by heating together equal parts of chalk and alumina to a high heat. By taking about four parts calcined alumina and one of aluminate of lime well pulverized and sieved, moistening with a little water, there is obtained a paste with which the inside of an earthen crucible is quickly and easily coated. The paste is spread evenly with a porcelain spatula, and compressed strongly until its surface has become well polished. It is allowed to dry, and then heated to bright redness to season the coating; which does not melt, and protects the crucible completely against the action of the aluminium and fluorspar. A crucible will serve several times in succession provided that the new material is put in as soon as the previous

14

charge is cast. The advantages of doing this are that the mixture and the sodium are put into a crucible already heated up, and so lose less by volatilization because the heating is done more quickly, and the crucible is drier than if a new one had been used or than if it had been let cool. A new crucible should be heated to at least 300° or 400° before being used. The saline slag contains a large quantity of calcium chloride, which can be washed away by water, and an insoluble material from which aluminium fluoride can be volatilized.

"Yet the operation just described, which was a great improvement on previous ones, requires many precautions and a certain skill of manipulation to succeed every time. But nothing is more easy or simple than to substitute cryolite for the fluorspar. Then the operation is much easier. The amount of metal produced is not much larger, although the button often weighs 22 grammes, yet if cryolite can only be obtained in abundance in a continuous supply, the process which I will describe will become most economical. The charge is made up as before, except introducing cryolite for fluorspar. In one of our operations we obtained, with 76 grms. of sodium, a button weighing 22 grms. and 4 grms. in globules, giving a yield of one part aluminium to two and eight-tenths parts sodium, which is very near to that indicated by theory. The metal obtained was of excellent quality. However, it contained a little iron coming from the aluminium chloride, which had not been purified perfectly. But iron does not injure the properties of the metal as copper does; and, save a little bluish coloration, it does not alter its appearance or its resistance to physical and chemical agencies.

"After these attempts we tried performing the reduction simply on the bed of a reverberatory furnace, relying on the immediate reaction of sodium on the double chloride to use up these materials before they could be perceptibly wasted by the furnace gases. This condition was realized in practice with unlooked-for success. The reduction is now made on a somewhat considerable scale at the Nanterre works, and never, since commencing to operate in this way, has a reduction failed, the results obtained being always uniform. The furnace now used has all the relative dimensions of a soda furnace. In fact, almost the temperature of an ordinary soda furnace is required that the operation may succeed perfectly. The absolute dimensions of the furnace, however, may vary with the quantity of aluminium to be made in one opera-tion, and are not limited. With a bed of one square metre surface, 6 to 10 kilos of aluminium can be reduced at once; and since each operation lasts about four hours and the furnace may be recharged immediately after emptying it of the materials just treated, it is seen that, with so small a bed, 60 to 100 kilos of aluminium can be made in twenty-four hours without any difficulty. In this respect, I think the industrial problem perfectly solved. The proportions which we employed at first were—

Aluminium-sodium chloride (crushed) . .	10 parts.
Fluorspar	5 "
Sodium (in ingots)	2 "

"As aluminium is still very dear, it is necessary to direct great attention to the return from the materials used, and on this point there is yet much progress to be made. We ascertained many times that the return was always a little better, and the reunion of the metal to a single ingot a little easier, when cryolite was substituted for fluorspar, the price of the former, after having been high, being now lowered to 350 francs per tonne (about $75 per long ton). For this reason we can now use cryolite instead of fluorspar, and in the same proportions. We can also recover alumina from this cryolite by treating the slags (see p. 119). The double chloride and pulverized cryolite are now mixed with the sodium in small ingots, and the mixture thrown on to the bed of the heated-up furnace. The dampers are then shut, to prevent as much as possible access of air. Very soon a lively reaction begins, with the production of such heat that the brick sides of the furnace, as well as the materials on the hearth, are made bright red-hot. At this heat the mixture is almost completely fused. Then it is necessary to open the damper and direct the flame on the bed in such manner as to heat the bath equally all over and unite the reduced aluminium. When the operation is considered ended, a casting is made by an opening in the back of the furnace and the slag is received in cast-iron pots. At the end of the cast the aluminium arrives in a single jet, which unites into a single lump at the bottom of the still-liquid slag. The gray slag flowing out last should be pulverized and sieved, to extract the divided globules of aluminium, 200 to 300 grammes of which can sometimes be extracted from one kilo of gray slag. The pulverization of the slag is in all cases indispensable in its subsequent treatment for extracting its alumina. The slag is of two kinds, one fluid and light, which covered the bath, and is rich in sodium chloride; the other less fusible and pasty, gray in color, which is more dense and lies in contact with the aluminium. The coloring material producing the grayness is carbon, coming either from the sodium or from the oil which impregnated it, or finally from the vapor of the oil. I attribute the slight pastiness of this slag to a little alumina dissolved by the fluorides. This slag contains about—

Sodium chloride	60 parts.
Aluminium fluoride	40 "

and on washing it the former dissolves while the latter remains, mixed with a little cryolite or alumina. This is the alumina which had been dissolved or retained in the bath of fluoride. It will be remarked that the bath of slag contains no other fluoride than aluminium fluoride, which does not attack earthen crucibles or siliceous materials in general except at a very high temperature. It is for this reason that the hearth and other parts of the furnace resist easily a fluoride slag containing only aluminium fluoride, which has not the property of combining with silicon fluoride at the expense of the silica of the bricks—as sodium fluoride does in like circumstances. In our operations, cryolite is used only as a flux. In the process of reduction based on cryolite alone, the sodium fluoride resulting is, on the contrary, very dangerous to crucibles, and it is due to that fact especially that the aluminium absorbs a large quantity of silicon, which always happens with this method. In fact, it is well known that metallic silicon can be prepared in this way by prolonging the operation a little."

Deville closes his account of the aluminium industry in 1859 with these words: "Many things yet remain to us to do, and we can scarcely say now that we know the true qualities of the substances we employ. But the matter is so new, is harassed with so many difficulties even after all that has been done, that our young industry may hope everything from the future when it shall have acquired experience. I ought to say, however, that the aluminium industry is now at such a point that if the uses of the metal are rapidly extended it may change its aspect with great rapidity. One may ask to-day how much a kilo of iron would cost if a works made only 60 to 100 kilos of it a month, if large apparatus were excluded from this industry, and iron obtained by laboratory processes which would permit it to become useful only by tedious after-treatment. Such will not be the case with aluminium, at least with the processes just described. In fact, in all I undertook, either alone or with my friends, I have always been guided by this thought—that we ought to adopt only such apparatus as is susceptible of being immediately enlarged, and to use only materials almost as common as clay itself for the source of the aluminium."

The Deville Process (1882).

The process just described reached a fair degree of perfection at Nanterre, under the direction of M. Paul Morin. Afterwards, some of the chemical operations incidental to the process were carried on at the works of the Chemical Manufacturing Company of Alais and Carmargue, at Salindres (Gard), owned by H. Merle & Co. At a later date the whole manufacture was removed to this place, while the Société Anonyme de l'Aluminium, at Nanterre, worked up the metal and placed it on the market. The Salindres works, about 1880, went under the management of A. R. Pechiney & Co., and under the personal attention of M. Pechiney the Deville process has reached its present state of perfection. The following account is taken mostly from M. Margottet's article on aluminium in Fremy's Encyclopédie Chimique.

An outline of the process, as it now stands, may very appropriately be given at this place, although detailed descriptions of the preliminary processes for preparing the materials for reduction are given under the appropriate headings (see pp. 109, 127).

The primary material to furnish the aluminium is beauxite.

To obtain the metal it is necessary to proceed successively through the following operations :—

I. Preparation of the aluminate of soda, and solution of this salt to separate it from the ferric oxide contained in the beauxite.

II. Precipitation of hydrated alumina from the aluminate of soda by a current of carbon dioxide; washing the precipitate.

III. Preparation of a mixture of alumina, carbon, and salt, drying it, and then treating with gaseous chlorine to obtain the double chloride of aluminium and sodium.

IV. Lastly, treatment of this chloride by sodium to obtain aluminium.

The principal chemical reactions on which this process rests are the following :—

Formation of aluminate of soda by calcining beauxite with sodium carbonate—

$$(AlFe)^2O^3.2H^2O + 3Na^2CO^3 = Al^2O^3.3Na^2O + Fe^2O^3 \\ + 2H^2O + 3CO^2.$$

Formation of alumina by precipitating the aluminate of soda with a current of carbon dioxide—

$$Al^2O^3.3Na^2O + 3CO^2 + 3H^2O = Al^2O^3.3H^2O + 3Na^2CO^3.$$

Formation of aluminium sodium chloride by the action of chlorine on a mixture of alumina, carbon, and sodium chloride—

$$Al^2O^3 + 3C + 2NaCl + 6Cl = Al^2Cl^6.2NaCl + 3CO.$$

Reduction of this double chloride by sodium—

$$Al^2Cl^6.2NaCl + 6Na = 2Al + 8NaCl.$$

As observed before, we will here consider only the last operation. The advances made since 1859 are mostly in matters of detail, which every one knows are generally the most important part of a process; and so, although a few of the details may be repeated, yet we think it best not to break the continuity of this description by excising those few sentences which are nearly identical in the two accounts.

The difficulty of this operation, at least from an industrial point of view, is to get a slag fusible enough and light enough to let the reduced metal easily sink through it and unite. This result has been reached by using cryolite, a white or grayish mineral originally from Greenland, very easy to melt, formula $Al^2F^6.6NaF$. This material forms with the sodium chloride resulting from the reaction a very fusible slag, in the midst of which the aluminium collects well, and falls to the bottom. In one operation the charge is now composed of—

100 kilos . . . . Aluminium-sodium chloride.
45 " . . . . Cryolite.
35 " . . . . Sodium.

The double chloride and cryolite are pulverized, the sodium, cut into small pieces a little larger than the thumb, is divided into three equal parts, each part being put into a sheet-iron basket. The mixture of double chloride and cryolite, being pulverized, is divided into four equal parts, three of these are respectively put in each basket with the sodium, the fourth being placed in a

Fig. 21.

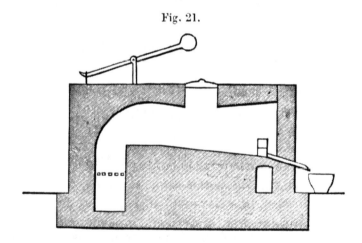

basket by itself. The reduction furnace (see Fig. 21) is a little furnace of refractory brick, with an inclined hearth and a vaulted roof. This furnace is strongly braced by iron tie-rods, because of the concussions caused by the reaction. The flame may at any given moment be directed into a flue outside of the hearth. At the back part of the furnace, that is to say, on that side towards which the bed slopes, is a little brick wall which is built up for each reduction and is taken away in operating the running out of the metal and slag. A gutter of cast-iron is placed immediately in front of the wall to facilitate running out the materials. All this side of the furnace ought to be opened or closed at pleasure by means of a damper. Lastly, there is an opening for charging in the roof, closed by a lid. At the time of an operation the furnace should be heated to low redness, then are introduced in rapid succession the contents of the three baskets containing sodium, etc., and lastly the fourth containing only double chloride and no sodium. Then all the openings of the furnace are closed and a very vivid reaction accompanied by dull concussions immediately takes place. At the end of fifteen minutes, the action subsides, the dampers are opened, and the heat continued, meanwhile stirring the mass from time to time with an iron poker. At the end of three hours the reduction is ended, and the metal collects at the bottom of the liquid bath. Then the running out is proceeded with in three phases : First.—Running off the upper part of the bath, which consists of a fluid material completely free from reduced aluminium and constituting the white slag. To run this out a brick is taken away from the upper course of the little wall which terminates the hearth. These slags are received in an iron wagon. Second.—Running out the aluminium. This is done by opening a small orifice left in the bottom of the brick wall, which was temporarily plugged up. The liquid metal is received in a cast-iron melting pot, the bottom òf which has been previously heated to redness. This aluminium is immediately cast in a series of small rectangular cast-iron moulds. Third.— Running out of the rest of the bath, which constitutes the gray slags. These were, like the white slags, formed by the sodium chloride and cryolite, but they contain, in addition, isolated globules of aluminium. To run these out, all the bricks of the little wall are taken away. This slag is received in the same melting pot into which the aluminium was run, the latter having been already moulded. Here it cools gradually, and after cooling there are always found at the bottom of the pot several grains of metal. In a good operation there are taken from one casting 10.5 kilos of aluminium, which is sold directly as commercial metal.

The following data as to the expense of this process may be very appropriately inserted here, giving the cost at Salindres in 1872, in which year 3600 kilos are said to have been made.

*Manufacture of one kilo of aluminium.

Sodium . . . 3.44 kilos @ 11.32 fr. per kilo			= 38 fr. 90 cent.
Aluminium-sodium				
chloride . . 10.04	" 2.48	" "	= 24 " 90 "
Cryolite . . . 3.87	" 61.0	" 100 kilos	= 2 " 36 "
Coal 29.17	" 1.40	" "	= 0 " 41 "
		Wages	1 " 80 "
		Costs	0 " 88 "
		Total	.69 " 25 "

This must be increased ten per cent. for losses and other expenses, making the cost of aluminium 80 fr. per kilo, and it is sold for 100. ($9.00 per lb.)

According to a statement in the ' Bull. de la Soc. de l'Industrie Minérale,' ii., 451, made in 1882, Salindres was then the only place in which aluminium was being manfactured.

Niewerth's Process (1883).

This method can be regarded as little more than a suggestion, since it follows exactly the lines of some of Deville's earlier experiments. Although theoretically very advantageous, yet in practice it has probably been found far inferior in point of yield of metal and expense to the ordinary sodium processes. The patent is said to be taken out in the United States and other countries in the name of H. Niewerth, of Hanover, and is thus summarized :—

†A compound of aluminium, with chlorine or fluorine, is brought by any means into the form of vapor, and conducted, strongly heated, into contact with a mixture of 62 parts sodium carbonate, 28 coal, and 10 chalk, which is also in a highly heated condition. This mixture disengages sodium, which reduces the gaseous chloride or fluoride of aluminium, the nascent sodium being the reducing agent. In place of the above mixture other suitable mixtures which generate sodium may be employed, or mixtures may also advantageously be used from which potassium is generated.

* A. Wurtz, Wagner's Jaresb., 1874, vol. xxi.
† Sci. Am. Suppl., Nov. 17, 1883.

Gadsden's Patent (1883).

H. A. Gadsden, of London, and E. Foote,* of New York, were granted a patent based on the principle of heating in a retort sodium carbonate and carbonaceous matter, or any suitable mixture for generating sodium, and conducting the vapor of sodium produced into another retort, lined with carbon, in which aluminium chloride, or aluminium-sodium chloride or cryolite has been placed and heated. The second English patent claims to heat a mixture which will generate sodium, in one retort, and pass chlorine over a mixture of carbon and alumina, thus generating aluminium chloride, in another retort, and then mixing the two vapors in a third retort or reaction chamber.

Frishmuth's Process (1884).

This was patented in the United States in 1884 (U. S. Pat. 308,152, Nov. 1884). In what the originality of the process consists, in view of Deville's publications and even in view of the processes just mentioned, we cannot see, and we simply acquiesce blindly to the mysterious penetration of our Patent Office Board. However, Col. Frishmuth himself admits, in 1887, having abandoned the sodium process; it is therefore probable that the difficulties of the method did not permit its competing with the more roundabout but more easily-conducted operation with solid sodium. A simple transcript of the claims in his patent will give a sufficiently extended idea of the reactions proposed to be used.

1. The simultaneous generation of sodium vapor and a volatile compound of aluminium in two separate vessels or retorts, and mingling the vapors thus obtained in a nascent (!) state in a third vessel, wherein they react on each other.

2. The sodium vapor is produced from a mixture of a sodium compound and carbon, or some other reducing agent; and the aluminous vapor from aluminous material.

3. The simultaneous generation of sodium vapor and vapor of

---

* English patents 1995 and 4930 (1883); German patent 27,572 (1884).

aluminium chloride or aluminium fluoride; or of sodium vapor and aluminium-sodium chloride.

4. Converting the aluminous material to a vapor by heating it in a retort with sodium chloride, and subjecting it at the same time to chlorine gas; mingling the vapor of aluminium-sodium chloride thus obtained with vapor simultaneously generated from sodium carbonate and carbon.

H. von Grousillier's Improvement (1885).

This suggestion as to the way of performing the reduction by sodium is the subject of the English patent 7858, June 29, 1885. Dr. Fischer remarks on it, in 'Wagner's Jahresbericht' for 1885, that "it is apparently wholly worked out at the writing-table." The patentee, Hector von Grousillier, Springe, Hanover, thus describes his invention :—

"In order to avoid the difficulties ordinarily met with in the use of aluminium-sodium chloride to obtain aluminium, I raise the volatilizing point of aluminium chloride by performing its reduction, either chemically or electrolytically, under pressure in a strong, hermetically-closed vessel lined with clay or magnesia and provided with a safety valve."

The Deville-Castner Process (1886).

This latest development of the old Deville process is now operated by the Aluminium Company, Limited, at their large new works at Oldbury, near Birmingham, England. The plant covers nearly five acres of ground, and adjoins Chance Bros.' large chemical works, from which the hydrochloric acid used is obtained and the waste soda-liquors returned, by means of large pipes connecting the two plants. The company is thus in position to obtain acid and dispose of its by-products to very good advantage. The principle on which the process works is similar to its predecessor, in being the reduction of aluminium-sodium chloride by sodium, but it improves on the other in the cheaper production of both these materials. For instance, the alumina used is obtained and converted into double chloride by Mr. Webster's processes, by which it is probable that this salt does not cost over 3d. per lb. (see p. 133), as against 12d., the cost at Salindres; further, by Mr. Castner's sodium process it is acknowledged that the sodium costs only about 9d. per lb., as against 48d., or $1, as formerly. Since 10 lbs. of the chloride and 3 lbs. of sodium are required to produce 1 lb. of aluminium, the average saving in these two items, over the old process, is somewhere about 75 per cent.

The works contain a sodium building, in which are four large sodium furnaces, each capable of producing over 500 lbs. of that metal in twenty-four hours; the sodium is also remelted and stored in the same building (see p. 166). The double chloride furnaces are in a building 250 feet by 50 feet wide, there being 12 furnaces each containing 5 retorts. The total output of double chloride is an average of 5000 lbs. per day. (See p. 129.) Connected with this building is a chlorine plant of the largest size, capable of supplying about a ton and a half of chlorine a day. In a separate building are two reverberatory furnaces in which the final reduction takes place and the aluminium is produced. Besides these, there are a rolling mill, wire mill, and foundry on the grounds. From the quantity of sodium and double chloride produced, we can see that the works can produce about 500 lbs. of aluminium a day or 150,000 lbs. a year, with some sodium left over for sale or other purposes.

The mode of conducting the reduction is not very different from that practised at Salindres. There are two regenerative reverberatory furnaces used, one about twice as large as the other. The larger furnace has a bed about six feet square, sloping towards the front of the furnace through which are several openings at different heights. The charge for this furnace consists of 1200 lbs. of double chloride, 350 lbs. of sodium, and 600 lbs. of cryolite for a flux. The chloride is in small pieces, the cryolite is in powder, and the sodium is cut into thin slices by a machine. These ingredients are put into a revolving wooden drum placed on a staging over the furnace, and are there thoroughly mixed. The drum is then opened and turned, when the contents fall into a small wagon beneath. The furnace having been raised to the required temperature, all the dampers are shut and the car is moved on a track immediately over a large hopper placed in the roof of the furnace. The hopper being opened the charge is dumped in and drops on to the centre of the hearth. The reaction is immediate and the whole charge becomes liquid in a very short time. After a few minutes, heating gas is again turned on and the furnace kept moderately hot for two or three hours. The reaction has been

$$Al^2Cl^6.2NaCl + 6Na = Al^2 + 8NaCl$$

and the aluminium gathers under the bath of cryolite and sodium chloride. One of the lower tap holes is then opened with a bar, and the aluminium run out into moulds. When the metal has all run out it is followed by slag, which flows into iron wagons. The openings are then plugged up and the furnace is ready for another charge. The charge given produces usually 115 to 120 lbs. of aluminium, the whole operation lasting about 4 hours. The large furnace could thus produce 840 lbs. in 24 hours, and the smaller one half that quantity. The first portion of metal running out is the purest, the latter portions and especially that entangled in the slag on the hearth, and which has to be scraped out, containing more foreign substances. This impure metal is about one-fourth of all the aluminium in the charge.

The purity of the metal run out depends directly on the purity of the chloride used. If the double chloride contains 0.2 per cent. of iron the metal produced will very probably contain all of it, or 2 per cent. Using the double chloride purified by Mr. Castner's new method (see p. 132), by which the content of iron is reduced to 0.05 per cent. or less, aluminium can be made containing less than 0.5 per cent. of iron and from 99 to 99.5 per cent. of aluminium. Professor Roscoe exhibited at one of his lectures a mass of metal weighing 116 lbs., being one single running from the furnace, and which contained only 0.3 per cent. silicon and 0.5 per cent. iron. In practice, the metal from 8 or 10 runnings is melted down together to make a uniform quality.

Taking the figures given, it appears that the metal run out represents 70 per cent. of the aluminium in the charge, and 80 per cent. of the weight which the sodium put in should reduce, but since an indeterminate weight is sifted and picked from the slag, it is probable that the utilization of the materials is more perfect than the above percentages. However, this seems to be the part of the old Deville process least improved upon in these new works, for there seems to be plenty of room for improvement in perfecting the utilization of materials especially in regard to loss of sodium by volatilization, which undoubtedly takes place and which can possibly be altogether prevented.

CHAPTER. X.

REDUCTION OF ALUMINIUM COMPOUNDS BY MEANS OF POTASSIUM OR SODIUM (*continued*).

II.

The methods based on the reduction of cryolite can be most conveniently presented in chronological order.

Rose's Experiments (1855).

We will here give H. Rose's entire paper, as an account of this eminent chemist's investigations written out by himself with great detail, describing failures as well as successes, cannot but be of value to all interested in the production of aluminium.[*]

"Since the discovery of aluminium by Wöhler, Deville has recently devised the means of procuring the metal in large, solid masses, in which condition it exhibits properties with which we were previously unacquainted in its more pulverulent form as procured by Wöhler's method. While, for instance, in the latter state it burns vividly to white earthy alumina on being ignited, the fused globules may be heated to redness without perceptibly oxidizing. These differences may be ascribed to the greater amount of division on the one hand and of density on the other.

---

[*] Pogg. Annalen, Sept. 1855.

According to Deville, however, Wöhler's metal contains platinum, by which he explains its difficulty of fusion, although it affords white alumina by combustion. Upon the publication of Deville's researches I also tried to produce aluminium by the decomposition of aluminium-sodium chloride by means of sodium. I did not, however, obtain satisfactory results. Moreover, Prof. Rammelsberg, who followed exactly the method of Deville, obtained but a very small product, and found it very difficult to prevent the cracking of the glass-tube in which the experiment was conducted by the action of the vapor of sodium on aluminium chloride. It appeared to me that a great amount of time, trouble, and expense, as well as long practice, was necessary to obtain even small quantities of this remarkable metal.

"The employment of aluminium chloride and its compounds with alkali chlorides is particularly inconvenient, owing to their volatility, deliquescence, and to the necessity of preventing all access of air during their treatment with sodium. It very soon occurred to me that it would be better to use the fluoride of aluminium instead of the chloride; or rather the combination of the fluoride with alkaline fluorides, such as we know them through the investigations of Berzelius, who pointed out the strong affinity of aluminium fluoride for sodium fluoride and potassium fluoride, and that the mineral occurring in nature under the name of Cryolite was a pure compound of aluminium fluoride and sodium fluoride.

"This compound is as well fitted for the preparation of aluminium by means of sodium as aluminium chloride or aluminium sodium chloride. Moreover, as cryolite is not volatile, is readily reduced to the most minute state of division, is free from water and does not attract moisture from the air, it affords peculiar advantages over the above-mentioned compounds. In fact, I succeeded with much less trouble in preparing aluminium by exposing cryolite together with sodium to a strong red heat in an iron crucible, than by using aluminium chloride and its compounds. But the scarcity of cryolite prevented my pursuing the experiments. In consequence of receiving, however, from Prof. Krantz, of Bonn, a considerable quantity of the purest cryolite at a very moderate price ($2 per kilo), I was enabled to renew the investigation.

"I was particularly stimulated by finding, most unexpectedly, that cryolite was to be obtained here in Berlin commercially at an inconceivably low price. Prof. Krantz had already informed me that cryolite occurred in commerce in bulk, but could not learn where. Shortly after, M. Rudel, the manager of the chemical works of H. Kunheim, gave me a sample of a coarse white powder large quantities of which were brought from Greenland, by way of Copenhagen, to Stettin, under the name of mineral soda, and at the price of $3 per centner. Samples had been sent to the soap boilers, and a soda-lye had been extracted from it by means of quicklime, especially adapted to the preparation of many kinds of soap, probably from its containing alumina. It is a fact, that powdered cryolite is completely decomposed by quicklime and water. The fluoride of lime formed contains no alumina, which is all dissolved by the caustic soda solution; and this, on its side, is free from fluorine, or only contains a minute trace. I found this powder to be of equal purity to that received from Prof. Krantz. It dissolved without residue in hydrochloric acid (in platinum vessels); the solution evaporated to dryness with sulphuric acid, and heated till excess of acid was dissipated, gave a residue which dissolved completely in water, with the aid of a little hydrochloric acid. From this solution, ammonia precipitated a considerable quantity of alumina. The solution filtered from the precipitate furnished, on evaporation, a residue of sulphate of soda, free from potash. Moreover, the powder gave the well-known reactions of fluorine in a marked degree. This powder was cryolite of great purity: therefore the coarse powder I first obtained was not the form in which it was originally produced. It is now obtainable in Berlin in great masses; for the preparation of aluminium it must, however, be reduced to a very fine powder.

"In my experiments on the preparation of aluminium, which were performed in company with M. Weber, and with his most zealous assistance, I made use of small iron crucibles, 1¾ inches high and 1¾ inches upper diameter, which I had cast here. In these I placed the finely divided cryolite between thin layers of sodium, pressed it down tight, covered with a good layer of potassium chloride (KCl), and closed the crucible with a well-fitting porcelain cover. I found potassium chloride the most advantageous flux to employ; it has the lowest specific gravity of any which could be used, an important point when the slight density of the metal is taken into consideration. It also increases the fusibility of the sodium fluoride. I usually employed equal weights of cryolite and potassium chloride, and for every five parts of cryolite two parts of sodium. The most fitting quantity for the crucible was found to be ten grammes of powdered cryolite. The whole was raised to a strong red heat by means of a gas-air blowpipe. It was found most advantageous to maintain the heat for about half an hour, and not longer, the crucible being kept closely covered the whole time; the contents were then found to be well fused. When quite cold the melted mass is removed from the crucible by means of a spatula, this is facilitated by striking the outside with a hammer. The crucible may be employed several times, at last it is broken by the hammer blows. The melted mass is treated with water, when at times only a very minute evolution of hydrogen gas is observed, which has the same unpleasant odor as the gas evolved during solution of iron in hydrochloric acid. The carbon contained in this gas is derived from a very slight trace of naphtha adhering to the sodium after drying it. On account of the difficult solubility of sodium fluoride, the mass is very slowly acted on by water, although the insolubility is somewhat diminished by the presence of the potassium chloride. After twelve hours the mass is softened so far that it may be removed from the liquid and broken down in a porcelain mortar. Large globules of aluminium are then discovered, weighing from 0.3 to 0.4 or even 0.5 grammes, which may be separated out. The smaller globules cannot well be separated from the undecomposed cryolite and the alumina always produced by washing, owing to their being specifically lighter than the latter. The whole is treated with nitric acid in the cold. The alumina is not dissolved thereby, but the little globules then first assume their true metallic lustre. They are dried and rubbed on fine silk muslin; the finely powdered, undecomposed cryolite and alumina pass through, while the globules remain on the gauze.

15

The mass should be treated in a platinum or silver vessel, a porcelain vessel would be powerfully acted on by the sodium fluoride. The solution, after standing till clear, may be evaporated to dryness in a platinum capsule, in order to obtain the sodium fluoride, mixed however with much potassium chloride. The small globules may be united by fusion in a small well-covered porcelain crucible, under a layer of potassium chloride. They cannot be united without a flux. They cannot be united by mere fusion, like globules of silver, for instance, for though they do not appear to oxidize on ignition in the air, yet they become coated with a scarcely perceptible film of oxide, which prevents their running together into a mass. This fusion with potassium chloride is always attended with a loss of aluminium. Buttons weighing 0.85 gramme lost, when so treated, 0.05 gramme. The potassium chloride when dissolved in water left a small quantity of alumina undissolved, but the solution contained none. Another portion of the metal had undoubtedly decomposed the potassium chloride; and a portion of this salt and aluminium chloride must have been volatilized during fusion (other metals, as copper and silver, behave in a similar manner—Pogg. lxviii. 287). I therefore followed the instructions of Deville, and melted the globules under a stratum of aluminium-sodium chloride in a covered porcelain crucible. The salt was melted first, and then the globules of metal added to the melted mass. There is no loss, or a very trifling one of a few milligrammes of metal, by this proceeding. When the aluminium is fused under potassium chloride its surface is not perfectly smooth, but exhibits minute concavities; with aluminium-sodium chloride this is not the case. The readiest method of preparing the double chloride for this purpose is by placing a mixture of alumina and carbon in a glass tube, as wide as possible, and inside this a tube of less diameter, open at both ends, and containing sodium chloride. If the spot where the mixture is placed be very strongly heated, and that where the sodium chloride is situated, more moderately, while a current of chlorine is passed through the tube, the vapor of aluminium chloride is so eagerly absorbed by the sodium chloride that none or at most a trace is deposited in any other part of the tube. If the smaller tube be weighed before the operation, the amount absorbed is readily determined. It is not uniformly combined with the sodium chloride, for that part which is nearest to the mixture of charcoal and alumina will be found to have absorbed the most.

"I have varied in many ways the process for the preparation of aluminium, but in the end have returned to the one just described. I often placed the sodium in the bottom of the crucible, the powdered cryolite above it, and the potassium chloride above all. On proceeding in this manner, it was observed that much sodium was volatilized, burning with a strong yellow flame, which never occurred when it was cut into thin slices and placed in alternate layers with the cryolite, in which case the process goes on quietly. When the crucible begins to get red hot, the temperature suddenly rises, owing to the commencement of the decomposition of the compound; no lowering of the temperature should be allowed, but the heat should be steadily maintained, not longer, however, than half an hour. By prolonging the process a loss would be sustained, owing to the action of the potassium chloride on the aluminium. Nor does the size of the globules increase on extending the time even to two hours; this effect can only be produced by obtaining the highest possible temperature. If the process be stopped, however, after five or ten minutes of very strong heat, the production is very small, as the metal has not had sufficient time to conglomerate into globules, but is in a pulverulent form and burns to alumina during the cooling of the crucible. No advantage is gained by mixing the cryolite with a portion of chloride before placing it between the layers of sodium, neither did I increase the production by using aluminium-sodium chloride to cover the mixture instead of potassium chloride. I repeatedly employed decrepitated sodium chloride as a flux in the absence of potassium chloride, without remarking any important difference in the amount of metal produced, although a higher temperature is in this case required. The operations may also be conducted in refractory unglazed crucibles made of stoneware, and of the same dimensions, although they do not resist so well the action of the sodium fluoride at any high heats, but fuse in one or more places. The iron crucibles fuse, however, when exposed to a very high temperature in a charcoal fire. The product of metal was found to vary very much, even when operating exactly in the manner recommended and with the same quantities of materials. I never succeeded in reducing the whole amount of metal contained in the cryolite (which contains only 13 per cent of aluminium). By operating on 10 grammes of cryolite, the quantity I always employed in the small iron crucible, the most successful result was 0.8 grm. But 0.6 or even 0.4 grm. may be considered favorable; many times I obtained only 0.3 grm., or even less. These very different results depend on various causes, more particularly, however, on the degree of heat obtained. The greater the heat the greater the amount of large globules, and the less amount of minutely divided metal to oxidize during the cooling of the crucible. I succeeded once or twice in reducing nearly the whole of the metal to one single button weighing 0.5 grm., at a very high heat in a stoneware crucible. I could not always obtain the same heat with the blowpipe, as it depended in some degree on the pressure in the gasometer in the gas-works, which varies at different hours of the day. The following experiment will show how great the loss of metal may be owing to oxidation during the slow cooling of the crucible and its contents: In a large iron crucible were placed 35 grms. of cryolite in alternate layers with 14 grms. of sodium and the whole covered with a thick stratum of potassium chloride. The crucible, covered by a porcelain cover, was placed in a larger earthen one also covered, and the whole exposed to a good heat in a draft furnace for one hour and cooled as slowly as possible. The product in this case was remarkably small, for 0.135 grm. of aluminium was all that could be obtained in globules. The differences in the amounts reduced depend also in some degree on the more or less successful stratification of the sodium with the powdered cryolite, as much of the latter sometimes escapes decomposition. The greater the amount of sodium employed, the less likely is this to be the case; however, owing to the great difference in their prices, I never employed more than 4 grms. of sodium to 10 grms. of cryolite. In order to avoid this loss by oxidation I tried another method of preparation: Twenty grms. of cryolite were heated intensely in a gun-barrel in a current of hydrogen, and then the vapor of 8 grms. of sodium passed over it. This was effected simply by placing the sodium in a little iron tray in a part of the gun-barrel without the fire, and pushing it forward when the cryolite had attained a maximum temperature. The operation went on very well, the whole being allowed to cool in a current of hydrogen. After the treatment with water, in which the sodium fluoride dissolved very slowly, I obtained a black powder consisting for the most part of iron. Its solution in hydrochloric acid gave small evidence of aluminium. The small amounts I obtained, however, should not deter others from making these experiments. These are the results of first experiments on which I have not been able to expend much time. Now that cryolite can be procured at so moderate a price, and sodium by Deville's improvements will in future become so much cheaper, it is in the power of every chemist to engage in the preparation of aluminium, and I have no doubt that in a short time methods will be found affording a much more profitable result.

"To conclude, I am of opinion that cryolite is the best adapted of all the compounds of aluminium for the preparation of this metal. It deserves the preference over aluminium-sodium chloride or aluminium chloride, and it might still be employed with great advantage even if its price were to rise considerably. The attempts at preparing aluminium direct from alumina have as yet been unattended with success. Potassium and sodium appear only to reduce metallic oxides when the potash and soda produced are capable of forming compounds with a portion of the oxide remaining as such. Pure potash and soda, with whose properties we are very slightly acquainted, do not appear to be formed in this case. Since, however, alumina combines so readily with the alkalies to form aluminates, one would be inclined to believe that the reduction of alumina by the alkali metals should succeed. But even were it possible to obtain the metal directly from alumina, it is very probable that cryolite would long be preferred should it remain at a moderate price, for it is furnished by nature in a rare state of purity, and the aluminium is combined in it with sodium and fluorine only, which exercise no prejudicial influence on the properties of the metal, whereas alumina is rarely found in nature in a pure state and in a dense, compact condition, and to prepare it on a large scale, freeing it from those substances which would act injuriously on the properties of the metal, would be attended with great difficulty.

"The buttons of aluminium which I have prepared are so malleable that they may be beaten and rolled out into the finest foil without cracking on the edges. They have a strong metallic lustre. Some small pieces, not globular, however, were found in the bottom of the crucible, and occasionally adhering to it, which cracked on being hammered, and were different in color and lustre from the others. They were evidently not so pure as the greater number of globules, and contained iron. On sawing through a large button weighing 3.8 grammes, it could readily be observed that the metal for about half a line from the exterior was brittle, while in the interior it was soft and malleable. Sometimes the interior of a globule contained cavities. With Deville, I have occasionally observed aluminium crystallized. A large button became striated and crystalline on cooling. Deville believes he has observed regular octahedra, but does not state this positively. According to my brother's examination, the crystals do not belong to any of the regular forms. As I chanced on one occasion to attempt the fusion of a large, flattened-out button of rather impure aluminium, without a flux, I observed before the heat was sufficient to fuse the mass, small globules sweating out from the surface. The impure metal being less fusible than pure metal, the latter expands in fusing and comes to the surface."

Experiments of Percy and Dick (1855).

After the publication of Rose's results, widespread attention was directed toward this field, and it was discovered that some six months previously Dr. Percy, in England, had accomplished almost similar results, and had even shown a specimen of the metal to the Royal Institution, but with the singular fact of exciting very little attention. These facts are stated at length in the following paper written by Allan Dick, Esq., which appeared in November, 1855, two months after the publication of H. Rose's paper :—*

---

* Phil. Mag., Nov. 1855.

"In the last number of this magazine was the translation of a paper by H. Rose, of Berlin, describing a method of preparing aluminium from cryolite. Previously, at the suggestion of Dr. Percy, I had made some experiments on the same subject in the metallurgical laboratory of the School of Mines, and as the results obtained agree very closely with those of Mr. Rose, it may be interesting to give a short account of them now, though no detailed description was published at the time, a small piece of metal prepared from cryolite having simply been shown at the weekly meeting of the Royal Institution, March 30, 1855, accompanied by a few words of explanation by Faraday.

"Shortly after the publication of Mr. Deville's process for preparing aluminium from aluminium chloride, I tried along with Mr. Smith to make a specimen of the metal, but we found it much more difficult to do than Deville's paper had led us to anticipate, and had to remain contented with a much smaller piece of metal than we had hoped to obtain. It is, however, undoubtedly only a matter of time, skill, and expense to join successful practice with the details given by Deville. Whilst making these experiments, Dr. Percy had often requested us to try whether cryolite could be used instead of the chlorides, but some time elapsed before we could obtain a specimen of the mineral. The first experiments were made in glass tubes sealed at one end, into which alternate layers of finely powdered cryolite and sodium cut into small pieces were introduced, and covered in some instances with a layer of cryolite, in others by sodium chloride. The tube was then heated over a gas blowpipe for a few minutes till decomposition had taken place and the product was melted. When cold, on breaking the tube, it was found that the mass was full of small globules of aluminium, but owing to the specific gravity of the metal and flux being nearly alike, the globules had not collected into a button at the bottom. To effect this, long-continued heat would be required, which cannot be given in glass tubes owing to the powerful action of the melted fluoride on them. To obviate this difficulty, a platinum crucible was lined with magnesia by ramming it in hard, and subsequently cutting out all but a lining. In this, alternate layers of cryolite and sodium were placed, with a thickish layer of cryolite on top. The crucible was covered with a tight-fitting lid, and heated to redness for about half an hour over a gas blowpipe. When cold it was placed in water, and after soaking for some time the contents were dug out, gently crushed in a mortar, and washed by decantation. Two or three globules of aluminium, tolerably large considering the size of the experiment, were obtained along with a large number of very small ones. The larger ones were melted together under potassium chloride. Some experiments made in iron crucibles were not attended with the same success as those of Rose, no globules of any considerable size remained in the melted fluorides; the metal seemed to alloy on the sides of the crucible, which acquired a color like zinc. It is possible that this difference may have arisen from using a higher temperature than Rose, as we made these experiments in a furnace, not over the blowpipe. Porcelain and clay crucibles were also tried, but laid aside after a few experiments, owing to the action of the fluorides upon them, which in most cases was sufficient to perforate them completely."

Deville's Methods (1856–8).

* "I have repeated and confirmed all the experiments of Dr. Percy and H. Rose, using the specimens of cryolite which I obtained from London through the kindness of MM. Rose and Hofmann. I have, furthermore, reduced cryolite mixed with sodium chloride by the battery, and I believe that this will be an excellent method of covering with aluminium all the other metals, copper in particular. Anyhow, its fusibility is considerably increased by mixing it with aluminium-sodium chloride. Cryolite is a double fluoride of aluminium and sodium, containing 13 per cent. of aluminium and having the formula $Al^2F^6.6NaF$. I have verified these facts myself by many analyses.

"In reducing the cryolite I placed the finely-pulverized mixture of cryolite and sodium chloride in alternate layers with sodium in a porcelain crucible. The uppermost layer is of pure cryolite, covered with salt. The mixture is heated just to complete fusion, and, after stirring with a pipe-stem, is let cool. On

* Ann. de Chem. et de Phys. [3], xlvi. 451.

breaking the crucible, the aluminium is often found united in large globules easy to separate from the mass. The metal always contains silicon, which increases the depth of its natural blue tint and hinders the whitening of metal by nitric acid, because of the insolubility of the silicon in that acid. M. Rose's metal is very ferruginous. I have verified all M. Rose's observations, and I agree with him concerning the return of metal, which I have always found very small. There are always produced in these operations brilliant flames, which are observed in the scoria floating on the aluminium, and which are due to gas burning and exhaling a very marked odor of phosphorus. In fact, phosphoric acid exists in cryolite, as one may find by treating a solution of the mineral in sulphuric acid with molybdate of ammonia, according to H. Rose's reaction.

"M. Rose has recommended iron vessels for this operation, because of the rapidity with which alkaline fluorides attack earthen crucibles and so introduce considerable silicon into the metal. Unfortunately, these iron crucibles introduce iron into the metal. This is an evil inherent in this method, at least in the present state of the industry. The inconveniences of this method result in part from the high temperature required to complete the operation, and from the crucible being in direct contact with the fire, by which its sides are heated hotter than the metal in the crucible. The metal itself, placed in the lower part of the fire, is hotter than the slag. This, according to my observations, is an essentially injurious condition. The slag ought to be cool, the metal still less heated, and the sides of the vessel where the fusion occurs ought to be as cold as possible. The yield from cryolite, according to Rose's and my own observations, is also very small. M. Rose obtained from 10 of cryolite and 4 of sodium about 0.5 of aluminium. This is due to the affinity of fluorine for aluminium, which must be very strong not only with relation to its affinity for sodium but even for calcium, and this affinity appears to increase with the temperature, as was found in my laboratory. Cryolite is most convenient to employ as a flux to add to the mixture which is fused, especially when operating on a small scale.

"The argument which decided the company at Nanterre not to

adopt the method of manufacture exclusively from cryolite was
the report of M. de Chancourtois, mining engineer, who had just
returned from a voyage to Greenland. According to the verbal
statements of this gentleman, the gîte at Evigtok is accessible only
during a very short interval of time each year, and, because of
the ice fields, can only be reached then by a steamboat. The
workmen sent from Europe to blast and load up the rock have
scarcely one or two months of work possible. The local workmen remain almost a whole year deprived of all communication
with the rest of the world, without fresh provisions or fuel other
than that brought from Europe in the short interval that naviga-
tion is open. The deposit itself, which is scarcely above sea-
level, can be easily worked with open roof, but the neighborhood
of the sea in direct contact with the vein, the unorganized manner of working, and the lack of care in keeping separate the
metalliferous portions of the ore—all combine to render the
mineral very costly and further developments underground almost
impossible.

" It is therefore fortunate that cryolite is not indispensable, for
no one would wish to establish an industry based on the employment of a material which is of uncertain supply."

Tissier Bros.' Method (1857).

The process adopted in the works at Amfreville, near Rouen,
directed by Tissier Bros., is essentially that described by Percy
and Rose. The method of operating is given by the Tissier Bros.
themselves in their book as follows :—

" After having finely powdered the cryolite, it is mixed with a
certain quantity of sodium chloride (sea salt), then placed between
layers of sodium used in the proportions given by M. Rose, in
large refractory crucibles. These are heated either in a reverberatory furnace or in a wind furnace capable of giving a temperature high enough to melt the fluoride of sodium produced by the
reaction. As the sodium fluoride requires a pretty high temperature to fuse it, the heat will necessarily be higher than that required in the reduction of the double chloride of aluminium and
sodium. When the contents of the crucible are melted, so as to

be quite liquid, the fusion is poured into cast-iron pots at the bottom of which the aluminium collects in one or several lumps."

Tissier Bros. claimed the following advantages for the use of cryolite:—

"Cryolite comes to us of a purity difficult to obtain with the double chloride of aluminium and sodium, to which it exactly corresponds; and since, thanks to the perfection we have attained in using it, the return of aluminium is exactly correspondent to the amount of sodium used in reduction, it is easily seen what immense advantages result from its employment. The double chloride deteriorates in the air, it gives rise in the works to vapors more or less deleterious and corrosive, and its price is always high. Cryolite can be imported into France at a price so low that we have utilized it economically for making commercial carbonate of soda; it remains unaltered in the air, emits no deleterious vapors, and its management is much more easy than that of the double chloride. Moreover, on comparing the residues of the two methods of reduction, the manufacture from double chloride leaves sodium chloride, almost without value, while the manufacture from cryolite leaves sodium fluoride, which may be converted for almost nothing into caustic soda or carbonate, and so completely cancels the cost of the cryolite from the cost of the aluminium. The most serious objection which can be made to using cryolite is that the sources of the mineral being up to the present very limited, the future prospect of aluminium lies necessarily in the utilization of clays and their transformation into aluminium chloride; but, admitting that other sources of cryolite may not be discovered hereafter, the abundance of those which exist in Greenland will for a long time to come give this mineral the preference in the manufacture of aluminium."

The most serious difficulty which this process had to meet, and which it could not overcome, was the high content of silicon in the metal produced. A specimen of their aluminium made in 1859 contained 4.4 per cent. of silicon alone (see p. 54, Analysis 7).* The firm at Rouen went out of business about 1863 or 1865, I am unable to give the exact date. From that time

---

* The analysis should read 0.8 iron and 4.4 silicon, not 0.8 silicon and 4.4 iron.

until quite recently, it has been considered that the best use of cryolite is as a flux in the preparation of aluminium from aluminium-sodium chloride, in which case the slag is not sodium fluoride but aluminium fluoride, which acts but slightly on the containing vessel.

Wöhler's Modifications (1856).

Wöhler suggested the following modifications of Deville's process of reducing cryolite in crucibles, by means of which the reduction can be performed in an earthen crucible without the metal produced taking up silicon.

* " The finely pulverized cryolite is mixed with an equal weight of a flux containing 7 parts sodium chloride to 9 parts potassium chloride. This mixture is then placed in alternate layers with sodium in the crucible, 50 parts of the mixture to 10 of sodium, and heated gradually just to its fusing point. The metal thus obtained is free from silicon, but only one-third of the aluminium in the cryolite is obtained." In spite of the small yield, this method was used for some time by Tissier Bros.

Gerhard's Furnace (1858).

This furnace was devised for the reduction of aluminium either from aluminium-sodium chloride or from cryolite, the object being to prevent loss of sodium by ignition. It was invented and patented by W. F. Gerhard.† " It consists of a reverberatory furnace having two hearths, or of two crucibles, or of two reverberatory furnaces, placed one above the other and communicating by an iron pipe. In the lower is placed a mixture of sodium with the aluminium compound, and in the upper a stratum of sodium chloride, or of a mixture of this salt and cryolite, or of the slag obtained in a previous operation. This charge, when melted, is made to run into the lower furnace in quantity sufficient to completely cover the mixture contained therein, and so to protect it from the air. The mixture thus covered is reduced as by the usual operation."

* Ann. der Chem. und Pharm. 99, 255.
† Eng. Pat. 1858, No. 2247.

Whether a furnace was ever put up and operated on this principle the author cannot say. It is possible that it may have been used in the English manufactories started in 1859 and 1860 at Battersea and Newcastle-on-Tyne.

Thompson and White's Patent (1887).

*J. B. Thompson and W. White recommend heating a mixture of 3 parts sodium and 4 parts of cryolite to 100°, whereby the sodium becomes pasty and the whole can be well kneaded together with an iron spatula. When cold, 4 parts of aluminium chloride are added, and the mixture put into a hopper on top of a well-heated reverberatory furnace, with a cup-shaped hearth. The charge is dropped into the furnace and the reaction takes place at once. To produce alloys, this patent claims that 16 parts of cryolite are mixed with 5 parts of sodium, the metal added before reduction and the mixture treated as above, by which means explosions are avoided. The preliminary heating to 100° is effected in a jacketed cast-iron pot connected with a circulating boiler.

Hampe's Experiment (1888).

†Dr. W. Hampe failed to produce aluminium bronze by treating cryolite with sodium in the presence of copper. A mixture of

44 grammes finely divided copper,
15 " sodium, in small pieces,
100 " finely powdered cryolite,

was melted rapidly in a carbon-lined crucible. There were no sounds given out such as usually accompany other reductions by sodium, but much sodium vapor was given off. The copper button contained only traces of aluminium.

Netto's Process (1887).

Dr. Curt Netto, of Dresden, patented in England and Germany, in spring and autumn of 1887, processes for producing sodium

* English Patent 8427, June 11, 1887.
† Chemiker Zeitung (Cöthen), xii. p. 391.

and potassium and methods of using them in producing aluminium. His experiments were made in conjunction with Dr. Salomon, of Essen, and the fact that the experimental apparatus was put up in Krupp's large steel works at Essen gave rise to reports that the latter had taken up the manufacture of aluminium by some new and very successful process, intending to use it for alloys in making cannon.*

In the latter part of 1888 we hear of the formation of the Alliance Aluminium Co. of London, England, capitalized at £500,000, purposing to manufacture potassium, sodium and aluminium, and owning the English, French, German, and Belgian patents of Dr. Netto for the production of those metals, also the processes of a Mr. Cunningham for the same purpose, also a process for the production of artificial cryolite by the regeneration of slags (provisionally protected by its inventor, Mr. Forster, of the Lonesome Chemical Works, Streatham), and, lastly, a process invented by Drs. Netto and Salomon by which aluminium can be raised to the highest standards of purity on a commercial scale. A note accompanying the above announcement stated that the exhaustive experiments made at Essen had satisfactorily demonstrated the practicability of the processes, and that the company had already contracted with the cryolite mines of Greenland for all the cryolite the company would need.

In June, 1888,† we learn that the Alliance Aluminium Company had in operation a small aluminium plant at King's Head Yard, London, E. C., and that when the process was in continuous operation the cost of the metal was set down at 6 shillings per pound. It is probable that the metal exhibited in the Paris Exposition of 1889 was produced at this place.

In April, 1889,‡ it was stated in the scientific journals that ten acres of ground had been leased at Hepburn on which to produce sodium by Capt. Cunningham's process. The sodium produced is to be sent to Wallsend to be used by the Alliance Aluminium Company, who are erecting a large works at that place.

* American Register, Paris, August, 1888.

† Engineering, June 1, 1888.

‡ E. and M. J., April 27, 1889.

As for Capt. Cunningham's sodium processes, they are apparently identical with Dr. Netto's. Cunningham's aluminium process* consists in melting the sodium to be used with lead, in order to facilitate the submerging of the sodium under the molten aluminium salt. The alloy is cast into bars and added piece by piece to the bath of molten aluminium salt on the hearth of a reverberatory furnace. After the reaction the mixture separates by specific gravity into lead, containing a little aluminium, and aluminium containing a little lead, the slag floating on top of all. Aluminium is known to have so small an attraction for lead that this result becomes possible.

Dr. Netto recommends several processes, the one used at London being the following :—†

One hundred parts of cryolite and 30 to 100 parts of sintered sodium chloride are melted at a red heat in a well-covered clay crucible. (Another arrangement, and apparently a better, is to melt this mixture on the hearth of a reverberatory furnace and to tap it into a deep, conical ladle, in which the succeeding operations proceed as about to be described.) As soon as the bath is well fused, 35 parts of sodium at the end of a rod, and covered over by a perforated concave plate, is quickly pushed down to the bottom of the crucible. The plate mentioned fits across the whole section of the crucible at its lower part, so that the fusible, easily volatile sodium, being vaporized, is divided into very fine streams as it passes upwards through the bath, and is all utilized before it reaches the surface. In this way the reaction is almost instantaneous, and the contents can be poured out at once into iron pots, where, on cooling, the metal is found as a large lump at the bottom.

It is further observed that to avoid explosions on introducing the sodium it should have in it no cavities which might contain moisture or hydrocarbons. In consequence of the reaction being over so quickly, and the heat set free in the reduction, the syrupy fusion becomes thin as water, and the aluminium disseminated through the mass collects together completely, so that the slag

* English Patent, 16727, Dec. 5, 1887.
† German Patent (D. R. P.) 45198, March 26, 1887.

contains no particles visible to the eye. Since the reduction, pouring, and cooling take place so quickly, the aluminium is not noticeably redissolved by the bath, thus insuring a high return of metal. By using 35 parts of sodium to 100 parts of cryolite, 10 parts of aluminium are obtained. Since the cryolite contains 13 per cent. of aluminium, the return is 77 per cent. of the amount of metal in the cryolite; since 35 parts of sodium should theoretically displace 14 parts of aluminium, the return is 71 per cent. of the amount which the sodium should produce. Dr. Netto claims that this is double the return formerly obtained from cryolite. The metal produced is said to be from 98.5 to 99 per cent. pure.

The apparatus erected at Krupp's works at Essen, which was described by the newspapers as similar to a Bessemer converter, was constructed and operated as follows: A large iron cylinder is pivoted at the centre in a manner similar to a Bessemer converter. Passing through the centre of the cylinder, longitudinally, is a large iron tube in which generator gas is burnt to heat the vessel. To heat it up, it is placed erect, connection made with the gas-main, while a hood above connects with the chimney. On top of the cylinder, a close valve communicates with the interior, for charging, and at the other end is a tap-hole. The charge of cryolite being put in, the flame is passed through the central tube until the mineral is well fused. Then solid or melted sodium is passed in at the top, the valve is screwed tight, the gas shut off, and the whole cylinder is rotated several times until reduction is complete, when it is brought upright, the taphole opened and slag and metal tapped into a deep iron pot, where they separate and cool. Aluminium thus made could not but contain much iron, even up to 14 per cent., it is said, which would prevent its use for any purpose except alloying with iron. To procure pure aluminium, the vessel would have to be properly fettled.

Dr. Netto also devised an arrangement similar to Heaton's apparatus for making steel. It consisted of a large, well-lined vessel on trunnions, the bottom of which was filled to a certain depth with sodium, then a perforated aluminium plate placed like a false bottom over it. On pouring molten cryolite into the vessel the aluminium plate prevented the sodium from rising *en masse* to the surface of the cryolite. After the reaction was over, the vessel was tilted and the slag and metal poured out into iron pots.

The modification of the crucible method appears to be the most feasible of Netto's processes, and is probably now being used at Wallsend by the Alliance Aluminium Company. Outside estimates of the cost of aluminium to this company place it at $1.50 to $2 per pound. They were selling in the latter part of 1889 at 11, 13, and 15 shillings per pound, according to quality.

III.

There is only one patentee claiming particularly the reduction of aluminium fluoride by sodium—Ludwig Grabau, of Hannover, Germany. His patents on this subject are immediately preceded by others on a method of producing the aluminium fluoride cheaply, which are described on p. 139, and the inventor is at present engaged on a process which will furnish him with cheap sodium. Mr. Alexander Siemens is authority for the statement that a plant was in operation in the Spring of 1889, in Hannover, producing aluminium by this process on a commercial scale. The principal object of Mr. Grabau's endeavors has been to produce metal of a very high degree of purity. To this end every precaution is taken to procure pure materials and to prevent contamination during reduction. We will quote from a paper written by Mr. Grabau* and also from his patent specifications,† the following explanation of the process:—

"The purifying of impure aluminium is accompanied by so many difficulties that it appears almost impossible. It is therefore of the greatest importance to so conduct the operation that every impurity is excluded from the start. Molten aluminium compounds, whether a flux is added or not, attack any kind of refractory vessels and become siliceous, if these vessels are made

---

* Zeitschrift für angewandte Chemie, 1889, vol. 6.

† German Pat. (D. R. P.) 47031, Nov. 15, 1887. English Pat. 15593, Nov. 14, 1887. U. S. Patents 386704, July 24, 1888; and 400449, April 2, 1889.

16

of chamotte or like materials, or if made of iron they become ferruginous. These impurities are reduced in the further processes and pass immediately into the aluminium as iron, silicon, etc. Evidently the case is altered if an aluminium compound which is infusible can be used advantageously. Aluminium fluoride is infusible and also retains its pulverized condition when heated up to the temperature needed for its use; it can therefore be heated in a vessel of any kind of refractory material or even in a metallic retort without danger of taking up any impurity.

"Further, it is necessary for succeeding in producing aluminium that the reduced metal shall unite to a large body after the reduction. For this purpose all previous processes use fluxes, and usually cryolite. But cryolite is impure and therefore here is a source of many of the impurities in commercial aluminium. Dr. K. Kraut, of Hannover, has observed that, according to the recent analyses of Fresenius and Hintz, commercial cryolite contains 0.80 to 1.39 per cent. of silicon and 0.11 to 0.88 per cent. of iron, and that these impurities inter-penetrate the mineral in such a manner as to be often only visible under the microscope and therefore totally impossible of removal by mechanical means. It is thus seen that the avoidance of the use of any flux is of great importance as far as producing pure metal is concerned, as well as from an economic standpoint.

"By the following process it is also possible to reduce aluminium fluoride by sodium without the vessel in which reduction takes place being attacked either by the aluminium-sodium fluoride formed or by the reduced aluminium. For this purpose the aluminium fluoride and sodium are brought together in such proportions that after the reaction there is still sufficient aluminium fluoride present to form with the sodium fluoride resulting from the reaction a compound having the composition of cryolite. The reaction, therefore, will be

$$2Al^2F^6 + 6Na = 2Al + Al^2F^6.6NaF.$$

"Using these proportions, the aluminium fluoride must be previously warmed up to about 600°, in order that when it is showered down upon the melted sodium the reaction may commence without further application of heat. The aluminium fluoride

Fig. 22.

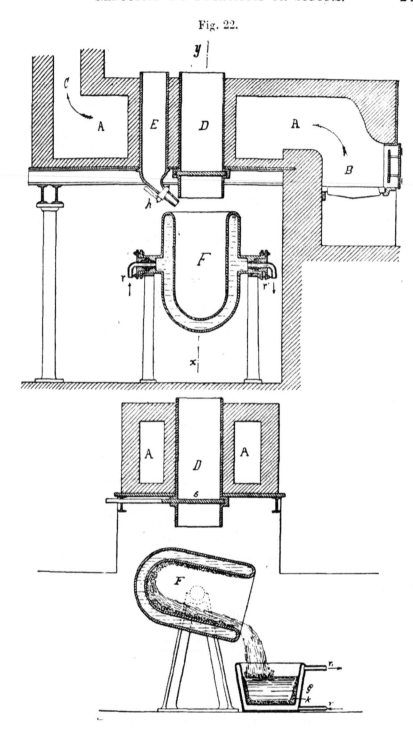

remains granular at this temperature and therefore remains on top of the melted sodium, like saw-dust or meal upon water, and under its protection the reaction proceeds from below upwards— an important advantage over the usual method of pouring molten aluminium compounds on to sodium, in which the lighter sodium floats to the top and burns to waste. If solid sodium is used in my process the aluminium fluoride must be somewhat hotter on being poured into the reduction vessel, or about 700°. For carrying out the process the reduction vessel must be artificially cooled, so as to form a lining by chilling some of the aluminium-sodium chloride formed by the reaction, on the inner walls. This lining is in no wise further attacked by the contents of the vessel, nor can it evidently supply to them any impurity.

"The furnace A (Fig. 22) with grate B and chimney C serves for heating the iron retorts D and E, which are coated with chamotte and protected from the direct action of the flame by brick work. The vessel D serves for heating the aluminium fluoride, and is provided with a damper or sliding valve beneath. The sodium is melted in E, and can be emptied out by turning the cock h. The water-jacketed reduction vessel is mounted on trunnions to facilitate emptying it. The retorts are first heated dark red-hot, and D is filled with the convenient quantity of aluminium fluoride. When this has become red hot, as is shown by a small quantity of white vapor issuing from it, the required quantity of sodium is put into E. This melts very quickly, and is then immediately run into the reduction vessel by opening the stop-cock h. As soon as it is transferred, the slide at the base of the retort D is pulled out and the whole quantity of aluminium fluoride falls at once upon the sodium and the reaction begins. As before remarked, the granular form of the aluminium fluoride keeps it on top of the sodium, so that the latter is completely covered during the whole reaction. This prevents almost altogether any waste of sodium by volatilization. Dr. K. Kraut testifies to an operation which he witnessed in which the return showed 83 per cent. of the sodium to have been utilized. An efficiency in this respect of over 90 per cent. has been occasionally reached, while the average is 80 to 90. Ad. Wurtz states that the average of several years' working of the Deville process showed only 74.3 per cent. of the quantity of aluminium produced which the sodium used could have given.

"During the reaction a very high temperature is developed, so that the cryolite formed becomes very fluid but is chilled against the sides of the vessel to a thickness of a centimetre or more. This crust is a poor conductor of heat, and is neither attacked by the fluid cryolite nor by the aluminium. In consequence of the great fluidity of the bath, it is possible for the aluminium to unite into a body without the use of any flux. The reaction being over, which is accomplished with the above proportions of materials in a few seconds, and the vessel having been shaken briskly backwards and forwards a few times to facilitate the settling of the aluminium, the whole is turned on the trunnions and emptied into a water-jacketed iron pot where it cools. The crust of cryolite inside the reduction vessel is left there, and the apparatus is ready for another operation."

M. Grabau, in a private communication to the author, sums up the advantages of his process, including the production of the aluminium fluoride, as follows :—

1. The process is not dependent on natural cryolite, which is expensive, impure and not easily purified.

2. The raw material—aluminium sulphate—can be procured in large quantities and of perfect purity.

3. The aluminium fluoride is produced by a wet process, which offers no difficulties to production on a large scale.

4. The fluorspar may be completely freed from foreign metals by washing with dilute acid ; any silica present is not injurious, as it remains undissolved in the residue during the reactions.

5. The cryolite formed in each reduction contains no impurities, and an excess of it is produced which can be sold.

6. The reduction of aluminium fluoride by my method gives a utilization of 80 to 90 per cent. of the sodium used, which is much more than can be obtained by other processes.

7. Aluminium fluoride is infusible, and can therefore be heated in a vessel of any refractory material without taking up any impurities. It is also unchanged in the air, and can be kept unsealed for any length of time without deteriorating in the least.

8. No flux has to be added for reduction, the use of impure flux being a frequent cause of impurity of the metal.

In point of fact, M. Grabau has succeeded in producing several hundred pounds of aluminium averaging over $99\frac{1}{2}$ per cent. pure. Dr. Kraut reports an analysis of an average specimen with 99.62 per cent. of aluminium (see Analysis 20, p. 54), and metal has been made as pure as 99.8 per cent., a piece of which has been kindly forwarded the author by M. Grabau, and I freely admit it to be the finest specimen of aluminium I have ever seen. If M. Grabau's statement that he can produce metal of this purity without difficulty on a commercial scale and at a price low enough to compete with the other commercial brands be realized, we will freely accord that gentleman the prize—not for *cheap*, but for *pure* aluminium; *cheap* aluminium is yet to come.

CHAPTER XI.

REDUCTION OF ALUMINIUM COMPOUNDS BY THE USE OF ELECTRICITY.

As preliminary to the presentation of the various electrolytic methods which have been proposed or used, it may be profitable to review briefly the principles of electro-metallurgy as they apply to the decomposition of aluminium compounds.

The atomic weight of aluminium being 27, its chemical equivalent, or the weight of it equal in combining power to one part of hydrogen, is 9. Therefore a current of quantity sufficient to liberate 1 part of hydrogen in a certain time would produce 9 parts of aluminium in the same time, according to the fundamental law of electric decomposition. It has been determined that a current of 1 ampere acting for one second, liberates 0.00001035 grammes of hydrogen; therefore it will produce or set free from combination in the same time, 0.00009315 grammes of aluminium. This is the electro-chemical equivalent of aluminium. Now, from thermo-chemical data we know that the amount of energy required to set free a certain weight of aluminium will vary with the compound from which it is produced; but the above equivalent is independent of the compound decomposed, therefore there must be some varying factor connected with the quantity of the current to account for the different amounts of work which the current does in decomposing different compounds of the same element. This is exactly in accordance with the principles of the mechanical or thermal equivalent of the electric current, for the statement "a current of one ampere," while it expresses a definite quantity of electricity, yet carries no idea of the energy represented by that current; we must know against what resistance or with what force that quantity is moved, and then we can calculate its mechanical equivalent. Now, a current of 1 ampere flowing against a resistance of 1 ohm, or in other words, with a moving force or intensity of 1 volt, represents a quantity of energy in one second equal to 0.00024 calories of heat or to 0.1 kilogrammetres of work, and is therefore nearly $\frac{1}{750}$ of a horse power. Therefore we can calculate the theoretical intensity of current necessary to overcome the affinities of any aluminium compound for which we know the appropriate thermal data. For instance, when aluminium forms its chloride (see p. 190) $\frac{321,960}{54} = 5960$ calories are developed per kilo of aluminium combining; consequently the liberating of 0.00009315 grammes of aluminium (its electrochemical equivalent), requires the expenditure of an amount of energy equal to $0.00009315 \times 5.960 = 0.000555$ calories. Since a current of 1 ampere at an intensity of 1 volt represents only 0.00024 calories, the intensity of current necessary to decompose aluminium chloride is theoretically $\frac{0.000555}{0.00024} = 2.3$ volts. In a similar manner we can calculate that to decompose alumina would require an electro-motive force of $\frac{391600}{54} \times \frac{0.00000009315}{0.00024} =$ 2.8 volts. These data would apply only to the substances named in a fused anhydrous state; with hydrated aluminium chloride in solution, a far greater electro-motive force would be necessary. If we had the thermal data we could also calculate the intensity of current necessary to decompose the sulphate, nitrate, acetate, etc., in aqueous solution; but, failing these, we can reason from analogy that it would be several volts in each case.

To utilize such calculations, we must bear in mind exactly what they represent. To decompose fused aluminium chloride, for instance, not only must the current possess an intensity of 2.3 volts but it must in addition have power enough above this to overcome the transfer resistance of the electrolyte; *i. e.*, to force the current through the bath from one pole to the other. So, then, 2.3 volts would be the absolute minimum of intensity which would produce decomposition, and the actual intensity practically required would be greater than this, varying with the distance of the poles apart and the temperature of the bath as far as it affects the conducting power of the electrolyte. From this it would immediately follow that if the substance to be decomposed is an absolute non-conductor of electricity, no intensity of current will be able to decompose it. If, on the other hand, the substance is a conductor and the poles are within reasonable distance, a current of a certain intensity will always produce decomposition. The objection is immediately made that in most cases no metal is obtained at all, which is true not because none is produced but because it is often dissolved by secondary actions as quickly as it is produced. I need but refer to the historic explanation of the decomposition of caustic soda in aqueous solution, although we have cases hardly parallel to this in which the electrolyte itself dissolves the separated metal.

How about the case of aqueous solutions? Water requires a minimum electro-motive force of 1.5 volts to decompose it, and hence a prominent electrician remarked of a compound which theoretically required over 2 volts that its decomposition in aqueous solution would involve the decomposition of the water and therefore was impossible. This remark is only partly true; for, caustic soda requires over 2 volts, yet if mercury is present to absorb the sodium as it is set free and protect it from the water, we will obtain sodium while the water is decomposed at the same time. The truth seems to be that if two substances are present which require different electro-motive force to decompose them, a current of a certain intensity will decompose the one requiring least force without affecting the other at all; but, if it is of an intensity sufficient to decompose the higher compound, then the current will be divided in some ratio between the two, decomposing them both.   This theory would render theoretically possible the decomposition of aluminium salts in aqueous solution, with a waste of power proportional to the amount of water decomposed at the same time; but whether any aluminium would be obtained would be contingent on the secondary action of the water on the aluminium.   Pure aluminium in mass is not acted on by water, but the foil is rapidly eaten away by boiling water.   The state of division of the metal, then, determines the action of water on it, and it is altogether probable that the reason why aluminium has not been easily and beyond question deposited from aqueous solution is that, like sodium, it is attacked as soon as isolated, the acidity of the solution converting the hydrate formed back into the salt, or else simply the hydrate remaining.   Unfortunately, mercury does not exercise the same function with aluminium as with sodium, for water attacks its amalgam with aluminium, and so destroys the metal.   It is possible that if some analogous solvent could be found which protected the aluminium from the action of water, the deposition from aqueous solution could be made immediately successful.   Perhaps some of the devices about to be described have successfully overcome these difficulties, but if so the proof of this has never been verified by any good authority, nor has the author seen any so-called aluminium plating (from aqueous solution) which really was so.

Further remarks as to the amount of aluminium theoretically obtainable per horse-power, etc. etc., will come up in connection with the various processes.

The consideration of these processes falls naturally under two heads :—

I. Deposition from aqueous solution.
II. Decomposition of fused aluminium compounds.

I.

DEPOSITION OF ALUMINIUM FROM AQUEOUS SOLUTION.

The status of this question is one of the curiosities of electro-metallurgic science.   Evidently attracted by the great reward to be earned by success, many experimenters have labored in this field, have recommended all sorts of processes, and patented all kinds of methods. We have inventors affirming in the strongest manner the successful working of their methods, while other experimenters have followed these recipes, and tried almost every conceivable arrangement, yet report negative results. To show that it is quite possible that many strong affirmations may be made in good faith, I have only to mention the fact that in March, 1863, Mr. George Gore described in the Philosophical Magazine some experiments by which he deposited coatings of aluminium from aqueous solutions, and afterwards, in his text book of Electro-metallurgy, asserts that he knows of no successful method of doing this thing. Mr. Gore found that he was in error the first time and was manly enough to acknowledge it. So, if we take the position of many eminent authorities that aluminium cannot by any methods so far advanced be deposited from aqueous solution, we will have to admit that the proposers of the following processes are probably misled by their enthusiasm in affirming so strongly that they can do this thing. Yet the problem is not impossible of solution, and I will simply assert again my previous statement, that no good authority testifies to the success of any process so far advanced, neither have I seen any so-called aluminium plating (from aqueous solution) which really was aluminium.

Messrs. Thomas and Tilly* coat metals with aluminium and its alloys by using a galvanic current and a solution of freshly precipitated alumina dissolved in boiling water containing potassium cyanide, or a solution of freshly calcined alum in aqueous potassium cyanide; also from several other liquids. Their patent covers the deposition of the alloys of aluminium with silver, tin, copper, iron, silver and copper, silver and tin, etc. etc., the positive electrode being of this metal or alloy.

M. Corbelli, of Florence,† deposits aluminium by electrolyzing a mixture of rock alum or sulphate of alumina (2 parts) with calcium chloride or sodium chloride (1 part) in aqueous solution (7 parts), the anode being mercury placed at the bottom of the

---

* English Patent, 1855, No. 2756.

† English Patent, 1858, No. 507.

solution and connected to the battery by an iron wire coated with insulating material and dipping its uncovered end into the mercury. The zinc cathode is immersed in the solution. Aluminium is deposited on the zinc, as a blackish powder or as a thin, compact sheet, and the chlorine which is liberated at the anode unites with the mercury, forming calomel.

J. B. Thompson* reports that he has for over two years been depositing aluminium on iron, steel, and other metals, and driving it into their surfaces at a heat of 500° F., and also depositing aluminium bronze of various tints, but declines to state his process.

George Gore,† the noted electrician, recommended the following procedure for depositing aluminium on copper, brass, or German silver:—

"Take equal measures of sulphuric acid and water, or one part sulphuric acid, one part hydrochloric acid and two parts of water, put into it half an ounce of pipe clay to the pint of dilute acid and boil for an hour. Take the clear, hot liquid and immerse in it an earthen porous cell containing sulphuric acid diluted with ten times its bulk of water, together with a rod or plate of amalgamated zinc. Connect the zinc with the positive wire of a Smee battery of three or four elements connected for intensity. The article to be coated, well cleaned, is connected with the negative pole and immersed in the hot clay solution. In a few minutes a fine, white deposit of aluminium will appear all over its surface. It may then be taken out, washed quickly in clean water, wiped dry, and polished. If a thicker coating is required, it must be taken out as soon as the deposit becomes dull, washed, dried, polished, and re-immersed, and this must be repeated at intervals as often as it becomes dull, until the required thickness is obtained. It is necessary to have the acid well saturated by boiling, or no deposit will be obtained."

Mierzinski asserts that Dr. Gore was mistaken when he supposed this deposit to be aluminium, and in Gore's Text Book of Electro-metallurgy no mention is made of these experiments, the

---

* Chem. News, xxiv. 194 (1871).
† Philosophical Magazine, March, 1863.

author thereby acknowledging the error. As to what the deposit could have been, we are left to conjecture, since no explanation has been advanced by Dr. Gore; it may possibly have been silicon, mercury, or zinc, as all three of these were present besides aluminium.

J. A. Jeancon* has patented a process for depositing aluminium from an aqueous solution of a double salt of aluminium and potassium of specific gravity 1.161; or from any solution of an aluminium salt, such as sulphate, nitrate, cyanide, etc., concentrated to 20° B. at 50° F. He uses a battery of four pairs of Smee's or three Bunsen's cells, with elements arranged for intensity, and electrolyzes the solutions at 140° F. The first solution will decompose without an aluminium anode, but the others require such an anode on the negative pole. The solution must be acidulated slightly with acid corresponding to the salt used, the temperature being kept at 140° F. constantly.

M. A. Bertrand† states that he deposited aluminium on a plate of copper from a solution of double chloride of aluminium and ammonia, by using a strong current, and the deposit was capable of receiving a brilliant polish.

Jas. S. Haurd,‡ of Springfield, Mass., patented the electrolysis of an aqueous solution formed by dissolving cryolite in a solution of magnesium and manganous chlorides.

John Braun§ decomposes a solution of alum, of specific gravity 1.03 to 1.07, at the usual temperature, using an insoluble anode. In the course of the operation, the sulphuric acid set free is neutralized by the continual addition of alkali; and, afterwards, to avoid the precipitation of alumina, a non-volatile organic acid, such as tartaric, is added to the solution. The intensity of the current is to be so regulated that for a bath of 10 to 20 litres two Bunsen elements (about 20 centimetres high) are used.

Dr. Fred. Fischer‖ stated that Braun's proposition was contrary to his experience. By passing a current of 8 to 9 volts and

---

* Annual Record of Science and Industry, 1875.

† Chem. News, xxxiv. 227.

‡ U. S. Patent, 228,900, June 15, 1880.

§ German Patent, No. 28,760 (1883).

‖ Zeitschrift des Vereins Deutsche Ingenieurs, 1884, p. 557.

50 amperes, using from 0.1 to 10 amperes per sq. centimetre of cathode, with various neutral and basic aluminium sulphate solutions, with and without organic acids, he obtained no aluminium. He obtained a black deposit of copper sulphide on the copper anode, which had apparently been mistaken by Braun for aluminium.

Moses G. Farmer[*] has patented an apparatus for obtaining aluminium electrically consisting of a series of conducting cells in the form of ladles, each ladle having a handle of conducting material extending upwards above the bowl of the next succeeding ladle; each ladle can be heated separately from the rest; the anodes are hung in the ladles, being suspended from the handles of the preceding ladles, the ladles themselves being the cathodes.

M. L. Senet[†] electrolyzes a saturated solution of aluminium sulphate, separated by a porous septum from a solution of sodium chloride. A current is used of 6 to 7 volts and 4 amperes. The double chloride, $Al^2Cl^6.2NaCl$, is formed, then decomposed, and the aluminium liberated deposited on the negative electrode. It has later been remarked of this process that it has not had the wished-for success on a large scale.

Col. Frismuth, Philadelphia, purports to plate an alloy of nickel and aluminium. He uses an ammoniacal solution, probably of their sulphates. The plating certainly resembles nickel, but whether it contains aluminium the author has not been able to determine.

Baron Overbeck and H. Neiwerth, of Hannover,[‡] have patented the following process: An aqueous or other solution of an organic salt of aluminium is used, or a mixture of solutions which by double decomposition will yield such salt. Or a mixture of a metallic chloride and aluminium sulphate may be used, this yielding nascent aluminium chloride, which the current splits up immediately into aluminium and chlorine.

Herman Rienbold[§] gives the following recipe, stating that it furnishes excellent results : 50 parts of potash alum are dissolved

---

[*] U. S. Patent, No. 315,266, April, 1885.
[†] Cosmos les Mondes, Aug. 10, 1885.
[‡] English Pat., Dec. 15, 1883, No. 5756.
[§] Jeweller's Journal, September, 1887.

in 300 parts of water, and to this are added 10 parts of aluminium chloride. The whole is then heated to 200° F., cooled, and then 39 parts of potassium cyanide added. A weak current should be used. It is stated that the plating, when polished, will be found equal to the best silver plating. "Iron," noticing this process, remarks, "there are a number of formulæ for electro-plating with aluminium, but few appear to have attained to practical utility in the arts, for the reason that there is no special demand for such processes. All the qualities that are possessed by an electro-deposit of aluminium are possessed to an equal or superior degree by other metals, silver, nickel, platinum, etc. Furthermore, it obstinately refuses to take and to retain a high lustre." This criticism is a little overdrawn, since the one quality in which aluminium is superior to silver—not blackening by contact with sulphurous vapor—is not mentioned.

Under the name of Count R. de Montegelas, of Philadelphia, several patents have been taken out in England for the electrolysis of aqueous solutions, which may be summarized as follows:—

*Alumina is treated with hydrochloric acid, and aluminium chloride obtained in solution. The liquid is then placed in a vessel into which dip a suitable anode and a cathode of brass or copper.. On passing an electric current through the bath the iron present in the liquid is first deposited, and as soon as this deposition ceases (as is apparent by the change of color of the deposit) the liquid is decanted into another similar bath, and to it is added about fifty per cent. by weight of the oxide of either lead, tin or zinc. On sending a current through this bath, aluminium together with the metal of the added oxide is said to be deposited on the cathode.

†A rectangular vessel is divided into two unequal compartments by a vertical porous partition, into the smaller of which is placed a saturated solution of common salt, in which is immersed a brass or copper electrode, into the larger is put a solution of aluminium chloride, immersed in which is an aluminium electrode. On passing the current the latter solution, which is normally yellow, is

---

* English Patent, Aug. 18, 1886, No. 10607.
† English Patent, Feb. 3, 1887, No. 1751.

gradually decolorized and converted into a solution of aluminium-sodium chloride.   When colorless, this solution is taken out and the aluminium deposited in a similarly arranged vessel.   The double chloride solution is placed in the larger compartment, with an electrode of brass, copper, or a thin plate of aluminium, while the smaller compartment contains a carbon electrode dipping into a solution of salt and surrounded by fragments of a mixture of salt and double chloride, fused together in equal parts.

· The author has been given several ounces of a very fine, metallic powder said to have been made by these processes, and which is certainly aluminium.   As I am not satisfied, however, that the specimen is really authentic, I feel justified in suspending a final expression of opinion on the process.

A. Walker, of Tarnowitz, has patented the following methods of procedure :—*

a. Pure commercial hydrate is dissolved in nitric acid free from chlorine, in slight excess, and tartaric acid added.   The liquid is let clear for some time, any potassium bi-tartrate, which may be formed from small quantities of potassium adhering to the hydrate, filtered out, and the clear solution electrolyzed. There is added to the solution during electrolysis organic acid— as formic, acetic, citric, oxalic—or, better, absolute alcohol.

· b. A solution of aluminium nitrate, as far as possible free from alkalies and sulphuric acid, is decomposed by a strong dynamic current in baths arranged in series, using platinized plates as anode and cathode.   With a weak current of 0.02 to 0.05 amperes to a square centimetre, the aluminium separates out on the cathode as a deep black deposit, sticking close to the copper.. The cathode is lifted from the solution, freed from small quantities of alumina coating it by gentle rinsing, and then the deposit washed off by a strong jet of water.   The powder obtained is washed further with clear, cold water particularly free from sodium chloride, and dried by gentle heating in the air.

H. C. Bull† proposes to manufacture aluminium alloys by using the metal to be alloyed with aluminium as a cathode in a

---

* German Pat. (D. R. P.) 40,626 (1887.)
† English Pat., 10199 A. (1887).

bath of aluminium sulphate, the anode being either of aluminium or of an insoluble substance. When enough aluminium is deposited, the cathode is taken out and melted down.

C. A. Burghardt and W. J. Twining, of Manchester, England, have patented the following methods :—

*To a solution of sodium or potassium aluminate containing about 7.2 oz. of aluminium per gallon are added 4 pounds of 95 per cent. potassium cyanide dissolved in a quart of water, and then gradually 2¼ pounds of potassium bi-carbonate. The whole is boiled 12 hours and made up to a gallon. The bath is used at 175° F. with aluminium or platinum anode and a carbon or copper cathode. The addition of a little free hydrocyanic acid insures a bright deposit when articles are being plated.

†Two and one-half kilos of aluminium sulphate in solution is precipitated by ammonia, and then re-dissolved by adding 1⅛ kilos of caustic soda dissolved in a litre of water ; the alumina is thus slightly in excess. Then hydrocyanic acid is added until a slight precipitate appears. This solution, warmed to 80°, is used as a bath from which aluminium is to be deposited.

‡The bath is prepared by dissolving alumina in a solution of chloride of copper, and treating further with caustic soda or potash for the purpose of causing the aluminium and copper to combine together. The precipitate, dissolved in hydrocyanic acid and diluted, forms a bath of double cyanide, which when electrolyzed deposits an alloy of aluminium and copper.

Besides the processes so far described, patents have been taken out in England by Gerhard and Smith,§ Taylor,‖ and Coulson,¶ the details of which have not been accessible to the author.

Over against all these statements and claims of enthusiastic inventors, let me place a few cool statements from authorities who have given much time and attention to elucidating the subject.

Sprague** states his inability to deposit aluminium electrically from solution.

---

* English Pat., July 2, 1887, No. 9389.
† German Pat. (D. R. P.), 45,020 (1887).
‡ English Pat., Oct. 28, 1887, No. 2602.
§ No. 16,653 (1884).   ‖ No. 1991 (1855).
¶ No. 2075 (1857).   ** Sprague's Electricity, p. 309.

Dr. Clemens Winckler* states that he has spent much time in trying all methods so far proposed, and comes to the conclusion that aluminium cannot be deposited by electricity in the wet way.

Dr. Geo. Gore† although having once proposed a method which he said attained this end, yet in his later work on Electro-metallurgy does not mention his former proposition, and quotes apparently as coinciding with his own opinion, the words of Sprague and Winckler given above.

Dr. S. Mierzinski‡ states, in 1883, that "the deposition of aluminium from an aqueous solution of its salt has not yet been accomplished."

Dr. W. Hampe§ claims to have shown that the electrolysis of aqueous aluminous solutions, although frequently patented, is not to be expected. From which we would infer that he could not testify to it ever having been done.

Alexander Watt‖ holds that the electrolytic production of aluminium from solution is very improbable. He tried acid solutions, alkaline solutions, cyanide combinations, etc., under most varied conditions, without any result.

Finally, I will quote from a letter of my good friend Dr. Justin D. Lisle, of Springfield, O., who with ample means at his disposal, an enthusiasm bred of love for scientific truth and talent to guide him in his work, has reached the following results : " I have tried in almost every conceivable way to deposit it (aluminium) from aqueous solution by electricity, using from 1 pint cells to 60 gallon cells successively ; the cells were connected for quantity and for intensity ; acid and neutral solutions were used ; carbon, platinum, and copper electrodes ; porous cups and diaphragms, were all thoroughly tried without the slightest deposit of metal. In some cases alumina was deposited, which has led me to think that aluminium was primarily deposited, and owing to the fine state in which it existed was promptly oxidized."

* Journal of the Chem. Soc., X. 1134.
† Text book of Electro-metallurgy.
‡ Die Fabrikation des Aluminiums.
§ Chem. Zeit. (Cöthen), XI. 935.
‖ London Electrical Review, July, 1887.

II.

The Electric Decomposition of Fused Aluminium Compounds.

This subdivision of the electrolytic methods includes all the electric processes which have given practical results. Under this head come Davy's first attempts to decompose alumina, in 1807, Deville's first success in producing pure aluminium, in 1854, and Grätzel's application of the dynamo-electric machine, in 1883, which introduced the first radical improvement the aluminium industry had known for twenty-five years. It is hardly too much to say that if the long sought for method of turning the potential energy of coal directly into electric energy ever be accomplished, these electrolytic methods will be beyond doubt the future means of bringing aluminium in price among the common metals.

There seem to be two ways of operating, as mentioned on p. 34, in the first of which the liquid compound is decomposed at moderate temperatures, such that the containing vessel can be heated to, in an ordinary fire, and in which almost all the current is utilized in decomposing the electrolyte; in the other enormous temperatures are reached by means of interrupting a powerful current, and a large part of the electric energy is converted into heat, while the decomposition may be partly electrolytic and partly a chemical reaction made possible by this extreme temperature. As it is impossible in one or two cases to draw this line, and since the practical requirements of the two methods of procedure are in most respects identical, we will consider the following processes in their chronological order, except in one or two cases where very similar ones are placed together.

Davy's Experiment (1810).

Sir Humphry Davy, in his Brompton Lecture before the Royal Philosophical Society,* described the following attempt to decompose alumina and obtain the metal of this earth. He

---

* Philosophical Transactions, 1810.

connected an iron wire with the negative pole of a battery consisting of 1000 double plates. The wire was heated to whiteness and then fused in contact with some moistened alumina, the operation being performed in an atmosphere of hydrogen. The iron became brittle, whiter, and on being dissolved in acid gave a solution from which was precipitated alumina, identical with that used.

Duvivier's Experiment (1854).

M. Duvivier* states that by passing an electric current from eighty Bunsen cells through a small piece of laminated disthene between two carbon points, the disthene melted entirely in two or three minutes, the elements which composed it were partly disunited by the power of the electric current, and some aluminium freed from its oxygen. Several globules of the metal separated, one of which was as white and as hard as silver.

Bunsen's and Deville's Methods (1854).

A method of decomposing aluminium-sodium chloride by the battery was discovered simultaneously by Deville in France and Bunsen in Germany, in 1854, and is nothing else but an application of the process already announced by Bunsen of decomposing magnesium chloride by the battery. Deville gives the more minute account, and we therefore quote his description of the process.

†"It appears to me impossible to obtain aluminium by the battery in aqueous solutions. I should believe this to be an absolute impossibility if the brilliant experiments of M. Bunsen in the preparation of barium, chromium and manganese did not shake my convictions. Still I must say that all the processes of this description which have recently been published for the preparation of aluminium have failed to give me any results. Every one knows the elegant process by means of which M. Bunsen has lately produced magnesium, decomposing fused mag-

---

* The Chemist, Aug. 1854.
† Ann. de Chem. et de Phys. [3], 46, 452 ; Deville's de l'Aluminium.

nesium chloride by an electric current. The illustrious professor
at Heidelberg has opened up a method which may lead to very
interesting results. However, the battery cannot be used for decomposing aluminium chloride directly, which does not melt, but
volatilizes at a low temperature; it is, therefore, necessary to use
some other material which is fusible and in which aluminium
alone will be displaced by the current. I have found this salt
in the double chloride of aluminium and sodium, which melts
towards 185°, is fixed at a somewhat high temperature, although
volatile below the fusing point of aluminium, and thus unites all
the desirable conditions.

"I put some of this double chloride into a porcelain crucible
separated imperfectly into two compartments by a thin leaf of
porcelain, and decomposed it by means of a battery of five elements and carbon electrodes. The crucible was heated more and
more as the operation progressed, for the contents became less and
less fusible, but the heat was not carried past the melting point of
aluminium. Arrived at this point, after having lifted out the
diaphragm and electrodes, I heated the crucible to bright redness
and found at the bottom a button of aluminium, which was
flattened out and shown to the Academy in the Séance of March
20, 1854. The button was accompanied by a considerable quantity of carbon, which prevented the union of a considerable mass
of shot-metal. This carbon came from the disintegration of the
very dense gas-retort carbon electrodes; in fact, the positive electrode was entirely eaten away in spite of its considerable thickness. It was evident, then, that this apparatus, although similar
to that adopted by Bunsen for manufacturing magnesium, would
not suit here, and the following is the process which after many
experiments I hold as best.

"To prepare the bath for decomposition, I heat a mixture of 2
parts aluminium chloride and 1 part sodium chloride, dry and
pulverized, to about 200° in a porcelain capsule. They combine
with disengagement of heat, and the resulting bath is very fluid.
The apparatus which I use for the decomposition comprises a
glazed porcelain crucible, which as a precaution is placed inside
a larger one of clay. The whole is covered by a porcelain
cover pierced by a slit to give passage to a large thick leaf of

platinum, which serves as the negative electrode; the lid has also a hole through which is introduced, fitting closely, a well-dried porous cylinder, the bottom of which is kept at some distance from the inside of the porcelain crucible. This porous vessel encloses a pencil of retort carbon, which serves as the positive electrode. Melted double chloride is poured into the porous jar and into the crucible so as to stand at the same height in both vessels; the whole is heated just enough to keep the bath in fusion, and there is passed through it the current from several Bunsen cells, two cells being strictly sufficient. The annexed diagram shows the crucibles in section.

Fig. 23.

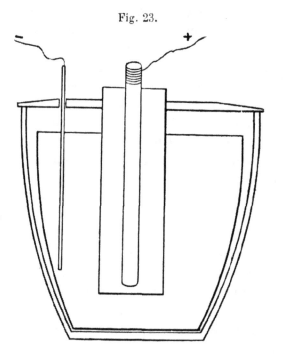

"The aluminium deposits with some sodium chloride on the platinum leaf; the chlorine, with a little aluminium chloride, is disengaged in the porous jar and forms white fumes, which are prevented from rising by throwing into the jar from time to time some dry, pulverized sodium chloride. To collect the aluminium, the platinum leaf is removed when sufficiently charged with the saline and metallic deposit; after letting it cool, the deposit is rubbed off and the leaf placed in its former position. The material thus detached, melted in a porcelain crucible, and after cooling washed with water, yields a gray, metallic powder, which by melting several times under a layer of the double chloride is reunited into a button."

Bunsen* adopted a similar arrangement. The porcelain crucible containing the bath of aluminium-sodium chloride kept in fusion was divided into two compartments in its upper part by a partition, in order to separate the chlorine liberated from the aluminium reduced. He made the two electrodes of retort carbon. To reunite the pulverulent aluminium, Bunsen melted it in a bath of the double chloride, continually throwing in enough sodium chloride to keep the temperature of the bath about the fusing point of silver.

As we have seen, Deville, without being acquainted with Bunsen's investigations, employed the same arrangement, but he abandoned it because the retort carbon slowly disintegrated in the bath, and a considerable quantity of double chloride was lost by the higher heat necessary to reunite the globules of aluminium after the electrolysis. Deville also observed that by working at a higher temperature, as Bunsen has done, he obtained purer metal, but in less quantity. The effect of the high heat is that silicon chloride is formed and volatilizes, and the iron which would have been reduced with the aluminium is transformed into ferrous chloride by the aluminium chloride, and thus the aluminium is purified of silicon and iron.

Plating aluminium on copper.—The same bath of double chloride of aluminium and sodium may be used for plating aluminium in particular on copper, on which Capt. Caron experimented with Deville. Deville says : "To succeed well, it is necessary to use a bath of double chloride which has been entirely purified from foreign metallic matter by the action of the battery itself. When aluminium is being deposited at the negative pole, the first portions of metal obtained are always brittle, the impurities in the bath being removed in the first metal thrown down ; so, when the metal deposited appears pure, the piece of copper to be plated is attached to this pole and a bar of pure aluminium to the posi-

---

* Pogg. Annalen, **97**, 648.

tive pole. However, a compact mixture of carbon and alumina can be used instead of the aluminium anode, which acts similarly to it and keeps the composition of the bath constant. The temperature ought to be kept a little lower than the fusing point of aluminium. The deposit takes place readily and is very adherent, but it is difficult to prevent it being impregnated with double chloride, which attacks it the moment the piece is washed. The washing ought to be done in a large quantity of water. Cryolite might equally as well be used for this operation, but its fusibility should be increased by mixing with it a little double chloride of aluminium and sodium and some potassium chloride."

Le Chatellier's Method (1861).

The subject of this patent* was the decomposition of the fused double chloride of aluminium and sodium, with the particular object of coating or plating other metals, the articles being attached to the negative pole. About the only novelty claimed in this patent was the use of a mixture of alumina and carbon for the anode, but we see from the previous paragraph that this was suggested by Deville several years before; the only real improvement was the placing of this anode inside a porous cup, in order to prevent the disintegrated carbon from falling into the bath.

Monckton's Patent (1862).

Monckton† proposes to pass an electric current through a reduction chamber, and in this way to raise the temperature to such a point that alumina will be reduced by the carbon present. We clearly see in this the germ of several more-recently patented processes.

Gaudin's Process (1869).

Gaudin‡ reduces aluminium by a process to which he applies the somewhat doubtful title of economic. He melts together

---

* English Patent, 1861, No. 1214.   † English Patent, 1862, No. 264.
‡ Moniteur Scientifique, xi. 62.

equal parts of cryolite and sodium chloride, and traverses the fused mass by a galvanic current. Fluorine is evolved at the positive pole, while aluminium accumulates at the negative.

Kagensbusch's Process (1872).

Kagensbusch,* of Leeds, proposes to melt clay with fluxes, then adding zinc or a like metal to pass an electric current through the fused mass, isolating an alloy of aluminium and the metal, from which the foreign metal may be removed by distillation, sublimation, or cupellation.

Berthaut's Proposition (1879).

Up to this time, all the proposed electric processes were confined to the use of a galvanic current, the cost of obtaining which was a summary bar to all ideas of economical production. About this period dynamo-electric machines were being introduced into metallurgical practice, and Berthaut is the first we can find who proposes their use in producing aluminium. The process which he patented† is otherwise almost identical with Le Chatellier's.

Grätzel's Process (1883).

This process‡ has little claim to originality, except in the details of the apparatus. A dynamo-electric current is used, the electrolyte is fused cryolite or double chloride of aluminium and sodium, and the anodes are of pressed carbon and alumina—none of which points are new. However, the use of melting pots of porcelain, alumina, or aluminium, and making them the negative electrode are points in which innovations are made.

In a furnace are put two to five pots, according to the power of the dynamo used, each pot having a separate grate. The pots are preferably of metal, cast-steel is used, and form the negative electrodes. The positive electrode, K (Fig. 24), can be made of

---

* English Patent, 1872, No. 4811.
† English Patent, 1879, No. 4087.
‡ German Patent (D. R. P.), No. 26962 (1883).

a mixture of anhydrous alumina and carbon pressed into shape and ignited. A mixture of alumina and gas-tar answers very well; or it can even be made of gas-tar and gas-retort carbon.

Fig. 24.

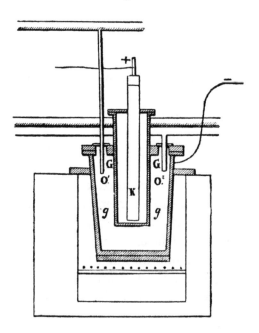

During the operation little pieces of carbon fall from it and would contaminate the bath, but are kept from doing so by the mantle, G. This isolating vessel, G, is perforated around the lower part at g, so that the molten electrolyte may circulate through. The tube O^1 conducts reducing gas into the crucible, which leaves by the tube O^2. This reducing atmosphere is important, in order to protect from burning any metal rising to the surface of the bath. The chlorine set free at the electrode, K, partly combines with the alumina in it, regenerating the bath, but some escapes, and, collecting in the upper part of the surrounding mantel, G, is led away by a tube connecting with it. Instead of making the electrode, K, of carbon and alumina, it may simply be of carbon, and then plates of pressed alumina and carbon are placed in the bath close to the electrode, K, but not connected with it. Also, in place of making the crucible of metal

and connecting it with the negative pole, it may be made of a
non-conducting material, clay or the like, and a metallic electrode
—as, for instance, of aluminium—plunged into the bath.

In a later patent,* Grätzel states that the bath is decomposed
by a current of comparatively low tension if magnesium chloride
be present; the chlorides of barium, strontium, or calcium act
similarly.

Prof. F. Fischer† maintains as impracticable the use of plates
of pressed alumina and carbon, which can, further, only be opera-
tive when they are made the positive electrode, and then their
electric resistance is too great. The incorporation into them of
copper filings, saturation with mercury, etc., gives no more practical results. There are also volatilized at the anodes considerable
quantities of aluminium chloride, varying in amount with the
strength of the current.

Large works were erected near Bremen by the Aluminium
und Magnesiumfabrik Pt. Grätzel, zu Hemelingen, in which
this process was installed. License was also granted to the large
chemical works of Schering, at Berlin, to operate it. R. Biedermann, in commenting on the process in 1886,‡ stated that the
results obtained so far were not fully satisfactory, but the difficulties which had been met were of a kind which would certainly
be overcome. They were principally in the polarization of the
cathode, by which a large part of the current was neutralized.
By using proper depolarizing substances this difficulty would be
removed. The utilization of the chlorine evolved would also
very much decrease the expenses. A more suitable slag, which
collected the aluminium together better, was also desirable.
Finally, the metal produced was somewhat impure, taking up
iron from the iron pots and silicon from the clay ones, to obviate
which Biedermann recommended the use of lime or magnesia
vessels.

Prof. Fischer, as we have seen, maintained the uselessness of
Grätzel's patent claims, and his later expression of this opinion

* English Patent, 14325, Nov. 23, 1885. U. S. Patent, 362441, May 3,
1887.
† Wagner's Jahresbericht, 1884, p. 1319; 1887, p. 376.
‡ Kerl und Stohman, 4th ed., p. 725.

in 1887 drew a reply from A. Saarburger,* director of the works at Hemelingen, to the effect that since October, 1887, they had abandoned the Grätzel process and were making aluminium at present by methods devised by Herr Saarburger; in consequence of which fact the directors of the company decided in January, 1888, to drop the addition Pt. Grätzel from the firm name. The methods now in use at Hemelingen are kept secret, but the author is informed by a friend in Hamburg that they are using a modified Deville sodium process. Herr Saarburger informed me in October, 1888, that they were producing pure aluminium at the rate of 12 tons a year, besides a large quantity sold in alloys. An attractive pamphlet issued by this firm sets forth precautions to be used in making aluminium alloys, together with a digest of their most important properties, which we shall have occasion to quote from later in considering those alloys.

Kleiner's Process (1886).

This was devised by Dr. Ed. Kleiner of Zurich, Switzerland, and has presumably been patented in most of the European States. The English patent is dated 1886.† The first attempts to operate it were at the Rhine Falls, Schaffhausen, and were promising enough to induce Messrs. J. G. Nethers, Sons & Co., proprietors of an iron works there, to try to obtain water rights for 1500 horse power, announcing that a company (the Kleiner Gesellschaft) with a capital of 12,000,000 francs was prepared to undertake the enterprise and build large works. The proposition is said to have met with strong opposition from the hotel-keepers and those interested in the Falls as an attraction for tourists, and the government declined the grant, considering that the picturesqueness of the falls would be seriously affected. This is the reason given by those interested in the process for it not being carried out in Switzerland, it being then determined to start a works in some part of England where cheap coal could be obtained, and test the process on a large scale. A small experi-

---

* Verein der Deutsche Ingenieure, Jan. 26, 1889.

† English Patents, 8531, June 29, 1886, and 15322, Nov. 24, 1886.

mental plant was then set up in the early part of 1887 on Far-rington Road, London, where it was inspected by many scientific men, among them Dr. John Hopkinson, F.R.S., who reported on the quantitative results obtained; a description of the process as here operated was also written up for " Engineering." With the co-operation of Major Ricarde-Seaver a larger plant was put up at Hope Mills, Tydesley, in Lancashire, where the process was inspected and reported on by Dr. George Gore, the electrician. After his report we learn that the patents have been acquired by the Aluminium Syndicate, Limited, of London, a combination of capitalists among whom are said to be the Rothschilds. The latest reports state that the process is still in the experimental stage, although Dr. Kleiner considers that the present results will justify working on a commercial scale in the near future.

The aluminium compound used is commercial cryolite. It is stated that the native mineral from Greenland contains on an average, according to Dr. Kleiner's analysis, 96 per cent. of pure cryolite, the remainder being moisture, silica, oxides of iron and manganese. As pure cryolite contains 13 per cent. of aluminium, the native mineral will contain 12½ per cent., all of which Dr. Kleiner claims to be able to extract. It is further remarked that as soon as sufficient demand arises, an artificial cryolite can be made at much less cost than that of the native mineral, which now sells at £18 to £20 a ton. The rationale of the process consists in applying the electric current in such a way that a small quantity of it generates heat and keeps the electrolyte in fusion, while the larger quantity acts electrolytically. Dry, powdered cryolite is packed around and between carbon electrodes in a bauxite-lined cavity; on passing a current of high tension (80 to 100 volts) through the electrodes, the cryolite is quickly fused by the heat of the arc and becomes a conductor. As soon as the electrolyte is in good fusion the tension is lowered to 50 volts, the quantity being about 150 amperes, the arc ceases and the decomposition proceeds regularly for two or three hours until the bath is nearly exhausted. The evolved fluorine is said to attack the bauxite and by thus supplying aluminium to the bath extends the time of an operation. In the first patent the negative carbon was inserted through the bottom of the melting cavity, the positive dipping into the bath from above, but it was found that while the ends of the positive carbon immersed in the cryolite were unattacked, the part immediately over the bath was rapidly corroded. In the second patent, therefore, the positive electrode was circular and entirely immersed in the cryolite, connection being made by ears which projected through the side of the vessel. As the carbons are thus fixed, the preliminary fusion is accomplished by a movable carbon rod suspended from above, passing through the circular anode and used only for this purpose. The bath being well fused and the current flowing freely between the fixed carbons, the rod is withdrawn. The carbons are said to be thus perfectly protected from corrosion, and able to serve almost indefinitely. The melting pots finally used were ordinary black-lead crucibles, which are not usually injured at all, since the fused part of the cryolite does not touch them, and they last as many as 300 fusions. After the operation, the carbons are lifted out of the bath and the contents cooled. When solid, the crucibles are inverted and the contents fall out. This residue is broken to coarse powder, the nodules of aluminium picked out, melted in a crucible and cast into bars. The coarse powder is then ground to fine dust. This powder is more or less alkaline and contains a greater or less excess of fluoride of sodium in proportion to the amount of aluminium which has been taken out. If only a small proportion of the metal has been extracted and the powder contains only a small excess of sodium fluoride, it is used again without any preparation in charging the crucibles; but if as much as 5 or 6 per cent. of aluminium has been removed and the powder, therefore, contains a large excess of sodium fluoride, it is washed with water for a long time to remove that salt, which slowly dissolves. The solution is reserved, while the powder remaining is unchanged cryolite, and is used over. Dr. Gore states that if the powder, electrodes and crucible are perfectly dry, there is no escape of gas or vapor during the process; but if moisture is present, a small amount only of fumes of hydrofluoric acid appear, and that there is no escape of fluorine gas at any time. If this is so, it is rather difficult to see where the fluorine with which the aluminium is combined goes to. If the vessel were lined with beauxite, it might be retained by this lining, but in the experiments seen by Dr. Gore, a plumbago crucible was used (which remained unattacked) and cryolite only.   It is certainly a mystery how any aluminium could be produced without fluorine vapors being liberated.   Dr. Kleiner hopes to soon dispense with the interruption of the process, washing, etc., by regenerating cryolite in the crucible itself and so making the process continuous.   One of the great advantages claimed is that the aluminium is obtained in nodules, and not in fine powder ; if it was, it could not all be collected because it is so light, some of it would float upon the water during the washing process and be lost, and even when collected it could not be dried and melted without considerable loss.

It has been found impossible in practice to obtain all the aluminium from a given quantity of cryolite in less than two fusions, for the sodium fluoride collecting in the bath hinders the production of the metal.   The proportion extracted by a single fusion depends upon its duration.   In the operations at Tydesley, a fusion lasting 24 hours separated only $2\frac{1}{2}$ per cent. of aluminium, whereas the cryolite contained $12\frac{1}{2}$ per cent.   At this rate, to extract the whole in two operations would require two fusions of 60 hours each.   As to the output, on an average a current of 38 electric horse power deposited 150 grammes of aluminium per hour, being a little over 3 grammes per horse-power.   Since a current of 50 volts and 150 amperes, such as was stated above as the current in each pot, is equal to $\dfrac{50 \times 150}{750}$ or 10 electric H. P., it is probable that the 38 H. P. current mentioned must have been used for four crucibles.   Now, the output of four crucibles, each with a current of 150 amperes, should have been 0.00009135 × 150 × 4 = 0.0548 grammes per second or 197.3 grammes per hour; the difference between this and the amount actually obtained, or 47.3 grammes, is the amount of aluminium which was produced and then afterwards lost either as fine shot-metal or powder or dissolved again by corroding elements in the bath. To calculate how the output of 3 grammes per electric H. P. per hour compares with the quantity of metal which this amount of energy should be able to produce, we need to know the heat of formation of cryolite, or we could form some idea if we knew even that of aluminium fluoride, but thermal data with regard to fluorides are entirely lacking, since free fluorine is needed as the basis of their experimental determination. As fluorine has lately been isolated, it may not be long before some of these figures are determined, and then the calculations referred to can be made. In the mean time we can observe no further than that about 75 per cent. of the metal produced in the bath is obtained and weighed, and that the high potential of the current seems to indicate a considerable loss of power in overcoming resistances other than that of decomposition, a loss greater than is met with in other somewhat similar operations.

As to the purity of the metal obtained, the process is met at the outset by the silica and iron oxide in the cryolite, which are probably all reduced with the first few grammes of aluminium thrown down. This can possibly be remedied by using a purer artificial cryolite; the impurities cannot generally be separated from the natural mineral. Then there are impurities of a similar nature coming from the carbons used, and which are generally present if especial pains are not taken to get very pure materials for making them. Dr. Kleiner's early attempts produced metal of 85 to 95 per cent. purity, but he now states that it is uniformly 95 to 98 per cent., and being put on the market in competition with other commercial brands. It appears, however, from a consideration of the preceding data, that unless great improvements have since been made in several details, the process will not enable the metal produced to be sold at 16 shillings a pound (the present selling price), and pay expenses.

Lossier's Method.

*This is a device for decomposing the natural silicates by electricity and obtaining their aluminium. The bath is composed of pure aluminium fluoride or of a mixture of this salt and an alkaline chloride, and is kept molten in a round bottomed crucible placed in a furnace. The electrodes are of dense carbon and are separated in the crucible by a partition reaching beneath the surface of the bath. The positive electrode is furnished with

* German Patent (D. R. P.), No. 31089.

a jacket or thick coating of some aluminium silicate, plastered on moist and well dried before use. When the current is passed, the aluminium fluoride yields up its fluorine at this pole and its aluminium at the other. The fluorine combines with the aluminium silicate, forming on the one hand aluminium fluoride, which regenerates the bath, on the other silicon fluoride and carbonic oxide, which escape as gases. The metal liberated at the negative pole is lighter than the fused bath, and therefore rises to the surface.

M. Grabau cites as one of the recommendations of aluminium fluoride for use in his process (p. 242), that it is quite infusible, so it would appear that Lossier has made a mistake in supposing that it could be melted alone in a crucible. It would, however, make a very fusible bath when the alkali chloride was added. It is probable that the carrying out of this method would develop great trouble from the attacking of the crucible by the very corrosive bath, the disintegration of the carbons, which would cause much trouble at the negative pole especially, and the oxidation of the fluid aluminium on the surface of the bath. I cannot learn that the process has ever been attempted on a large scale.

Omholt's Furnace.

I. Omholt and the firm Böttiger and Seidler, of Gössnitz, have patented the following apparatus for the continuous electrolysis of aluminium chloride :—*

The bed of a reverberatory furnace is divided by transverse partitions into two compartments, in each of which are two retorts semi-circular in section, lying side by side horizontally across the furnace, with the circular part up. They are supported on refractory pillars so that their open side is a small distance above the floor of the furnace. The aluminium compound being melted on the hearth, it stands to the same depth in both retorts, and if the electrodes are passed through the bottom of the hearth they may remain entirely submerged in molten salt and each under its own retort cover. The metal therefore collects in a liquid state under

* German Patent (D. R. P.), No. 34728.

one retort and the chlorine under the other, both being preserved from contact or mixture with the furnace gases by the lock of molten salt. The chlorine can thus be led away by a pipe, and utilized, while the aluminium collects without loss, and is removed at convenient intervals.

Henderson's Process (1887).

A. C. Henderson, of Dublin,* patents the process of fluxing alumina with cryolite, the bath being put into a graphite crucible, which serves as the negative electrode, and which is put inside a larger crucible and the space between filled with graphite. The positive electrode is of carbon and dips into the fused material. A current of only 3 volts is used, and the dissolved alumina only is decomposed, the cryolite remaining unaltered. The aluminium collects in the bottom of the crucible, and as the operation proceeds alumina is added to renew the bath. To prepare alloys, a negative electrode is made of the metal and used in a similar position to the positive electrode, and as the current passes the alloy is formed and falls melted to the bottom of the crucible.

We must give Mr. Henderson credit for having introduced the idea of decomposing alumina held in solution in a fused bath, an idea which is, however, more fully developed by Hall (see p. 288); and also for hitting what Hampe designates as the best mode of procedure for obtaining alloys in such processes (see p. 286). It is to be regretted that, having such a good beginning, we have not heard more of Mr. Henderson's process in the three years since it was patented.

Bernard Bros.' Process (1887).

Messrs. M. and E. Bernard, of Paris, have patented a process† which consists in electrolyzing a mixture of sodium chloride with aluminium fluoride or with the separate or double fluorides of aluminium and sodium, melted in a non-metallic crucible or in a

---

* English Patent, No. 7426 (1887).
† English Patent, No. 10057, July 18, 1887.

metallic one inclosed in a thin refractory jacket to avoid filtration. The details of the apparatus and bath are as follows :—

Disposition of the Apparatus.—The pots or crucibles used may be of refractory earth, plumbago or of metal, and in cases where an alloy is required the crucible itself serves as an electrode. None of these, however, resist the corrosive power of the electrolyte and would under ordinary conditions be quickly destroyed. To overcome this difficulty two special devices are employed. When alloys are to be made directly, the pot is cast of the metal with which the aluminium is to be combined. It is shaped with a sloping bottom and provided with a tap hole. The pot is encased in thin brickwork and is then made the negative electrode, the positive being two carbon rods dipping into the bath. As soon as the current is passed aluminium is deposited on the walls of the pot, forming a rich alloy with the metal of which the pot is made (iron or copper). When this coating becomes sufficiently rich in aluminium, the heat of the bath melts it and it trickles down and collects at the bottom. After a certain time, the alloy can be tapped out regularly at intervals without interrupting the electrolysis. The metal thus obtained is principally aluminium containing a few per cent. of the metal of the pot, which is of no consequence since the end to be finally attained is the production of an alloy with a smaller quantity of aluminium. When pure aluminium is to be obtained, an ingenious device is used to protect the metal from contamination by the metal of the pot. Two carbon rods serve as anode and cathode, the cathode standing upright in a small crucible placed upon a plate resting on the bottom of the pot. This crucible and plate are made from carbon blocks or from fused alumina or fluorspar moulded into the shape desired. As the metal is set free it trickles down the cathode and is caught in the crucible or cup, thus being prevented from spreading out over the bottom of the pot. To prevent the bath from corroding the pot, a wire is passed from the latter to the negative pole of the battery. The pot is thus made part of the negative electrode, but it is not intended that much of the current should pass through it, so a resistance coil is interposed between it and the battery or dynamo, so that the derived current passing through the sides of the pot is only 5 to 10 per cent. of the whole current. The effect of this is that a small amount of aluminium is deposited on and alloys with the sides of the vessel, which protects the latter from corrosion and is only feebly acted upon by the bath. The metal deposited in the crucible is thus kept nearly pure, while a small amount of alloy falls to the bottom of the pot and is poured out after the crucible has been removed. When it is wished to obtain the purest aluminium, the intensity of the derived current passing through the pot is increased by removing part of the resistance interposed between it and the negative wire, thus also decreasing the intensity of the principal current. The nature of the electrodes proper may be varied. For producing pure aluminium the anode is carbon, the cathode carbon and the pot either of copper or iron; for producing copper alloys the anode may be either carbon or bright copper, and the cathode (pot) of carbon or, copper; for producing iron alloys the anode may be either carbon or iron, while the vessel used as cathode is either of cast-iron or plumbago.

Composition of the bath.—The proportions of the different salts used for the bath vary between 30 to 40 per cent. of fluorides of aluminium and of sodium and 60 to 70 per cent. of sodium chloride. Very good results are reported with—

Aluminium fluoride	40
Sodium chloride	60
	100

Pure cryolite may be used, mixed with varying quantities of sodium chloride. Moreover, the separate fluorides of aluminium and sodium can be used in different proportions to those in which they are found in cryolite; for instance—

Aluminium fluoride	35
Sodium fluoride	10
" chloride	55
	100

As aluminium is removed, the bath becomes poor in aluminium fluoride, and this salt must be added to keep up its strength. For each kilo of aluminium produced about 3 kilos of aluminium fluoride would need to be added, but only $1\frac{1}{2}$ kilos is added as such, the other $1\frac{1}{2}$ kilos being regenerated by causing the fluorine vapors evolved to act on alumina or beauxite placed somewhere about the anode. The materials used, then, for producing 100 kilos of aluminium are estimated as—

Aluminium fluoride	150 kilos.
Commercial alumina	200 "
Sodium chloride	100 "

Power required.—M. Ad. Minet, who has written a sketch of Bernard's process as carried out at their works at Creil (Oise), maintains that aluminium fluoride is the principal electrolyte. The bath is very fluid and the temperature and composition kept constant during the operation, the laws of electrolysis can therefore be applied easily to the discussion of the process. M. Minet states that the electro-motive force absorbed by the bath is from 4 to 5 volts, and that this is not much above the minimum potential necessary to decompose aluminium fluoride, deduced from its heat of formation, which is $3\frac{1}{2}$ volts. I confess that I do not know of any determination of this heat of formation referred to; we can probably draw the inference that it is greater than that of aluminium chloride, but I do not know that its exact value has been determined. Taking, however, Minet's figure of $3\frac{1}{2}$ volts, the current is certainly very economically applied if decomposition is produced with 4 volts. With $3\frac{1}{2}$ volts tension, a current of 1 horse-power should produce 72 grammes of aluminium per hour. It is stated that 25 grammes are produced per hour per indicated mechanical horse-power, which shows that the aluminium produced represents a quantity of energy equal to 35 per cent. of the power of the engine. Assuming that 20 per cent. of the engine power is lost in being converted into electric energy, we have only 56 per cent. of the electric current not productive, including the loss by transfer resistance of the bath. Since 5 volts are absorbed by the bath altogether, the amount lost by resistance is to the amount utilized in decomposition as $1\frac{1}{2}$ to $3\frac{1}{2}$, which would show the former item to be $(100-56) \times \frac{3}{7}$ or 19 per cent. of the energy of the current. This loss is unavoidable, and as small as can be well expected, so that the real loss in working is 56—19 or 37 per cent. This is caused principally by re-solution of aluminium in the bath. We might reach this conclusion by another way. One horse-power furnished by the engine would produce 0.8 electric horse-power, which at a tension of 5 volts would furnish 120 amperes. According to Faraday's law, 120 amperes would set free 40 grammes of aluminium per hour. As only 25 were obtained practically, it shows that 15 grammes have been produced and re-dissolved by the bath, making this loss 37.5 per cent. From the figures furnished, we see that to produce 100 kilos of aluminium in 20 hours would require an engine of 200 horse-power, and since each pot produces 4 kilos of pure metal or 6 kilos of aluminium in alloys, per hour, a plant of this output would require 25 pots for making pure aluminium or 18 for working on alloys.

Quality of metal.—When working for pure aluminium, about three-fourths of the metal produced is taken from the crucible in which the cathode stands, and is 98 to 99 per cent. pure; the other one-fourth has been deposited on the sides of the cast-iron pot, and contains 10 to 20 per cent. of iron. It is poured out and used for making ferro-aluminium.

Reactions in the process.—M. Minet claims that aluminium fluoride is the chief electrolyte, since the yield of aluminium increased with the proportion of this salt in the bath. However, on reviewing this gentleman's statements, we find that when the bath contains—

Aluminium fluoride	40
Sodium chloride	60
	100

the best results are obtained. It is conceivable that the yield of aluminium increases with the proportion of aluminium fluoride up to this point, but there is no reason for saying that a further increase would give a better yield if this has been found the best mixture. Mr. Rogers found that when the proportion of aluminium fluoride to sodium fluoride in an electrolytic bath was greater than 40 to 60 (the proportions in which they exist in cryolite) the resistance increased very materially, from which he concluded that pure aluminium fluoride is not an electrolyte (p. 283). Further, Mr. Hall has found that when making a bath of

Aluminium fluoride 67
Sodium fluoride 33
_____
100

it was hardly possible to pass a current through it, but on adding alumina the latter dissolved in the bath and was easily decomposed by a current of low tension; however, as soon as the alumina was exhausted, the resistance rose quickly. It seems probable that sodium chloride, in Bernard's process, is the chief electrolyte, or else its combination with aluminium fluoride in certain proportions, but that the aluminium fluoride is the electrolyte is hardly probable.

Messrs. Bernard exhibited at the recent Paris Exposition a collection of articles made of their metal, such as round tubes, medals, keys, opera-glasses, ingots, etc., for which they received the same reward as the other exhibitors of aluminium—a gold medal.

Feldman's Method (1887).

A. Feldman, of Linden, Hannover, patented the following electrolytic process :—*

A double fluoride of aluminium and an alkaline earth metal, mixed with an excess of a chloride of the latter group, is either electrolyzed or reduced by sodium. The proportions of these substances to be used are such as take place in the following reactions :—

1. $(Al^2F^6+2SrF^2)+6SrCl^2=2Al+5SrF^2+3SrCl^2+6Cl.$
2. $(Al^2F^6+2SrF^2)+6SrCl^2+6Na=2Al+5SrF^2+3SrCl^2+6NaCl.$

The three equivalents of strontium chloride are found in practice to be most suitable. Potassium chloride may also be added to increase the fluidity, but in this case the strontium chloride must be in still greater excess.

Even if the above reactions and transpositions do take place, the use of so much costly strontium salts would appear to render the process uneconomical.

* English Patent, No. 12575, Sept. 16, 1887.

Warren's Experiments (1887).

Mr. H. Warren, of the Everton Research Laboratory, has outlined the following methods or suggestions, some of which had already been carried out, and probably others have since given useful ideas to workers in this line. The principle can hardly be called new, since suggestions almost identical with Mr. Warren's were made previously to his, but the latter's results are the first recorded in this particular direction :* "This method of preparing alloys differs only slightly from the manner in which amalgams of different metals are prepared, substituting for mercury the metals iron, copper, or zinc made liquid by heat. These metals are melted, connected with the negative pole of a battery, and the positive pole immersed in a bath of molten salt floating on top of the melted metal. The apparatus used is a deep, conical crucible, through the bottom of which is inserted a graphite rod, projecting about one inch within, the part outside being protected by an iron tube coated with borax. As an example of the method, to prepare silicon bronze-copper is melted in the crucible, a bath of potassium silico-fluoride is fused on top to a depth of about two inches. A thick platinum wire dips into this salt, and on passing the electric current an instantaneous action is seen, dense white vapors are evolved and all the silicon, as it is produced, unites with the copper, forming a brittle alloy. Cryolite may be decomposed in like manner if melted over zinc, forming an alloy of zinc and aluminium from which the zinc can be distilled leaving pure aluminium."

Mr. Warren does not affirm that he has actually performed the decomposition of cryolite in the way recommended, but states that it *may* be done; from which we would infer that he simply supposed it could. A well-recorded experiment, then, is needed to establish the truth of this statement. Neither does he propose to make aluminium bronze in this way; it may be that it was attempted and did not succeed, for Hampe states that an experiment thus conducted did not furnish him aluminium bronze (p. 283).

* Chemical News, Oct. 7, 1887.

Bognski's Patent.

J. Bognski,* of Warsaw, Russia, appears to have patented the above principle in 1884, for in his patent he states that the metal to be alloyed with aluminium is melted in a crucible, covered with a fusible compound of aluminium for a flux (alumina and potassium carbonate may be used) and made the negative pole of an electric current, the positive pole being a carbon rod dipping in the flux.

Grabau's Apparatus.

Ludwig Grabau,† of Hanover, Germany, proposes to electro-lyze a molten bath of cryolite mixed with sodium chloride. The features of the apparatus used are an iron pot, in which the bath is melted, and water-cooled cylinders surrounding both electrodes, the jacket surrounding the negative one having a bottom, the other not. The object of these cylinders is, at the positive elec-trode, to keep the liberated fluorine from attacking the iron pot and so contaminating the bath, at the other pole the liberated aluminium is kept from dropping to the bottom of the pot, where it might take up iron, and can be removed from the bath by simply lifting out the water-cooled cylinders and carbon electrode. Mr. Grabau states that he has abandoned this process because the inseparable impurities in the cryolite produced impurities in the metal; it may be that with the pure artificial cryolite, which he makes by his other processes (see p. 139), this electrolytic process may again be taken up.

Rogers' Process (1887).

In July, 1887, the American Aluminium Company, of Mil-waukee, was incorporated, with a capital stock of $1,000,000, for the purpose of extracting aluminium by methods devised by Prof. A. J. Rogers, a professor of chemistry in that city. This gentleman had been working at the subject for three or four

---

* English Patent 3090, Feb. 11, 1884.
† German Patent (D. R. P.), No. 45012.

REDUCTION BY THE USE OF ELECTRICITY.   281

years previous to that time, but it has not been until quite recently that patents have been applied for, and they are still pending.

The principle made use of has already been suggested in connection with the production of sodium (p. 183). It is briefly, that if molten sodium chloride is electrolyzed using a molten lead cathode, a lead-sodium alloy is produced. This alloy is capable of reacting on molten cryolite, setting free aluminium, which does not combine with the lead remaining because of its small affinity for that metal. If, then, cryolite is placed in the bath with the sodium chloride, the two reactions take place at once, and aluminium is produced. In the early part of 1888, the company erected a small experimental plant, with a ten horse-power engine, with which the following experiments, among many others, were made :—

*1. A current of 60 to 80 amperes was passed for several hours through a bath of cryolite melted in a crucible lined with alumina, and using carbon rods $2\frac{1}{4}$ inches in diameter as electrodes, one dipping into the bath from above, the other passing through the bottom of the crucible into the bath. Only 1 or 2 grammes of aluminium were obtained, showing that the separated metal was almost all redissolved or reunited with fluorine. With the temperature very high, it was found that sodium passed away from the bath without reducing the cryolite.

2. A current averaging 54 amperes and 10 volts was passed for five and a half hours through a mixture of 1 part cryolite and 5 parts sodium chloride placed in a crucible with 370 grammes of molten lead in the bottom as the cathode. After the experiment, 25 grammes of aluminium were found in globules on top of the lead-sodium alloy. This latter alloy contained some aluminium. The globules were about as pure as ordinary commercial aluminium and contained no lead or sodium. From another experiment it was determined that the lead-sodium alloy must first acquire a certain richness in sodium before it will part with any of that metal to perform the reduction of the cryolite. It was also found that a certain temperature was necessary in order that aluminium be produced at all.

* Proceedings of the Wisconsin Nat. Hist. Soc., April, 1889.

3. A current of 75 amperes and about 5 volts sufficed to decompose the bath and to produce 105 grammes of aluminium in seven hours. This would be nearly 30 grammes per hour for each electric horse-power.

4. A current of 80 amperes and 24 volts was passed through four crucibles connected in series for six hours, using a bath of 1 part cryolite and 3 parts sodium chloride with 450 grammes of lead in each crucible. The crucibles were heated regularly to a moderate temperature. There were obtained altogether 250 grammes of quite pure aluminium. This would be equal to 16 grammes per electric horse-power-hour.

A large number of similar experiments afforded a return of $\frac{3}{4}$ to $1\frac{1}{2}$ lbs. of aluminium per electric horse-power per day. The experimental plant now in operation consists of a 40 volt—100 ampere dynamo, the current being sent through six pots connected in series. When the bath is completely electrolyzed the contents of the crucible are tapped off at the bottom and a fresh supply of melted salt poured in quickly. The lead-sodium alloy run off is put back into the crucibles, thus keeping approximately constant in composition and going the rounds continuously. With this apparatus, 3 to 4 lbs. of aluminium are produced regularly per day of 12 hours. As soon as patents are obtained, it is the intention of the company to put up a plant of 50 lbs. daily capacity which can be easily increased to any extent desired as the business expands.

Professor Rogers observes in regard to the apparatus that he has tried various basic linings for his clay crucibles, but a paste of hydrated alumina, well fired, has succeeded best. Some "shrunk" magnesia lining, such as is used in basic steel furnaces, answered well but could not be used because of the amount of iron in it. Lime could not be used, as it fluxed readily. The carbon rods lasted 48 hours without much corrosion if protected from the air during electrolysis. Carbon plates and cylinders were tried, but the solid rods gave the best results. About 8 to 10 per cent of aluminium can be extracted from cryolite containing 12.85 per cent. The mineral used was obtained from the Pennsylvania Salt Company, and was called pure, but it contained 2 per cent. of silica and 1 per cent. of iron. These impurities pass largely into the aluminium produced, but the company hope to be able to manufacture an artificial aluminium fluoride which will not only be purer but less costly than this commercial cryolite. Professor Rogers infers that pure aluminium fluoride would not be an electrolyte, since the resistance of the bath increases as the amount of other salts present decreases.

It is useless to base any accurate estimation of the cost of aluminium by this process on the data given above, since they are only for a small experimental plant. If, however, 75 per cent. of the aluminium in cryolite can be extracted at the rate of 1 lb. of metal per day per electric horse-power, and the metal is free from lead and sodium, (a sample sent me recently is of very fair quality) it would seem that the process is in a fair way to compete on an equal footing with the other electrolytic processes which are coming into prominence.

Dr. Hampe on the Electrolysis of Cryolite.

Prof. W. Hampe, of Clausthal, whose name is a guarantee of careful and exact observations, has written the following valuable information on this subject, in presenting which we will also give the remarks of Dr. O. Schmidt, called forth by Hampe's first article.

*"The electrolysis of a bath of cryolite mixed with sodium and potassium chlorides, using a layer of melted copper in the bottom of the crucible as cathode and a carbon rod as anode, gave balls of melted sodium which floated on the surface and burnt, but scarcely a trace of aluminium. Yet here the conditions were most favorable to the production of the bronze. The battery used consisted of twelve large zinc-iron elements."

†Dr. O. Schmidt, referring to this statement of Hampe's, quotes an opposite experience. He fused cryolite and sodium chloride together in a well-brasqued crucible in the proportions indicated by the reaction

$$Al^2F^6.6NaF + 6NaCl = Al^2Cl^6 + 12NaF.$$

At a clear red-heat the bath becomes perfectly fluid and transparent, and an anode of gas carbon and a cathode of sheet copper

---

* Chemiker Zeitung, xii. 391 (1888).   † Idem, xii. 457 (1888).

are introduced. On passing the current the copper did not melt
but became covered with a film of deposited aluminium, which in
part penetrated the electrode and in part adhered to the surface
as a rich alloy which utimately fused off and sank to the bottom
of the crucible. With a plate 1 to $1\frac{1}{2}$ millimetres thick, 10 per
cent. of its weight of aluminium could thus be deposited ; with
one 3 millimetres thick, about 5 per cent. The metal could be
made perfectly homogeneous by subsequent fusion in a graphite
crucible. Dr. Schmidt further remarks (evidently on the supposition that the reaction he gives actually takes place) that on
thermo-chemical grounds sodium would not here be reduced, because while the molecule of sodium chloride requires 97.3 calories
for its decomposition, that of aluminium chloride, $\dfrac{Al^2Cl^6}{2}$, requires
only 80.4, and the current would attack first the most easily decomposed. He also states that the calculated difference of potential for the dissociation of aluminium chloride, which is $\dfrac{80.4}{23} =$
3.5 volts, was actually observed, and the tension of the current
must have been increased to about 4.5 volts to bring about the
decomposition of the sodium chloride.*

Dr. Hampe's statement occasioned several other communications, which he considers and replies to in the following article :—†

---

* Aside from Hampe's subsequent remarks as to no aluminium chloride
being formed, we would further point out the fact that the decomposition of a
chemically equivalent quantity of aluminium chloride requires not $\dfrac{160.8}{2} =$
80.4 calories, but $\dfrac{321.600}{6}$ or 53.6 calories, and the calculated difference of
potential is properly $\dfrac{53.6}{23}$ or 2.3 volts. The fact that the observed tension
was 3.5 volts shows that the current was not strong enough to decompose the
sodium chloride, as Schmidt observes, and the fact that this current deposited
aluminium would show that the heat of formation of aluminium fluoride cannot be greater than $23 \times 3.5 \times 6 = 483$ (thousand) calories, while it is probably much less than this, for the 3.5 volts, besides decomposing the aluminium compound, were also partly expended in overcoming resistances, as
explained on p. 248.

† Chemiker Zeitung (Cöthen) xiii. 29 and 49.

"Dr. O—— (whose name I withhold at his own request) writes to me that by electrolyzing pure cryolite, using a negative pole of molten copper, he never obtained aluminium bronze; but, on the other hand, always obtained it if he used the mixture of cryolite and sodium chloride mentioned by Dr. Schmidt, and in place of the molten copper a thick stick of the unfused metal. A letter from R. Grätzel, Hannover, contains a similar confirmation of the latter observation. By electrolyzing a mixture of 100 parts cryolite with 150 of sodium chloride in a graphite crucible holding 30 kilogrammes, aluminium bronze dripped down from the ring-shaped copper cathode used, while chlorine was freely disengaged at the carbon anode. But after a time, long before the complete decomposition of the cryolite, the formation of bronze stopped—even an attacking of that already formed sometimes taking place. Pellets of an alloy of sodium and aluminium appear on the surface and burn with a white light.

"These comments excited me to further research in the matter. At first, it was necessary to consider or prove whether by melting sodium chloride with cryolite a true chemical decomposition took place, such as Dr. Schmidt supposed. If this were the case, the very volatile aluminium chloride must necessarily be mostly driven off on melting the mixture, and at a temperature of 700° to 1000° C. there could not be any left in it. But an experiment in a platinum retort showed that such a reaction positively does not occur; for neither was any aluminium chloride volatilized nor did the residue contain any, for on treatment with water it gave up no trace of a soluble aluminium compound. During the melting of the mixture acid vapors proceeded from the retort, and a small quantity of cryolite was volatilized into the neck of the retort. Dr. Klochman has shown that cryolite always contains quartz, even colorless, transparent pieces which to the naked eye appear perfectly homogeneous showing it when examined in thin sections under the microscope, and on melting the mineral opportunity is given for the following reactions :—

$$SiO^2 + 4NaF = SiF^4 + 2Na^2O,$$
$$3Na^2O + Al^2F^6 = 6NaF + Al^2O^3,$$

as is rendered probable by the appearance of delicate crystals of alumina on the inner surface of the retort just above the fusion.

The silicon fluoride probably passes away as silico-fluoride of sodium.

"If cryolite is fused with such metallic chlorides that really do bring about a decomposition, there is never any aluminium chloride formed in these cases, but the sodium of the cryolite is exchanged for the other metal. Dr. O———, to whom I owe this observation, fused cryolite with calcium chloride, hoping that aluminium chloride would distil, but obtained instead crystals of the calcium salt of alumino-fluoric acid; thus,

$$Na^6Al^2F^{12} + 3CaCl^2 = 6NaCl + Ca^3Al^2F^{12},$$

and in like manner can be obtained the analogous strontium or barium compounds.

"Just as erroneous as the supposed production of aluminium chloride are the other arguments advanced by Dr. Schmidt, regarding the reasons why sodium could not be set free. The self-evident premises for the propositions are lacking, viz: that the two bodies compared are conductors. On the contrary, I have previously shown* that aluminium chloride and bromide and more certainly its fluoride belong to the non-conductors. It follows, then, that there can remain no doubt that on electrolyzing pure cryolite, or a mixture of it with sodium chloride, only sodium will be set free at first, either from sodium fluoride or the more easily decomposable sodium chloride. The presence or absence of sodium chloride is consequently, chemically, without significance.

"Since the experiments with solid cathodes gave aluminium, while those with molten copper did not, these results being independent of the presence or absence of sodium chloride, the next attempt made was to seek for the cause of the different results in the differences of temperature. It was found that when the electrolysis takes place at a temperature about the melting point of copper, bubbles of sodium vapor rise and burn, and any aluminium set free is so finely divided that it is attacked and dissolved by the cryolite. To explain this action of the cryolite it is necessary to admit the formation of a lower fluoride of aluminium

---

* Chemiker Zeitung (Cöthen) xi. p. 934 (1887).

and sodium, such as I have recently proven the existence of.[*]
The solution of the aluminium takes place according to the following reaction

$$Al^2F^6.6NaF + Al = 3(AlF^2.2NaF.)$$

If the electrolysis takes place at a temperature so low that the sodium separates out as a liquid (its volatilizing point is about 900°), large globules of aluminium will be produced on which the cryolite seems to exert no appreciable action. Nevertheless, the yield of aluminium is much below the theoretical quantity set free. Since pure copper melts at 1050°, and aluminium bronze at 800°, the copper electrodes can remain unfused in the bath while the bronze melts off as it forms, while the temperature can be low enough to keep the sodium in the liquid state. By mixing sodium or potassium chlorides with the cryolite, the melting point is lowered, or at a given temperature the bath is more fluid and so, easier to work. When there is not enough aluminium fluoride present in the bath to utilize all the sodium liberated, the excess of sodium may form an alloy with some aluminium, and rising to the surface, burn to waste. Since cryolite always contains silica, as previously explained, the bronze thus obtained is always rendered hard with silicon, and is not of much value commercially."

Winkler's Patent.

[†]August Winkler, of Görlitz, proposes to electrolyze a fusible phosphate or borate of aluminium. This bath is made by melting alumina or kaolin with phosphoric or boracic acid, the proportions being such that the acid is saturated; the separation of aluminium will not be hindered if alumina is added continually to combine with the acid set free. Carbon electrodes are used.

---

* Chem. Zeit. (Cöthen) xiii. p. 1 (1889). Hampe melted together aluminium and sodium fluorides in the proportions of one molecule of the first to four of the second, and obtained what is apparently a lower fluoride than cryolite, in which aluminium cannot be otherwise than diatomic, since analysis gives it the formula $AlF^2.2NaF$. This salt is similar in appearance and properties to cryolite. As there are still some doubts, however, about this compound, the above explanation of the solution of aluminium by the cryolite need not be accepted as final.

† German Patent, 45824, May 15, 1888.

Faure's Proposition.

Camille A. Faure, whose process of making aluminium chloride is described on p. 134, proposes to obtain the metal therefrom by electrolysis, using carbon electrodes. M. Faure states that if the process is carried out on a large scale the chlorine set free can be utilized to form bleaching powder, and will thus nearly repay the whole cost of manufacturing the aluminium. Patents have been applied for covering the details of the electrolytic apparatus, but have not yet been granted. The inventor states, however, that he has determined on a large scale that anhydrous, molten aluminium chloride can be practically decomposed at 300° by an electro-motive force of 5 volts, which comprises the force required for actual decomposition and also that required to overcome the resistance of the bath. While, therefore, the reduction of 1 kilo of aluminium per hour theoretically requires a minimum expenditure of 9.2 electric horse-power, the actual resistance of 5 volts would increase this requirement to 20 horse-power or 9 horse-power per lb. produced per hour. Therefore, if each bath could be decomposed by 5 volts, the production of 2000 lbs. of aluminium in 20 hours would require the use of a 920 horse-power current, and could not be possibly achieved by a 400 horse-power dynamo, as calculated by M. Faure.

Hall's Process (1889).

Mr. Chas. M. Hall, a graduate of Oberlin College, has, since 1885, experimented with electrolytic aluminium processes, and has finally attained such success that a company has been formed to work by his methods. The Pittsburgh Reduction Company was organized about the middle of 1888, and since March, 1889, have had their metal on the market. They are located on Fifth Avenue, Pittsburgh, Pa. The plant is at present equal to a production of about 300 lbs. of aluminium a week, and their metal is quoted at $2 per lb. Contracts have recently been given out for the erection of a plant of 2500 lbs. weekly capacity; a plant of the same size is also being erected at Patricroft, Lanc., England.

*Mr. Hall claims the process of dissolving alumina in a fluid

* U. S. Patents, 400664 to 400667, and 400766, April 2, 1889.

bath composed of aluminium fluoride and potassium fluoride, or with also the addition of lithium fluoride, then electrolyzing this bath using an anode of non-carbonaceous material. The bath is formed by fusing a mixture of the required fluorides in certain proportions; thus, 169 parts of aluminium fluoride and 116 parts of potassium fluoride form proportions corresponding to the formula $Al^2F^6.2KF$. A slight variation from these proportions affects the process but little, but it is observed that a larger proportion of potassium fluoride increases the capacity of the bath for dissolving alumina, while a larger proportion of aluminium fluoride renders the bath more fusible but decreases the amount of alumina it can dissolve. However, the bath is rendered more fusible and its capacity for dissolving alumina increased also, if lithium fluoride is added to the above mixture or substituted for part of the potassium fluoride. Thus, the combinations in proportions represented by the formulæ $Al^2F^6.KF.LiF$ and $2Al^2F^6.3KF.3LiF$ are useful in both respects. These materials may be conveniently prepared by saturating aluminium hydrate and carbonates of potassium and lithium, mixed in the proportion required, with hydrofluoric acid. In electrolyzing the bath, the negative electrode is to be of carbon or a metal coated with carbon and the positive electrode of copper, platinum, or other suitable non-carbonaceous material. When of copper, it soon becomes coated with oxide of copper, which is a conductor at a red heat, and therefore does not affect the passage of the current, while it forms a protecting cover over all the surface of the anode and prevents further oxidation, the oxygen thereafter escaping at this electrode in a free state. The containing vessel is of metal protected by a carbon lining, which is preferably made the negative electrode. A low red heat is sufficient for carrying on the operation, and on account of the liability of reducing the solvent a current of low electro-motive force is used.

In the second patent, Hall claims the use of a bath composed of alumina dissolved in compound fluorides of aluminium with alkaline-earth metals, such as in proportions varying from $Al^2F^6.CaF^2$ to $Al^2F^6.3CaF^2$. Since this bath is of higher specific gravity than aluminium, that metal would rise to the surface and there be subject to loss by oxidation; to remedy which a quantity of the salt represented by $Al^2F^6.2KF$ may be added sufficient to lower the specific gravity of the bath below that of aluminium.* If it is desired to produce alloys, the metal to be alloyed may be made the negative electrode, in which case the addition of $Al^2F^6.2KF$ is unnecessary, because the alloy will be sufficiently heavy to sink. For making alloys the barium compound is especially recommended, for its high specific gravity is of no inconvenience, and it is more fusible than the compounds of calcium and strontium. These double fluorides are said not to be subject to a decrease in efficiency such as occurs with the double fluoride of potassium and aluminium when used alone.

In a third patent, the use of a bath formed of fluorides of calcium, sodium and aluminium, in which alumina is dissolved, is claimed; these materials being obtained by melting together cryolite, aluminium fluoride and fluorspar in the proportions represented by the formula $Al^2F^6.6NaF + Al^2F^6.CaF^2$. This bath is said not to become so readily clogged as the previous ones; but when it does become so it is cleared by the addition of three or four per cent. of calcium chloride, and this device is said to permit the use of a carbon anode without the bath being affected by its disintegration.

The plant now being operated in Pittsburgh consists of a 50 horse-power engine driving two dynamos connected in parallel, the current produced varying from 16 to 25 volts in tension and 1700 to 1800 amperes in quantity. Two reducing pots are used, coupled in series. Each pot is of cast-iron lined with carbon, the lining forming the negative electrode, while a number (6 to 10) of three-inch carbon cylinders are suspended in the bath and form the positive electrode. Each pot holds 200 to 300 pounds of the electrolyte, its dimensions being 24 inches long, 16 inches wide and 20 inches deep. These vessels are not heated from outside, as was done in the early stages of the process when a current of only 4 to 6 volts tension was employed, but the dis-

---

* A specimen of the salt represented by the formula $Al^2F^6.2KF$ was sent the author, who found its specific gravity to be 2.35. I should infer from analogy that its specific gravity when molten would be much less, probably not much over 2, since solid cryolite has a specific gravity of 2.9, and yet, when molten, a piece of aluminium of gravity 2.6 will sink beneath it.

tance between the electrodes is now increased until the electro-
motive force absorbed in each bath is 8 to 12 volts, and enough
heat is thus generated by this resistance to keep the bath at the
necessary temperature. The bath at present employed for producing pure aluminium is the last one described, and the temperature is continuously kept very near to the melting point of brass,
sometimes it is hot enough to melt copper, but the high temperature is not a disadvantage since the bath is more efficient—*i e.*,
conducts better and collects the aluminium better—the hotter it
is, within certain limits. During the operation, alumina (obtained
by calcining pure aluminium hydrate) is fed in small quantities
of 5 to 10 pounds as required. The exhaustion of the alumina
in the bath is immediately shown by a rise in the resistance, so
that the current can hardly be made to pass at all. Mr. Hall
has estimated that when the bath is saturated with alumina its
conducting power is at least 200 times that of copper sulphate
solution. In an experiment which he made, a copper anode of
about 30 to 40 square inches area transmitted as high as 150
amperes with an electro-motive force less than $3\frac{1}{2}$ volts. Only
the dissolved alumina is decomposed by the current, for the
fluoride solvents waste only very slightly and require replenishing
to the extent of a small fraction of the weight of metal made;
and as these materials cost only about 7 cents a pound, their
waste forms a very small item of expense. An accurate account
of the amount of calcined alumina used shows that a fraction over
50 per cent. of aluminium is extracted from it (theoretically pure
alumina contains 52.94 per cent.). It is thus seen that the process
is able to extract nearly all the aluminium from commercial alumina in one direct operation. When it is desired to form bronze,
a bath of different composition is used, as before mentioned, and
both electrodes are of copper. On working at a temperature just
below that of melting copper, the anode remains undissolved and
practically unattacked, while the bronze formed at the cathode
drips down melted to the bottom of the vessel. In either case
the metal is allowed to collect in the pots for one or two days,
and is then ladled out with cast-iron ladles, taking the metal from
the bottom as one might dip water out from under oil. The production is about one pound of aluminium an hour from each pot,

and the operations are kept up without stopping for several weeks at a time. The metal produced has varied from 94 to over 98 per cent. pure. Three analyses by Mr. Hall have given—

	I.	II.	III.
Aluminium	94.16	95.93	98.34
Silicon	4.36	2.01	1.34
Iron	1.48	2.06	0.32

The metal being made at present averages over 97 per cent., and a specimen kindly sent the author compares favorably with other commercial brands. A specimen sent me at an early stage of the process contained copper, probably from the copper anodes, but this has since been avoided by discontinuing the use of copper anodes in making the pure aluminium.

The essential features of Mr. Hall's process have been antedated; the principle of dissolving alumina in a fluid bath, and electrolyzing it without decomposing the solvent, is the matter of Henderson's patent issued in 1887, and Bernard Bros. claim the use of a copper anode in their patent of the same year (pp. 273 and 274). It is only just to Mr. Hall, however, to observe that his patent applications were dated in 1886, and that the use of these principles is clearly original with him. If we seek to find the efficiency of the process, basing our calculations on the data given, we reach the following results: The current is said to average 1700 amperes and 20 volts, being 10 volts to each pot. A current of this size would represent $\frac{1700 \times 20}{750}$ or a little over 45 electric horse-power, which would need about a fifty horse-power engine to drive the dynamos. If the energy of this current could be entirely utilized for dissociating alumina into aluminium and oxygen, it would be able to produce $\frac{1700 \times 20 \times 0.00024}{7250} = 1\frac{1}{9}$ grammes per second, or 4 kilos per hour. Since the output is stated as 1 lb. per hour for each pot, the production of aluminium represents $\frac{2}{8.8}$ or nearly one-quarter of the energy of the current, which it is almost needless to observe, is a high efficiency in this kind of work. Further, a current of 1700 ampères passing through two pots will set free $0.00009135 \times 1700 \times 60 \times 60 \times 2$ $=1008$ grammes of aluminium per hour, or about 1 kilogramme. This shows that $\frac{2}{2.2}$ or 91 per cent. of the aluminium set free is practically obtained and weighed, the other 9 per cent. being re-dissolved or otherwise lost. Since we have calculated that the decomposition of alumina requires an electro-motive force of 2.8 volts, the fact that something like 10 volts is required for each bath would show that about 7.2 volts, or 72 per cent. of the energy of the current is absorbed in other resistances, being principally converted into heat, and thus keeping the bath in fusion.

As to the probable cost of aluminium by this process, I have no official figures to present, but an approximate idea can easily be estimated from the data given. Pure hydrated alumina should not cost over 3 cents per lb., and since it contains about one-third water, the alumina produced costs $4\frac{1}{2}$ cents, with the cost of calcination to be added, which should not be over $1\frac{1}{2}$ cents per lb. The bath wastes very little, let us suppose 15 per cent. of the weight of aluminium produced. The cost of power, number of men, etc., will have to be guessed at. We might then put down for twenty-four hours' work—

100 lbs. calcined alumina @ 6 cts.	. . . .	= $6.00
fluorides for bath @ 7 cts.	. . . .	= .50
carbons	. . . .	= 1.00
50 horse-power engine	. . . . .	= 10.00
2 engineers @ $3.00 per diem .	. . .	= 6.00
6 workmen @ $2.00 "	. . . .	= 12.00
Superintendence, office expenses, etc.	. . .	= 10.00
Interest on plant, rent, etc.	. . . .	= 4.50
Cost of about 50 lbs. of aluminium	. . .	$50.00

When the large plant now being erected is in operation, the cost of aluminium by this process will not exceed $0.50 per lb.

Cowles Bros.' Process.

Messrs. E. H. and A. H. Cowles patented in the United States and Europe* an electric furnace and its application for producing

* U. S. Patents 324658, 324659, Aug. 18, 1885; English Patent 9781, same date; German Patent, 33672.

aluminium. Their patent claims "reducing an aluminium compound in company with a metal in presence of carbon in a furnace heated by electricity; the alloy of aluminium and the metal formed being further treated to separate out the aluminium." The history of the development of this process has already been sketched, we will proceed to describe the details of its operation. The first public description was given in two papers, one read before the American Association for the Advancement of Science* by Prof. Chas. F. Mabery, of the Case School of Applied Science, Cleveland, the other before the American Institute of Mining Engineers† by Dr. T. Sterry Hunt, of Montreal.

Prof. Mabery said in his paper: "Some time since, the Messrs. Cowles conceived the idea of obtaining a continuous high temperature on an extended scale by introducing into the path of an electric current some material that would afford the requisite resistance, thereby producing a corresponding increase in the temperature. After numerous experiments, coarsely pulverized carbon was selected as the best means for maintaining an invariable resistance, and at the same time as the most available substance for the reduction of oxides. When this material mixed with the oxide to be reduced was made a part of the electric circuit, inclosed in a fire-clay retort, and subjected to the action of a current from a powerful dynamo, not only was the oxide reduced, but the temperature increased to such an extent that the whole interior of the retort fused completely. In other experiments lumps of lime, sand, and corundum were fused, with a reduction of the corresponding metal; on cooling, the lime formed large, well-defined crystals, the corundum beautiful red-green and blue octahedral crystals. Following up these results, it was soon found that the intense heat thus produced could be utilized for the reduction of oxides in large quantities, and experiments were next tried on a large scale with the current from a fifty horse-power dynamo. For the protection of the walls of the furnace, which were of fire-brick, a mixture of ore and coarsely pulverized gas-carbon was made a central core, and was surrounded on the side and bottom by fine char-

* Ann Arbor Meeting, Aug. 28, 1885.
† Halifax Meeting, Sept. 16, 1885.

coal, the current following the lesser resistance of the core from carbon electrodes inserted in the ends of the furnace in contact with the core. The furnace was charged by first filling it with charcoal, making a trough in the centre, and filling this with the ore mixture, the whole being covered with a layer of coarse charcoal. The furnace was closed on top with fire-brick slabs containing two or three holes for the escape of the gaseous products of the reduction, and the whole furnace was made air tight by luting with fire-clay. Within a few minutes after starting the dynamo, a stream of carbonic oxide issued through the openings, burning usually with a flame eighteen inches high. The time required for complete reduction was ordinarily about an hour. Experience has already shown that aluminium, silicon, boron, manganese, sodium, and potassium can be reduced from their oxides with ease. In fact, there is no oxide that can withstand the temperature attainable in this furnace. Charcoal is changed to graphite; does this indicate fusion? As to what can be accomplished by converting enormous electrical energy into heat within narrow limits, it can only be said that it opens the way into an extensive field of pure and applied chemistry. It is not difficult to conceive of temperature limited only by the power of carbon to resist fusion.

"Since the motive power is the chief expense in accomplishing reductions by this method, its commercial success is closely connected with obtaining power cheaply. Realizing the importance of this point, Messrs. Cowles have purchased at Lockport, N. Y., a water power where they can utilize 1200 horse-power. An important feature in the use of these furnaces from a commercial standpoint is the slight technical skill required in their manipulation. The four furnaces operated in the experimental laboratory at Cleveland are in charge of two young men, who six months ago knew absolutely nothing of electricity. The products at present manufactured are the various grades of aluminium bronze, made from a rich furnace product obtained by adding copper to the charge of ore. Aluminium silver is also made; and a boron bronze may be prepared by the reduction of boracic acid in contact with copper, while silicon bronze is made by reducing silica in contact with copper. As commercial results may be mentioned the production in the experimental laboratory, which averages 50 lbs. of 10 per cent. aluminium bronze daily, which can be supplied to the trade in large quantities on the basis of $5 per lb. for the aluminium contained, the lowest market quotation of aluminium being now $15 per lb."

Dr. Hunt stated further that if the mixture consisted of alumina and carbon only, the reduced metal volatilized, part escaping into the air and burning to alumina, part condensing in the upper layer of charcoal, affording thus crystalline masses of nearly pure aluminium and yellow crystals supposed to be a compound of aluminium with carbon. Great loss was met in collecting this divided metal into an ingot, so that only small quantities were really obtained. To gather all the aluminium together, a metal such as copper was added, thus producing an alloy with 15 to 20 per cent. of aluminium ; on substituting this alloy for pure copper in another operation, an alloy with over 30 per cent. of aluminium was obtained.

Dr. Hunt, in a later paper,* stated that pure aluminium has been obtained in this process by first producing in the furnace an alloy of aluminium and tin, then melting this with lead, when the latter takes up the tin and sinks with it beneath the aluminium. He also stated that in the early experiments a dynamo driven by a 30 horse-power engine yielded a daily output of 50 lbs. of 10 per cent. aluminium bronze, but with a larger machine the output was proportionately much greater. In the latest practice, one-half cent per horse-power per hour is said to cover the expense of working, making the 10 per cent. bronze cost about 5 cents per lb. over the copper used.

Various shapes of furnaces have been used by the Cowles Bros., the first described being a rectangular box, lined with carbon, with the electrodes passing through the ends. Although two other forms have been patented, we understand that the kind now used, and which is described at length in Mr. Thompson's paper, is also of the oblong, horizontal style. Chas. S. Bradley and Francis B. Crocker, of New York, patented and assigned to the

* National Academy of Science, Washington Meeting, April 30, 1886.

Cowles Electric Smelting Company,* the use of a retort, composed of conducting material, surrounded by a substance which is a poor conductor of heat, and having inside a mixture of charcoal and the ore to be heated. Electric connection being made with the ends of the retort, the walls of the retort and the material in it are included in the circuit and constitute the greater part of the resistance. The retort may be stoppered at each end during the operation, and the heating thus performed in a reducing atmosphere. Mr. A. H. Cowles devised a style of furnace adapted for continuous working and utilizing the full current of a dynamo of the largest size.† The electrodes are tube-shaped and placed vertically. The positive pole is above, and is surmounted by a funnel in which the mixture for reduction is placed. The regular delivery of the mixture is facilitated by a carbon rod, passing through the cover of the funnel, which is serrated on the end and can be worked up and down. The melted alloy produced, with any slag, passes down through the negative electrode. The distance between the poles can be regulated by moving the upper one, and the whole is inclosed in a fire-brick chamber. The space between the electrodes and the walls is filled with an isolating material, which is compact around the lower electrode but coarse grained around the upper to facilitate the escape of the gases produced. The chamber is tightly closed excepting a small tube for the escape of gas.

A very complete description of the Cowles process was given by Mr. W. P. Thompson (agent for the Cowles Co. in England) in a paper read before the Liverpool Section of the Society of Chemical Industry.‡ He describes the process as then carried on in Lockport; the dynamo used being a large Brush machine weighing $2\frac{1}{2}$ tons and consuming about 100 horse-power in being driven at 900 revolutions per minute.

"Conduction of the current of the large dynamo to the furnace and back is accomplished by a complete metallic circuit, except where it is broken by the interposition of the carbon electrodes

* U. S. Patent 335499, Feb. 2, 1886.
† English Patent 4664 (1887).
‡ Journal of the Society of Chemical Industry, April 29, 1886.

and the mass of pulverized carbon in which the reduction takes place. The circuit is of 13 copper wires, each 0.3 inch in diameter. There is likewise in the circuit an ampère meter, or ammeter, through whose helix the whole current flows, indicating the total strength of the current being used. This is an important element in the management of the furnace, for, by the position of the finger on the dial, the furnace attendant can tell to a nicety what is being done by the current in the furnace. Between the ammeter and the furnace is a resistance coil of German silver kept in water, throwing more or less resistance into the circuit as desired. This is a safety appliance used in changing the current from one furnace to another, or to choke off the current before breaking it by a switch.

"The furnace (see Figs. 25, 26, 27) is simply a rectangular box, A, one foot wide, five feet long inside, and fifteen inches deep, made of firebrick. From the opposite ends through the pipes BB the two electrodes CC pass. The electrodes are immense electric-light carbons three inches in diameter and thirty inches long. If larger electrodes are required, a series this size must be used instead, as so far all attempts to make larger carbons that will not disintegrate on becoming incandescent have failed. The ends of the carbons are placed within a few inches of each other in the middle of the furnace, and the resistance coil and ammeter are placed in the circuit. The ammeter registers 50 to 2000 ampères. These connections made, the furnace is ready for charging.

"The walls of the furnace must first be protected, or the intense heat would melt the fire brick. The question arose, what would be the best substance to line the walls? Finely powdered charcoal is a poor conductor of electricity, is considered infusible and the best non-conductor of heat of all solids. From these properties it would seem the best material. As long as air is excluded it will not burn. But it is found that after using pure charcoal a few times it becomes valueless; it retains its woody structure, as is shown in larger pieces, but is changed to graphite, a good conductor of electricity, and thereby tends to diffuse the current through the lining, heating it and the walls. The fine charcoal is therefore washed in a solution of lime-water, and after drying, each particle is insulated by a fine coating of lime. The bottom of the furnace is now filled with this lining about two or three inches deep. A sheet-iron gauge is then placed along the sides of the electrodes, leaving about two inches between them

Fig. 25.

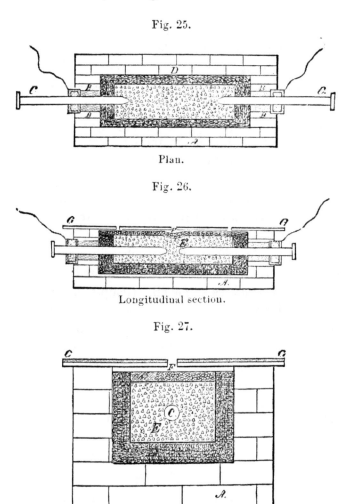

Plan.

Fig. 26.

Longitudinal section.

Fig. 27.

Transverse section.

and the side walls, in which space more of the charcoal is placed. The charge *E*, consisting of about 25 pounds of alumina, in its native form as corundum, 12 pounds of charcoal and carbon, and 50 pounds of granulated copper, is now placed within the gauge and spread around the electrodes to within a foot of each end of the furnace. For making iron alloy, where silicon also is not harmful, beauxite or various clays containing iron and silica may be used instead of the pure alumina or corundum. In place of granulated copper, a series of short copper wires or bars can be placed parallel to each other and transverse to the furnace, among the alumina and carbon, it being found that where grains are used they sometimes fuse together in such a way as to short-circuit the current. After this, a bed of charcoal, F, the granules of which vary in size from a chestnut to a hickory, is spread over all, and the gauge drawn out. This coarse bed of charcoal above the charge allows free escape of the carbonic oxide generated in the reduction. The charge being in place, an iron top, G, lined with fire brick, is placed over the whole furnace and the crevices luted to prevent access of air. The brick of the walls insulate the cover from the current.

"Now that the furnace is charged and the cover luted down, it is started. The ends of the electrodes were in the beginning placed close together, as shown in the longitudinal section, and for this cause the internal resistance of the furnace may be too low for the dynamo, and cause a short circuit. The operator, therefore, puts sufficient resistance into the circuit, and by watching the ammeter and now and then moving one of the electrodes out a trifle, he can prevent undue short circuiting in the beginning of the operation. In about ten minutes, the copper between the electrodes has been melted and the latter are moved far enough apart so that the current becomes steady. The current is now increased till 1300 ampères are going through, driven by 50 volts. Carbonic oxide has already commenced to escape through the two orifices in the top, where it burns with a white flame. By slight movements outward of the electrodes during the coming five hours, the internal resistance in the furnace is kept constant, and at the same time all the different parts of the charge are brought in turn into the zone of reduction. At the close of the run the electrodes are in the position shown in the plan, the furnace is shut down by placing a resistance in the circuit and then the current is switched into another furnace charged in a similar manner. It is found that the product is larger if the carbons are inclined at angles of 30° to the horizontal plane.

"This regulating of the furnace by hand is rather costly and unsatisfactory. Several experiments have therefore been tried to make it self-regulating, and on January 26, 1886, a British patent was applied for by Cowles Bros., covering an arrangement for operating the electrodes by means of a shunt circuit, electro-magnet, and vibrating armature. Moreover, if the electrodes were drawn back and exposed to the air in their highly heated state, they would be rapidly wasted away. To obviate this, Messrs. Cowles placed what may be called a stuffing-box around them, consisting of a copper box filled with copper shot. The wires are attached to the boxes instead of the electrodes. The hot electrodes as they emerge from the furnace first encounter the shot, which rapidly carry off the heat, and by the time they emerge from the box they are too cool to be oxidized by contact with the air.

"Ninety horse-power have been pumped into the furnace for five hours. At the beginning of the operation the copper first melted in the centre of the furnace. There was no escape for the heat continually generated, and the temperature increased until the refractory corundum melted, and being surrounded on all sides by carbon gave up its oxygen. This oxygen, uniting with the carbon to form carbonic oxide, has generated heat which certainly aids in the process. The copper has had nothing to do with the reaction, as it will take place in its absence. Whether the reaction is due to the intense heat or to electric action it is difficult to say. If it be electric, it is Messrs. Cowles' impression that we have here a case where electrolysis can be accomplished by an alternating current, although it has not been tried as yet. Were the copper absent, the aluminium set free would now absorb carbon and become a yellow, crystalline carbide of aluminium; but, instead of that, the copper has become a boiling, seething mass, and the bubblings of its vapors may distinctly be heard. The vapors probably rise an inch or two, condense and fall back, carrying with them the freed aluminium. This continues till the current is taken off the furnace, when we have the copper charged with 15 to 30 per cent., and in some cases as high as 40 per cent. of its weight of aluminium, and a little silicon. After cooling the furnace this rich alloy is removed. A valuable property of the fine charcoal is that the metal does not spread and run through its interstices, but remains as a liquid mass surrounded below and on the sides by fine charcoal, which sustains it just as flour or other fine dust will sustain drops of water for considerable periods, without allowing them to sink in. The alloy is white and brittle. This metal is then melted in an ordinary crucible furnace, poured into large ingots, the amount of aluminium in it determined by analysis, again melted, and the requisite amount of copper added to make the bronze desired.

"Two runs produce in ten hours' average work 100 pounds of white metal, from which it is estimated that Cowles Bros., at Lockport, are producing aluminium in its alloys at a cost of about 40 cents per lb. The Cowles Company will shortly have 1200 horse-power furnaces. With a larger furnace, there is no reason why it should not be made to run continuously like the ordinary blast furnace.

" In place of the copper any non-volatile metal may be used as a condenser to unite with any metal it may be desired to reduce, provided, of course, that the two metals are of such a nature that they will unite at this high temperature. In this way aluminium may be alloyed with iron, nickel, silver, tin, or cobalt. Messrs. Cowles have made alloys containing 50 aluminium to 50 of iron, 30 aluminium to 70 of copper, and 25 aluminium to 75 of nickel. Silicon or boron or other rare metals may be combined in the same way, or tertiary alloys may be produced; as, for instance, where fire-clay is reduced in presence of copper we obtain an alloy of aluminium, silicon, and copper."

Soon after Mr. Thompson's description, the plant at Lockport was increased by the addition of the largest dynamo yet constructed, built by the Brush Electric Company, and dubbed the " Colossus." This machine weighs almost ten tons, and when driven at 423 revolutions per minute, with 68 volts resistance in the external circuit, it produced a useful current of 3400 ampères, or at 405 revolutions produced a current of 3200 ampères with 83 volts electro-motive force, indicating 249000 Watts or 334 electric horse-power. The steam engine was, in the latter case, developing nearly 400 horse-power, and could not supply more; it was judged that the dynamo could have been driven to 300,000

Watts with safety. The first run with this machine was made in September, 1886. The furnaces used for this current are of the same style as that described by Mr. Thompson, but are larger, the charge being 60 lbs. of corundum, 60 lbs. of granulated copper, 30 lbs. of coarse charcoal besides the pulverized lime-coated charcoal used in packing. The operation of reducing this charge takes about two hours. As soon as the operation is finished the current is switched off into another furnace prepared and charged, so that the dynamo is kept working continuously. In 1888, the Cowles Company had two of these large dynamos in operation and eight furnaces in use. With two-hour runs a furnace is tapped every hour, producing about 80 lbs. of bronze averaging 18 per cent. of aluminium. The capacity of the plant is, therefore, about $1\frac{1}{2}$ tons of 10 per cent. bronze per day. Their alloys are now sold on the basis of $2.50 per lb. for the contained aluminium.

The Cowles Syndicate Company, of England, located at Stoke-on-Trent, have set up a large plant at Milton, where, profiting by the experience of the parent concern in America, still larger electric currents are used, these being found more economical. The dynamo in use at this works was built by Crompton, and supplies a current of 5000–6000 ampères at 50 to 60 volts. There are two furnace-rooms, each containing six furnaces, aluminium and silicon bronze being produced in one room and ferro-aluminium in the other. The furnaces used measure 60 by 20 by 36 inches, inside dimensions. The electrodes used are formed by bundling together 9 carbon rods, each $2\frac{1}{4}$ inches in diameter, each electrode weighing 20 lbs. More recently larger carbons have been obtained, 3 inches in diameter, and an electrode formed of five of these weighs 36 lbs. The furnace is charged as previously described.

The current is started at 3000 ampères, gradually increasing to 5000 during the first half hour, and then keeping steady until the run is ended, which is about one and a half hours from starting. The product of each run is about 100 lbs. of raw bronze containing 15 to 20 per cent. of aluminium. The return is said to average 1 lb. of contained aluminium per 18 electric horse-power per hour, or $1\frac{1}{3}$ lbs. per electric horse-power per day. The works produce about 200 lbs. of aluminium contained in alloys per day. The raw bronze is stacked until several runs have accumulated, then a large batch is melted at once in a reverberatory furnace, refined, and diluted to the proportion of aluminium required by adding pure copper.

The Cowles Company, both in England and America, produce six standard grades of bronze as follows :—

"Special" A	11 per cent. of aluminium.
A	10 " "
B	$7\frac{1}{2}$ " "
C	$5-5\frac{1}{2}$ " "
D	$2\frac{1}{2}$ " "
E	$1\frac{1}{4}$ " "

Their ferro-aluminium is sold with usually 5 to 7 per cent. of aluminium, but 10, 13, and 15 per cent. is furnished if asked for.

Products of the Cowles furnace.—Dr. W. Hampe obtained the following results on analyzing a sample of Cowles Bros.' 10 per cent. bronze :—

Copper	90.058
Aluminium	8.236
Silicon	1.596
Carbon	0.104
Magnesium	0.019
Iron	trace
	100.013

A sample of 10 per cent. bronze, made in the early part of 1886, and analyzed in the laboratory of the Stevens Institute, showed—

Copper	88.0
Aluminium	6.3
Silicon	6.5

but it is evident that the percentage of silicon has since then been lowered.

The ferro-aluminium used by Mr. Keep in his tests on cast-iron was furnished by the Cowles Company, and analyzed—

Aluminium	11.42
Silicon	3.86

A sample shipped to England in December, 1886, contained—

Iron	86.69
Combined carbon 1.01	
Graphitic " 1.91	
Total . . . . ——	2.92
Silicon	2.40
Manganese	0.31
Aluminium	6.50
Copper	1.05
Sulphur	0.00
Phosphorus	0.13
	100.00

The copper in this alloy was present by accident, the alloy regularly made containing none, but the rest of the analysis gives a correct idea of the constitution of the alloy. Prof. Mabery gives several analyses of Cowles' ferro-aluminium :—*

Iron . . .	85.17	85.46	86.04	86.00	84.00
Aluminium . .	8.02	8.65	9.00	9.25	10.50
Silicon . .	2.36	2.20	2.52	2.35	2.40
Carbon . .		3.77		2.41	2.50

The slags formed in the furnace in producing this alloy were analyzed as follows :—

Silica	0.78	4.10
Alumina (insoluble)	0.20	—
Lime	28.50	14.00
Iron	1.50	29.16
Alumina (soluble) + aluminium . .	38.00	• 48.70
Sulphur	0.50	—
Graphite	5.00	2.60
Combined carbon	0.90	0.48

The slags formed when producing bronze vary in composition, and are usually crystalline, with a shining, vitreous lustre. Their analysis shows—

Alumina (insoluble) . . .	55.30	66.84	—
Alumina (soluble) + aluminium .	21.80	14.20	—
Lime	3.70	1.44	6.77
Copper	—	3.32	1.00
Carbon	0.65		

* American Chemical Journal, 1887, p. 11.

20

The lime present probably existed as calcium aluminate. These slags contained only a small amount of aluminium, rarely any iron, and were usually free from silica.

The same chemist analyzed a peculiar product sometimes formed in the furnace when smelting for bronze, in the shape of crystalline masses, steel-gray to bright yellow in color, semi-transparent and with a resinous lustre. These all contained aluminium, copper, silicon and calcium in various proportions, and when exposed to the air fell to powder. Analyses gave

Copper	.	.	.	26.70	35.00	20.00	—
Aluminium	.	.	.	66.20	53.30	74.32	15.23
Silicon	.	.	.	5.00	12.30	2.86	20.55
Calcium	.	.	.	2.00	0.20	2.86	—
Tin	.	.	.	.	—	—	—	49.26
				99.90	100.80	100.04	85.04

The latter product was formed in smelting for aluminium-tin.

Prof. Mabery also found that the soot collecting at the orifices on top of the furnace contained 10 to 12 per cent. of aluminium; also that when alumina and carbon alone were heated and silica was present, the aluminium formed dissolved up to 10 per cent. of silicon, which, on dissolving the aluminium in hydrochloric acid, was left as crystalline or graphitic silicon.

Reactions in Cowles' process.—The inventors, themselves, claim " reduction in a furnace heated by electricity in presence of carbon and a metal." In their first pamphlet they say that " the Cowles process accomplishes the reduction of alumina by carbon and heat." Professor Mabery and Dr. Hunt, already quoted, and Dr. Kosman* look at the process in no other light than that the electric current is utilized simply by its conversion into heat by the resistance offered, and that pure electrolysis is either absent or occurs to so small an extent as to be inappreciable. Indeed, if we consider the arrangement of the parts in the Cowles furnace we see every effort made to oppose a uniform, high resistance to the passage of the current and so convert its energy into heat, and an entire absence of any of the usual arrangements for electrolysis. For instance, electroysis requires a fluid bath in

---

* Stahl und Eisen, Jan. 1889.

circulation, so that each element of the electrolyte may be continuously liberated at one of the poles and the presence of any foreign material, as bits of carbon, between the poles is to be avoided if possible, since they short-circuit the current and hinder electrolysis proper. I think the arrangement of the furnace shows no attempt to fulfil any of the usual conditions for electrolysis, and one of the best arrangements for converting the energy of the current entirely into heat. Dr. Hampe, however, in spite of these evident facts, draws the conclusion that because he was unable to reduce alumina by carbon in presence of copper at the temperature of a Deville lime-furnace that it was therefore to be assumed that even the somewhat higher temperature of the electric furnace alone would be insufficient to accomplish the desired reaction, and hence the effect of the electric arc must be not only electro-thermic in supplying heat, but afterwards electrolytic, in decomposing the fused alumina.

If we figure out the useful effect of the current, $i.\ e.$, the proportion of its energy utilized for the purpose of reducing alumina, we find a low figure, but it is well to note that although the power required is one of the main features of this way of reduction yet this item is so cheap at the firm's works that it becomes a secondary consideration in the economy of the process. A 300 horse-power current is equivalent to an expenditure of $\dfrac{300 \times 270000}{424} = 191000$ calories of heat per hour. Theoretically, this amount of heat would produce $\dfrac{191000}{7250} = 26\frac{1}{3}$ kilos or 58 pounds of aluminium. However, about 7 pounds are obtained in an hour's working, which would show a useful effect of 12 per cent. This should even be diminished, since no account has been taken of the combustion of carbon in the furnace to carbonic oxide. The remainder of the heat account, probably 90 per cent. of the whole, is partly accounted for by the heat contained in the gases escaping and the materials withdrawn from the furnace (of which no reasonable estimate can be made, since the question of temperatures is so uncertain) and the large remainder must be put down as lost by radiation and conduction. As before remarked, water power is obtained by this company very cheaply, and even this large loss does not make much show in the cost of the alloy, yet the figures show that a much larger useful effect should be possible, and it is not at all improbable that the prospect of getting double or triple the present output from the same plant is at present inciting the managers to fresh exertions in utilizing the power to better advantage.

Since writing the above, I have seen Mr. H. T. Dagger's paper* on the Cowles process in England, in which the product at their Milton works is said to be 1 lb. of aluminium to 18 electric horse-power per hour, which would show that the dissociation of the alumina represented nearly 30 per cent. of the energy of the current, but the data given in the body of this gentleman's paper (p. 303) do not seem to indicate so large a return as is stated above. Mr. Dagger, moreover, maintains the purely electro-thermic action of the current, denying that any electrolysis takes place at all.

In the discussion of Heroult's process (p. 314) it will be shown that in both it and Cowles' process the largest part of the reduction must necessarily be performed by chemical and not by electrolytic action. I do not introduce this discussion here, since the two processes resemble each other so closely in the reaction involved that they can best be considered together.

Menges' Patent.

†This inventor proposes to produce aluminium or aluminium bronze by mixing aluminous material with suitable conducting material, such as coal, and a cohesive material, then pressing into cylinders and baking hard. These strong, compact bars conduct electricity, and are to be used like the carbon electrodes of electric lamps in a suitably inclosed space.

Farmer's Patent.

M. G. Farmer‡ mixes aluminous material with molasses or pitch, making a paste which is moulded into sticks, burned, and

---

* Read before the British Association for Adv. Science, Newcastle, 1889.
† German Patent, 40354 (1887).
‡ English Patent, 10815, Aug. 6, 1887.

used as electrodes, inclosed in a furnace. Aluminium is produced by the arc, and drops into a crucible placed immediately beneath.

It appears that Messrs. Menges and Farmer hit upon the same idea at about the same time, but the practicability of the process as outlined has still to be demonstrated, and appears to be very improbable of attainment.

The Heroult Process (1887).

This is the invention of P. L. C. Heroult, of Paris, and has been patented in the United States, England, and most European countries.* As has already been outlined in Chapter I., the process was first put in operation at the works of the Societé Mettallurgique Suisse, at Neuhausen on the Rhine, where large water power is obtained from the Rhine-Falls.

The English patent is headed "an improved process for the production of aluminium, aluminium bronze, and other alloys of aluminium by electrolysis." It specifies that for producing pure aluminium a mixture of cryolite and alumina is fused in a carbon crucible contained within one of plumbago, and set in a wind furnace. The inner crucible serves as the cathode of an electric current, while it is provided with a lid having two holes through one of which connection is made by a carbon rod with the crucible, through the other another carbon rod dips into the middle of the bath. The cover is banked up with loam and garden mould; the two carbon rods being protected from oxidation by passing through earthen tubes which pass through the arch of the furnace above. By using a current of 3 volts electro-motive force the alumina is electrolyzed, aluminium being deposited on the walls of the crucible and a corresponding amount of oxygen set free at the carbon anode, which is gradually consumed, thereby producing carbonic oxide. The bath must be replenished with alumina and the anode renewed from time to time. To form alloys, as of copper, the metal is melted in a carbon crucible by a voltaic arc, the positive pole being a movable carbon rod above

---

* U. S. Patent, 387876, August 14, 1888; English Patent, 7426, June 21, 1887; French Patent, 170003, April 15, 1887.

and the copper serving as the negative pole, connection being made with the crucible. When the copper is melted, alumina is introduced by degrees, without any flux. The intense heat fuses the alumina, and it is electrolyzed between the copper and the carbon anode above. The electrolyte is then liquid alumina, and it, as well as the copper cathode, is kept melted solely by the heat developed by the electric current. The alloy is tapped out and fresh materials added at suitable intervals without interruption to the process. A convenient strength of current for a crucible 20 centimetres deep by 14 centimetres in diameter inside, with a carbon anode 5 centimetres in diameter, is found to be 400 ampères, with an electro-motive force of 20 to 25 volts. An ampère-meter introduced into the circuit indicates the progress of the operation and the necessity for tapping or adding new material.

Of the two processes described in the above specification the first is very similar to that of Henderson, and to Hall's process (pp. 273 and 288), and has not been exploited as the second one has been. For some time the province of the " Heroult process" has generally been considered to be in producing aluminium alloys, which is the second part of the English patent and the whole subject of the United States patent referred to. We must therefore conclude that the process for producing pure aluminium has been abandoned, and our subsequent remarks will be concerned solely with the process for producing the alloys.

The Societé Metallurgique Suisse, which owned the patents, put up a plant on a commercial scale in July, 1888. It is said that this firm experimented some time with Dr. Kleiner's process, but abandoned it, about the middle of 1887, to try Heroult's process, and with such successful results that the plant about to be described was decided on. The instalment consisted of two large dynamos constructed especially for this work by the Oerliken Engineering Company, and directly coupled to a 300 horse-power Jonval turbine situated between them and mounted on a horizontal shaft. A separate dynamo of 300 ampères and 65 volts, driven by a belt from a pulley upon the main shaft, is used to excite the field magnets of the two large machines. These large dynamos were originally intended to give each a current of 6000 ampères at 20 volts electro-motive force when running at 180 revolutions per minute, but sufficient margin was allowed in the strength of the field to be able to work to 30 volts. They have even worked up to 35 volts on unusual occasions without any undue heating. It happens sometimes that the end of the anode touches the molten cathode, producing a short circuit, when the current will suddenly rise to from 20,000 to 25,000 ampères without, however, damaging the machine.

. The main conductors are naked copper cables, about 10 centimetres diameter, and no special precautions are taken to insulate them, since the current is of comparatively low potential, and a leakage of 100 ampères more or less in such a large current is too insignificant to take the trouble to avoid. An ampèremeter is placed in the main circuit, its dial being traversed by an index about 1 metre long, which is closely watched by the workman controlling the furnace.

The furnace or crucible first used consisted of an iron box cast around a carbon block, the iron, on contracting by cooling, securely gripping the surface of the carbon on all sides, and thus insuring perfect contact and conduction of the current from the cathode inside. This method was found only suitable for small crucibles, and the next furnace was built up of carbon slabs held together by a strong wrought-iron casing. The interior depth of the crucible was 60 centimetres, length 50 and breadth 35 centimetres, which would permit the introduction of the carbon anode and leave a clear space of 4 centimetres all around it horizontally. At the botton of the cavity is a passage to a tap-hole, D (Fig. 28), closed by the plug E, which is withdrawn from time to time to run off the alloy. The carbon anode, F, is suspended vertically above the crucible by pulleys and chains, which permit it to be raised and lowered easily and quickly. This anode is built up of large carbon slabs laid so as to break joints, and securely fastened together by carbon pins. The whole bar is 250 centimetres long with a section of 43 by 25 centimetres, and weighs complete 255 kilos. The conductor is clamped to the anode by means of the copper plates, G. The crucible is covered on top by carbon slabs, H, H, 5 centimetres thick, leaving an opening just large enough for the anode to pass through. The openings, J, J, closed by the lids, K, K, serve for introducing fresh copper and alumina.

The materials used have just been mentioned. Electrolytic or Lake Superior copper is used, the former being perhaps preferred if from a good manufacturer. The alumina is bought as commercial hydrated alumina, costing in Europe 22 francs per 100 kilos

Fig. 28.

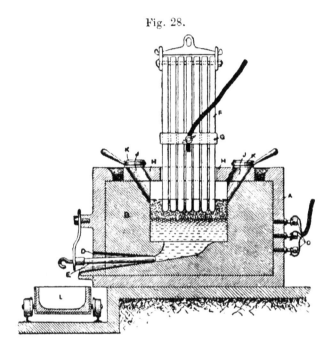

(2 cents per lb.). This is, of course, calcined before using, each 100 kilos furnishing about 65 kilos of alumina. Corundum can be substituted for the artificial alumina; some from North Carolina was tried, and is said to have given even more satisfactory results. Commercial beauxite has been used, but since it contains more or less iron its use is confined to the manufacture of ferro-aluminium. It is very cheap in Europe, and requires no other preparation than simple calcining.

The operation is begun by placing copper, broken into rather small pieces, in the crucible. The carbon anode is then approached to the copper, which is quickly melted by the current. The bath of fluid copper then becomes the negative pole, and ore is immediately fed into the crucible. It also is soon melted and floats on top of the copper. The electrolysis now proceeds, care being taken that while the anode dips into the molten ore, it does not

touch the molten cathode.   Particular stress is laid on the economy of keeping the distance between the electrodes small, the
reason given being that "the space between, being filled with a
layer of badly-conducting molten ore, offers a resistance which
increases with the distance; and although resistance is necessary,
in order that the current should produce heat, it is not economical
to have more heat than is necessary to melt the ore—the work of
separating the metal from the oxygen being chiefly done by the
electrolytic action of the current, and not by the high temperature."
In practice, this intervening space is not over 3 millimetres (one-
tenth of an inch).   The workman in charge, by watching the
indications of the ampèremeter, is enabled to maintain the anode
at its proper distance without any difficulty; it is proposed to do
this regulating automatically by means of an easily-constructed
electrical device.   The oxygen liberated gradually burns away
the anode, it being found that about 1 kilo of the anode is consumed for every kilo of aluminium produced.

After the operation commences, the alumina and metal are
introduced alternately in small quantities at frequent and regular
intervals, and the alloy is tapped out about every twelve hours.
The only wear to which the crucible is subjected is from the
accidental admission of small quantities of air; this waste is
scarcely appreciable.   All oxygen evolved from the bath is
evolved in contact with the anode and burns it.   When the anode
has worn down until too short for further use, it is replaced by
another, the pieces of carbon left being utilized for repairing,
covering or building up the crucible.   The operation is kept up
night and day, and it is generally more than a day after starting before the crucible is thoroughly up to its maximum heat
and work.   Two reliefs of five men each operate the plant, one
to superintend, one to prepare and dry the alumina, a third to
control the working of the crucible by working the anode, a
fourth to feed ore and metal into the crucible, and the fifth to
take care of the machinery, prepare anodes, crucibles, etc.   All
five work together to replace an anode or tap the crucible.   A
part of each tapping is analyzed, to determine its percentage of
aluminium.   It is the aim to produce as rich a bronze as possible
at the first operation (over 42 per cent. of aluminium has been

reached) and its subsequent dilution to any percentage desired is done in any ordinary smelting furnace.

The average current supplied the crucible is 8000 ampères and 28 volts, requiring an expenditure of a little over 300 horse-power in the turbine. Starting cold it required, in one instance, 36 hours to produce 670 kilos of aluminium bronze containing 18.3 per cent. or 122.67 kilos of aluminium. Taking the current as 300 electric horse-power, this would be a return of 11 grammes per hour or 0.264 kilos (0.6 lbs.) per day for each electric horse-power. It is claimed by Mr. Heroult that the furnace takes several days to attain its full efficiency, and that when it does so the above charge can be worked in 12 hours, which would triple the above production per horse-power. This claim is backed by figures as to 271 hours of actual operation, during which time the crucible cooled several times, but the average over the whole period was $22\frac{2}{3}$ grammes per hour or 0.544 kilos (1.2 lbs.) per day for each horse-power (163 kilos of aluminium per day, total production). During actual operation at full efficiency, Mr. Heroult claims to get 35 to 40 grammes of aluminium per horse-power-hour, which would mean 11 to 15 horse-power-hours per pound of aluminium, or 1.75 to 2.1 lbs. of aluminium per horse-power per day.

An idea of the percentage of useful effect derived from the current may be had very easily by considering that 1 electric horse-power = 750 Watts = 644.4 calories of heat per hour. (See p. 247.) As each gramme of aluminium evolves 7.25 calories in forming alumina, the production of 1 electric horse-power in 1 hour (if its energy were utilized solely for separating aluminium from oxygen) would be 88.88 grammes. Therefore the heat energy of the current is amply sufficient to account for all the alumina decomposed, leaving over the heat produced in the crucible by the union of oxygen with the carbon anode. Looking at the other side of the question, the electrolytic action, we can easily calculate from the strength of the current what it could perform. A current of 8000 ampères can liberate 2.68 kilos of aluminium per hour,* according to the fundamental law of electro-

---

* N. B. Only one furnace is used on the circuit.

deposition. If, then, from the figures given, there was actually produced 3.3 kilos and 6.8 kilos per hour, and 10.5 to 12 kilos are claimed when up to full efficiency, it is impossible that more than a fraction of the aluminium is produced by *electrolytic* decomposition of alumina, and the claim that the process is essentially electrolytic is without foundation. Similar calculations with the data given with regard to Cowles' process will lead to exactly similar conclusions, viz: that the absolute energy of the current, if converted into its heat equivalent, is many times more than sufficient to account for the decomposition of the alumina on thermal grounds, but the amount of current used will not suffice to explain the decomposition of the alumina as being electrolytic. Therefore, in both these processes the oxygen is abstracted from alumina by carbon, the condition allowing this to take place being primarily the extremely high temperature and secondarily the fluidity of the alumina. The presence of copper is immaterial, as is clearly shown in the Cowles process.

The Heroult process has been rapidly extended. In November, 1888, a syndicate was formed in Berlin, with a capital of $2,500,-000, which purchased the Heroult continental patents and has united with the former Swiss owners in forming the Aluminium Industrie Actien Gesellschaft, which has commenced to erect a very large plant in place of the former one, at Neuhausen. Dr. Kiliani has been made manager of the works, which are being rapidly completed, and will include, when finished, foundries and machine shops for casting and utilizing their product. The new plant will consist of 8 crucibles, capable of producing at least 10 tons of ten per cent. bronze in 24 hours.

In the beginning of 1889, the Societé Electro-Metallurgique of France, located at Froges (Isere), commenced to manufacture alloys by the Heroult process. Their plant consists of two turbines of 300 horse-power each, with two dynamos of 7000 ampères and 20 volts each. The output is estimated at 3000 kilos of alloys per day, which probably means 200 to 300 kilos of aluminium.

An experimental plant, under the direction of Mr. Heroult, was started in July, 1889, at Bridgeport, Conn., but the dynamo proved inadequate to the work required and was burnt out, stopping operations temporarily.   A dynamo was then ordered from the Oerliken Works, at Zurich, which arrived the following November, and another plant has been started at Boonton, N. J. The American company has not yet been incorporated.

CHAPTER XII.

REDUCTION OF ALUMINIUM COMPOUNDS BY OTHER MEANS THAN SODIUM OR ELECTRICITY.

No very exact classification of these numerous propositions can be made, since often many reducing agents are claimed in one general process.   Where such general statements are made, the method will be found under the most prominent reducing agent named, with cross references under the other headings.

REDUCTION BY CARBON WITHOUT THE PRESENCE OF OTHER METALS.

About the first attempt of this nature we can find record of, is the following article by M. Chapelle :—*

" When I heard of the experiments of Deville, I desired to repeat them, but having neither aluminium chloride nor sodium to use, I operated as follows : I put natural clay, pulverized and mixed with ground sodium chloride and charcoal, into an ordinary earthen crucible and heated it in a reverberatory furnace, with coke for fuel.   I was not able to get a white heat.   After cooling, the crucible was broken, and gave a dry pulverulent scoria in which were disseminated a considerable quantity of small globules about one-half a millimetre in diameter, and as white as silver. They were malleable, insoluble in nitric or cold hydrochloric acids, but at 60° dissolved rapidly in the latter with evolution

---

* Compt. Rendus, 1854, vol. xxxviii. p. 358.

of hydrogen; the solution was colorless and gave with ammonia a gelatinous precipitate of hydrated alumina. My numerous occupations did not permit me to assure myself of the purity of the metal. Moreover, the experiment was made under conditions which leave much to be desired, but my intention is to continue my experiments and especially to operate at a higher temperature. In addressing this note to the Academy I but desire to call the attention of chemists to a process which is very simple and susceptible of being improved. I hope before many days to be able to exhibit larger globules than those which my first experiment furnished."

M. Chapelle never did address any further communications to the Academy on this subject, and we must presume that further experiments did not confirm these first ones. The author was once called upon to examine a slag full of small, white, metallic globules, the result of fusing slate-dust in a similar manner to M. Chapelle's treatment of clay. They proved to be globules of siliceous iron reduced from the iron oxide present in the slate. It is not impossible that Chapelle's metallic globules were something similar in composition to these.

G. W. Reinar* states that the pyrophorous mass, which results from igniting potash or soda alum with carbon, contains a carboniferous alloy of aluminium with potassium or sodium, from which the alkaline metal can be removed by weak nitric acid.

The manager of an aluminium company in Kentucky claims to produce pure aluminium by a process which the newspapers state consists in smelting down clay and cryolite in a waterjacketed cupola reducing furnace, it being also stated that the aluminium is reduced so freely, and gathers under the slag so well, that it is tapped from the furnace by means of an ordinary syphon-tap. These are all the particulars which have been made public. As to whether this company really does make aluminium by any such process, I am unable to assert; about all that can be said further is that it has advertised its metal extensively at \$2 to \$3 per lb., and a sample of it sent to a friend of mine upon application was truly aluminium of fair quality.

* Wagner's Jahresb., 1859, p. 4.

O. M. Thowless,* of Newark, N. J., proposes to prepare a solution of aluminium chloride by dissolving precipitated aluminium hydrate in hydrochloric acid. The solution is concentrated and mixed with chalk, coal, soda, and cryolite, and the mass resulting heated in closed vessels to a strong, red heat. It is also stated that aluminium fluoride may be used instead of the chloride. The resulting fused mass is powdered and washed, when it is said that aluminium is obtained in the residue.

According to a patent granted to Messrs. Pearson, Liddon, and Pratt, of Birmingham,† an intimate mixture is made by grinding together—

> 100 parts cryolite.
> 50 " beauxite, kaolin or aluminium hydrate.
> 50 " calcium chloride, oxide or carbonate.
> 50 " coke or anthracite.

These are heated to incipient fusion in a carbon-lined furnace or crucible for two hours, when the aluminium is said to be produced and to exist finely disseminated through the mass. A mixture of 25 parts each of potassium and sodium chlorides is then to be added and the heat raised to bright redness, when the aluminium collects in the bottom of the crucible. A better utilization of the fine powder is effected by washing it, drying, and then pouring fused zinc upon it, which alloys with the aluminium and can be afterwards removed by distillation. If melted copper is used, a bronze is obtained.

REDUCTION BY CARBON AND CARBON DIOXIDE.

J. Morris,‡ of Uddington, claims to obtain aluminium by treating an intimate mixture of alumina and charcoal with carbon dioxide. For this purpose, a solution of aluminium chloride is mixed with powdered wood-charcoal or lampblack, then evaporated till it forms a viscous mass which is shaped into balls. During the evaporation hydrochloric acid is given off. The residue consists

---

* U. S. Patent, 370220, Sept. 20, 1887; English Patent, 14407 (1886).
† English Patent, 5316, April 10, 1888.
‡ Dingler, 1883, vol. 259, p. 86. German Pat., No. 22150, Aug. 30, 1882.

of alumina intimately mixed with carbon. The balls are dried, then treated with steam in appropriate vessels for the purpose of driving off all the chlorine, care being taken to keep the temperature so high that the steam is not condensed. The temperature is then raised so that the tubes are at a low red heat, and dry carbon dioxide, CO^2, is then passed through. This gas is reduced by the carbon to carbonic oxide, CO, which now, as affirmed by Mr. Morris, reduces the alumina. Although the quantity of carbonic oxide escaping is in general a good indication of the progress of the reduction, it is, nevertheless, not advisable to continue heating the tubes or vessels until the evolution of this gas has ceased, as in consequence of slight differences in the consistency of the balls some of them give up all their carbon sooner than others. The treatment with carbon dioxide lasts about thirty hours when the substances are mixed in the proportion of five parts carbon to four parts alumina. Morris states further that the metal appears as a porous spongy mass, and is freed from the residual alumina and particles of charcoal either by smelting it, technically "burning it out," with cryolite as a flux or by mechanical treatment.

REDUCTION BY HYDROGEN.

F. W. Gerhard* decomposes aluminium fluoride or cryolite by subjecting them to hydrogen at a red heat. The aluminium compound is placed in a number of shallow dishes of glazed earthenware, each of which is surrounded by a number of other dishes containing iron filings. These dishes are placed in an oven previously heated to redness, hydrogen gas is then admitted, and the heat increased. Aluminium then separates, hydrofluoric acid, HF, being formed, but immediately taken up by the iron filings and thereby prevented from reacting on the aluminium. To prevent the pressure of the gas from becoming too great, an exit tube is provided, which may be opened or closed at pleasure. This process, patented in England in 1856, No. 2920, is ingenious and

* Watts' Dictionary, article "Aluminium."

was said to yield good results. The inventor has, however, returned to the use of the more costly reducing agent, sodium,
which would seem to imply that the hydrogen method has not yet
quite fulfilled his expectations.

(See also Comenge's processes.)

Reduction by Carburetted Hydrogen.

Mr. A. L. Fleury,* of Boston, mixes pure alumina with gas-
tar, resin, petroleum, or some such substance, making it into a
stiff paste which may be divided into pellets and dried in an
oven. They are then placed in a strong retort or tube which is
lined with a coating of plumbago. In this they are exposed to
a cherry-red heat. The retort must be sufficiently strong to stand
a pressure of from 25 to 30 lbs. per square inch, and be so
arranged that by means of a safety valve the necessary amount
of some hydrocarbon may be introduced into the retort among
the heated mixture, and a pressure of 20 to 30 lbs. must be maintained. The gas is forced in by a force pump. By this process
the alumina is reduced, the metal remaining as a spongy mass
mixed with carbon. This mixture is remelted with metallic zinc,
and when the latter has collected the aluminium it is driven off
by heat. The hydrocarbon gas under pressure is the reducing
agent. The time required for reducing 100 lbs. of alumina, earth,
cryolite, or other compound of aluminium, should not be more
than four hours. When the gas can be applied in a previously
heated condition as well as being strongly compressed, the reduction takes place in a still shorter period.

Nothing is now heard of this process, and it has been presumably a failure. It is said that several thousand dollars were
expended by Mr. Fleury and his associates without making a
practical success of it. We should be glad to hear in the future
that their sacrifices have not been in vain, and that the process
still has possibilities in it which will some time be realized.

Petitjean† states that aluminium sulphide, or the double sul-

* Chemical News, June, 1869, p. 332.
† Polytechnisches Central Blatt., 1858, p. 888.

phide of aluminium and sodium (see p. 144), may be reduced by putting them into a crucible or retort, through the bottom of which is passed a stream of carburetted hydrogen. Some solid or liquid hydrocarbon may be placed in the bottom of the crucible. The aluminium is said to be thus separated from its combination with sulphur. The powder must be mixed with metallic filings, as iron, and melted, in order to collect the aluminium. Or, metallic vapor may be passed into the retort in place of carburetted hydrogen.

Messrs. Reillon, Montague, and Bourgerel* patent the production of aluminium sulphide (see p. 143) and its reduction by carburetted hydrogen exactly as above.

REDUCTION BY CYANOGEN.

According to Knowles' patent,[†] aluminium chloride is reduced by means of potassium or sodium cyanide, the former, either fused or in the form of vapor, being brought in contact with either the melted cyanide or its vapor. The patent further states that pure alumina may be added to increase the product. The proportions necessary are in general—

        3 equivalents of aluminium chloride.
        3-9   "   potassium or sodium cyanide.
        4-9   "   alumina.

Corbelli, of Florence,[‡] patented the following method in England: Common clay is freed from all foreign particles by washing, then well dried. One hundred grammes of it are mixed with six times its weight of concentrated sulphuric or hydrochloric acid; then the mixture is put in a crucible and heated to 400° or 500°. The mass resulting is mixed with 200 grammes of dry yellow prussiate of potash and 150 grammes of common salt, and this mixture heated in a crucible to whiteness. After cooling, the reduced aluminium is found in the bottom of the crucible as a button.

* English Patent 4576, March 28, 1887.
† Sir Francis C. Knowles, English Patent, 1857, No. 1742.
‡ English Patent, 1858, No. 142.

21

According to Deville's experiments, this process will not give any results. Watts remarks that any metal thus obtained must be very impure, consisting chiefly of iron.

Lowthian Bell* attempted to obtain aluminium in his laboratory by exposing to a high heat in a graphite crucible mixtures of alumina and potassium cyanide, with and without carbon. In no case was there a trace of the metal discovered.

REDUCTION BY DOUBLE REACTION.

M. Comenge,† of Paris, produces aluminium sulphide (see p. 143) and reduces it by heating it with alumina or aluminium sulphate in such proportions that sulphurous acid gas and aluminium may be the sole products. The mixture is heated to redness on the bed of a reverberatory furnace, in an unoxidizing atmosphere, the reaction being furthered by agitation. It is stated that the resulting mass may be treated in the way commonly used in puddling spongy iron and afterwards pressed or rolled together. The reactions involved would be, if they occurred,

$$Al^2S^3 + 2Al^2O^3 = 6Al + 3SO^2.$$
$$Al^2S^3 + Al^2(SO^4)^3 = 4Al + 6SO^2.$$

It is also claimed that metallic alloys may be prepared by the action of metallic sulphides on aluminium sulphate; as, for instance—

$$Al^2(SO^4)^3 + 3FeS = Al^2Fe^3 + 6SO^2.$$

The sulphide is also reduced by hydrogen, iron, copper or zinc, the reactions being

$$Al^2S^3 + 6H = 2Al + 3H^2S.$$
$$Al^2S^3 \quad 3Fe = 2Al + 3FeS.$$

In the case of reduction by a metal, alloys are formed.

Mr. Niewerth's‡ process may be operated in his newly invented furnace, but it may also be carried on in a crucible or other form

---

* Chemical Reactions in Iron Smelting, p. 230.
† English Patent, 1858, No. 461, under name of J. H. Johnson.
‡ Sci. Am. Suppl., Nov. 17, 1885.

of furnace. The furnace alluded to consists of three shaft furnaces, the outer ones well closed on top by iron covers, and connected beneath by tubes with the bottom of the middle one: the
tubes being provided with closing valves. These side shafts are
simply water-gas furnaces, delivering hot water-gas to the central
shaft, and by working the two alternately supplying it with a
continuous blast. The two producers are first blown very hot by
running a blast of air through them with their tops open, then
the cover of one is closed, the blast shut off, steam turned on just
under the cover, and water-gas immediately passes from the tube
at the bottom of the furnace into the central shaft. The middle
shaft has meanwhile been filled with these three mixtures in their
proper order :—

First. A mixture of sodium carbonate, carbon, sulphur and
alumina.

Second. Aluminium sulphate.

Third. A flux, preferably a mixture of sodium and potassium
chlorides.

This central shaft must be already strongly heated to commence
the operation, it is best to fill it with coke before charging, and as
soon as that is hot to put the charges in on the coke. Coke may
also be mixed with the charges, but it is not necessary. The
process then continues as follows : The water-gas enters the bottom
of the shaft at a very high temperature. These highly heated
gases, carbonic oxide and hydrogen, act upon the charges so that
the first breaks up into a combination of sodium sulphide and
aluminium sulphide, from which, by double reaction with the
second charge of aluminium sulphate, free aluminium is produced.
As the latter passes down the shaft, it is melted and the flux
assists in collecting it, but is not absolutely necessary. Instead
of producing this double sulphide, pure aluminium sulphide
might be used for the first charge, or a mixture which would generate it; or again pure sulphide of sodium, potassium, copper, or
any other metallic sulphide which will produce the effect alone, in
which case aluminium is obtained alloyed with the metal of the
sulphide. Instead of the first charge, a mixture of alumina, sulphur and carbon might be introduced. Or the aluminium sulphate of the second charge might be replaced by alumina. So

one charge may be sulphide of sodium, potassium or any other metallic sulphide, and the second charge may be either alumina or aluminium sulphate.

Messrs. Pearson, Turner, and Andrews* claim to produce aluminium by heating silicate of alumina or compound silicates of alumina and other bases with calcium fluoride and sodium or potassium carbonate or hydrate, or all of these together. If other metals are added, alloys are obtained.

REDUCTION IN PRESENCE OF OR BY COPPER.

Calvert and Johnson† obtained copper alloyed with aluminium by recourse to a similar chemical reaction to that employed to get their iron-aluminium alloy. Their mixture was composed of—

20 equivalents of copper	. . . .	. 640 parts.
8 (24) "	aluminium chloride .	. . 1076 "
10 "	lime	. . 280 "

" We mixed these substances intimately together, and after having subjected them to a high heat for one hour we found at the bottom of the crucible a melted mass covered with cuprous chloride, Cu^2Cl^2, and in this mass small globules, which on analysis contained 8.47 per cent. aluminium, corresponding to the formula—

5 equivalents of copper .	. .	160	91.96 per cent.
1 "	aluminium .	. 14	8.04 "
			100.00

" We made another mixture of aluminium chloride and copper in the same proportions as above, but left out the lime. We obtained an alloy in this case also, which contained 12.82 per cent. aluminium, corresponding to the formula—

3 equivalents of copper .	. .	96	87.27 per cent.
1 "	aluminium .	. 14	12.73 "
			100.00

---

* English Patent, 12332, Sept. 12, 1887.
† Phil. Mag. 1855, x. 242.

M. Evrard,* in order to make aluminium bronze, makes use of an aluminous pig-iron. (It is not stated how this aluminous pig-iron is made.) This is slowly heated to fusion, and copper is added to the melted mass. Aluminium, having more affinity for copper than for iron, abandons the latter and combines with the copper. After the entire mass has been well stirred, it is allowed to cool slowly so as to permit the bronze, which is heavier than iron, to find its way to the bottom of the crucible. M. Evrard makes silicon bronze in the same way by using siliceous iron.

Benzon† has patented the reduction of aluminium with copper, forming an aluminium-copper alloy. He mixes copper, or oxidized copper, or cupric oxide, in the finest possible state, with fine, powdered, pure alumina and charcoal, preferably animal charcoal. The alumina and copper or copper oxide are mixed in equivalent proportions, but an excess of charcoal is used. The mixture is put in a crucible such as is used for melting cast-steel, which is lined inside with charcoal. The charge is covered with charcoal, and the crucible subjected first to a temperature near the melting point of copper, until the alumina is reduced, and then the heat is raised high enough to melt down the alloy. In this way can be obtained a succession of alloys, whose hardness and other qualities depend on the percentage of aluminium in them. In order to obtain alloys of a certain composition, it is best to produce first an alloy of the highest attainable content of aluminium, to analyze it, and then melt it with the required quantity of copper. The same process can be used for the reduction of alumina with iron or ferric oxide, only the carbon must in this case be in greater excess, and a stronger heat kept up longer must be used than when producing the copper-aluminium alloy. In contact with ferric oxide the alumina is more easily reduced than with metallic iron.

Benzon further remarks that some of these alloys, as the ferro-aluminium, may be subsequently treated so as to separate out the metallic aluminium; also that the iron alloy may be mixed with steel in the melting pot, or suitable proportions of alumina and carbon may be put into the melting pot. The iron alloy may be

* Annales du Genie Civil, Mars, 1867, p. 189.
† Eng. Pat. 1858, No. 2753.

useful for many purposes, especially in the manufacture of cast-steel.

The question opened up by Benzon's statements is whether carbon reduces alumina in presence of copper. This has been the subject of many careful experiments, and the verdict of the most reliable observers is that at ordinary furnace temperatures it does not. This principle has been the subject of numerous patents, and before presenting the negative evidence on this point we will review the claims made in these patents.

G. A. Faurie[*] states that he has succeeded in obtaining aluminium bronze by taking two parts of pure, finely-powdered alumina, making it into a paste with one part of petroleum and then adding one part of sulphuric acid. When the yellow color is uniform and the mass homogeneous, sulphur dioxide begins to escape. The paste is then wrapped up in paper and thrown into a crucible heated to full redness, where the petroleum is decomposed. The calcined product is cooled, powdered and mixed with an equal weight of a metal in powder, e. g., copper. This mixture is put into a graphite crucible and heated to whiteness in a furnace supplied with blast. Amidst the black, metallic powder are found buttons of aluminium alloy. In an English patent,[†] by Mr. Faurie, it is further claimed that by making bricks out of the calcined alumina mixture and alloying metal, and using similar bricks of lixiviated soda ashes mixed with tar for flux, the reduction can be affected in a cupola.

Bolley,[‡] at his laboratory in Zurich, and List,[§] at the royal foundry at Augsburg, have shown that by following the process claimed by Benzon the resulting copper contained either no aluminium, or at most a trace. In an experiment made by the author to test this point—

                40 grammes of copper oxide and copper,
                 5   "   alumina,
                 5   "   charcoal,

---

[*] Comptes Rendue, 105, 494, Sept. 19, 1887.
[†] English Patent, 10043, Aug. 18, 1887.
[‡] Schweizer Polytechnisches Zeitschrift, 1860, p. 16.
[§] Wagner's Jahresbericht, 1865, p. 23.

were intimately mixed and finely powdered, put in a white-clay crucible and covered with cryolite. The whole was slowly heated to bright redness, and kept there for two hours. A bright button was found at the bottom of the crucible. This button was of the same specific gravity as pure copper, and a qualitative test showed no trace of aluminium in it. A friend of mine, Dr. Lisle, has repeated this experiment, taking the metal produced and returning it to another operation and repeating this four times, but the resulting button scarcely showed a trace of aluminium.

Dr. W. Hampe has lately made an exhaustive test of this subject with the following conclusions :—[*]

"The reduction of alumina by carbon, although often patented, is on thermo-chemical grounds highly improbable, but since aluminium in alloying with copper, especially in the proportions 9.7 parts of the former to 90.3 parts of the latter ($AlCu^4$), evolves much heat, it might be possible that the reaction

$$Al^2O^3 + 3C + 8Cu = 2AlCu^4 + 3CO$$

is exothermic. I therefore mixed alumina with the necessary quantity of lamp-black and copper, in other cases evaporated together to dryness solutions of aluminium and copper nitrates, afterwards igniting them to oxides and adding the necessary amount of carbon. These mixtures were put into gas-carbon crucibles contained within plumbago pots with well-luted covers, and heated in a Deville blast-furnace to a temperature sufficient to frit together the quartz sand with which the space between the two crucibles had been filled. In no case was there a trace of aluminium produced, nor did the addition of any flux for the alumina affect the result in any way."

The possibility of reducing aluminium sulphide by copper has been generally decided affirmatively. M. Comenge claimed that it was possible (see p. 322), Reichel[†] also stated unreservedly that copper filings performed the reduction at a high temperature. In an experiment by the author, copper foil was used instead of copper filings, the latter not being immediately at hand, and the

[*] Chemiker Zeitung (Cöthen), xii. p. 391 (1888).
[†] Journal für Pr. Chemie, xi. p. 55.

result was negative. As a similar test with iron filings gave a good result, it seems quite probable that copper would have performed the reduction under proper conditions.

Andrew Mann,* of Twickenham, patents a process which may be stated briefly as follows: Aluminium sulphate is mixed with sodium chloride and heated until a reaction begins to take place. The mass is mixed intimately with lime, and to this mixture aluminium sulphate and ground coke added. This is calcined, the powder mixed with a metal, as copper, and melted down. In this case the slags are calcium sulphide and copper chloride, while aluminium bronze is obtained.

L. Q. Brin,† of Paris, claims to produce aluminium bronze by the following process: Sheet copper is cleaned by pickling, and then covered with a mixture of 2 parts borax, 2 parts common salt, and 1 part sodium carbonate, made into a paste with water. The metal is then put into a reverberatory furnace, heated to bright redness and vapors of aluminium chloride led over it, carried in by a current of inert gas. (It is stated that the vapors of aluminium chloride are produced by heating in a retort a mixture of clay, salt, and fluorspar.) The aluminium compound is said to be decomposed, and the nascent aluminium to combine with the copper forming $1\frac{1}{2}$ to 2 per cent. bronze at one operation, and by using this over it may be enriched to any extent desired. In a modification of this method, the coating put on the metal contains clay or other earth rich in alumina. It is also stated that the metal thus coated can be put into a cupola with alternate layers of fuel and run down to an alloy.

REDUCTION BY OR IN PRESENCE OF IRON.

M. Comenge claims that aluminium sulphide is reduced by iron (see p. 322); the statement is repeated by a writer in the "Chemical News," 1860; F. Lauterborn‡ states that the reduction takes place at a red heat; Reichel§ also records his success

---

* English Patent, 9313, June 30, 1887; German Patent, 45755, Dec. 20, 1887.

† English Patents, 3547-8-9, March 7, 1888; U. S. Patent, 410574, Sept. 10, 1889.

‡ Dingler, 242, 70.   § Jrnl. für Pr. Chemie, xi. 55.

in this reaction; finally, the author has obtained encouraging results. I used a product containing 32.3 per cent. of aluminium sulphide. On mixing this intimately with fine iron filings, and subjecting to a high heat for one and a half hours, the product was a loose powder in which were small buttons of metal. They were bright, yellower than iron, and contained by analysis 9.66 per cent. of aluminium.

H. Niewerth* has patented the following process: "Ferro-silicum is mixed with aluminium fluoride in proper proportions and the mixture submitted to a suitable red or melting heat by which the charge is decomposed into volatile silicon fluoride (SiF^4), iron and aluminium, the two latter forming an alloy. In order to obtain the valuable alloy of aluminium and copper from this iron-aluminium alloy, the latter is melted with metallic copper, which will then by reason of greater affinity unite with the aluminium, while the iron will retain but an insignificant amount of it. On cooling the bath, the bronze and iron separate in such a manner that they can readily be kept apart. In place of pure aluminium fluoride, cryolite may advantageously be employed, or aluminium chloride may also be used, in which case silicon chloride volatilizes instead of the fluoride. Or, again, pure silicon may be used with aluminium fluoride, cryolite, or aluminium chloride, in which case pure aluminium is obtained."

Mr. W. P. Thompson† has taken out a patent in England‡ for the manufacture of aluminium and similar metals, which is carried out as follows: The inventor employs as a reducing agent iron, either alone or conjointly with carbon or hydrogen. The operation is effected in an apparatus similar to a Bessemer converter, divided into two compartments. In one of these compartments is placed melted iron, or an alloy of iron, which is made to run into the second by turning the converter. This last compartment has two tuyeres, one of which serves to introduce hydrogen, while by the other is introduced either aluminium chloride, fluoride, double chloride or double fluoride with sodium, in liquid or gaseous state. In presence of the hydrogen, the iron

---

* Sci. Am. Suppl., Nov. 17, 1883.
† Bull. de la Soc. Chem. de Paris, 1880, xxiv. 719.
‡ March 27, 1879, No. 2101.

takes up chlorine or fluorine, chloride or fluoride of iron is dis-engaged, and aluminium mixed with carbon remains as a residue. Then this mixture of iron, aluminium and carbon is returned to the other compartment where the carbon is burnt out by means of a current of air. The mass being then returned to the chamber of reduction, the operation described is repeated. When almost all the iron has been consumed, the reduction is terminated by hydrogen alone. There is thus obtained an alloy of iron and aluminium. (The preparation of sodium does not require the intervention of hydrogen. A mixture of iron with an excess of carbon and caustic soda (NaOH) is heated in the converter, when the sodium distils off. When all the carbon has been burnt, the iron remaining as a residue may be converted into Bessemer steel. As iron forms an alloy with potassium, the method would scarcely serve for the production of that metal.) To obtain the pure aluminium, sodium is first prepared by the process indicated, the chloride or fluoride of aluminium is introduced into the apparatus in the other chamber, when the metal is reduced by the vapor of sodium. The chambers ought to be slightly inclined, and an agitator favors the reaction. The inventor intends to apply his process to the manufacture of magnesium, strontium, calcium and barium.

Calvert and Johnson* made experiments on the reduction of aluminium by iron, and the production thereby of iron-aluminium alloys. We give the report in their own words :—

"We shall not describe all the fruitless efforts we made, but confine ourselves only to those which gave satisfactory results. The first alloy we obtained was by heating to a white heat for two hours the following mixture :—

8 equivalents of aluminium chloride	.		.		.	1076 parts.		
40 "	iron filings		.		.	.	1120 "	
8 "	lime	.	.	.	.	.	224 "	

"The lime was added to the mixture with the view of removing the chlorine from the aluminium chloride, so as to liberate the metal and form fusible calcium chloride, $CaCl^2$. Subtracting the lime from the above proportion, we ought to have obtained an

* Phil. Mag., 1855, x. 240.

alloy having the composition of 1 equivalent of aluminium to 5 equivalents of iron, or with 9.09 per cent. of aluminium. The alloy we obtained contained 12 per cent., which leads to the formula $AlFe^4$. This alloy, it will be noticed, has an analogous composition to the one we made of iron and potassium, and like it was extremely hard, and rusted when exposed to a damp atmosphere. Still it could be forged and welded. We obtained a similar alloy by adding to the above mixture some very finely pulverized charcoal and subjecting it to a high heat in a forge furnace for two hours. This alloy gave on analysis 12.09 per cent.[*] But, in the mass of calcium chloride and carbon remaining in the crucible there was a large amount of globules varying in size from a pin-head to a pea, as white as silver and extremely hard, which did not rust in the air or in hyponitric fumes. Its analysis gave 24.55 per cent. aluminium; the formula Al^2Fe^3 would give 25 per cent. Therefore this alloy has an analogous composition to alumina, iron replacing oxygen. We treated these globules with weak sulphuric acid, which removed the iron and left the aluminium, the globules retaining their form, and the metal thus obtained had all the properties of the pure aluminium.

"We have made trials with the following mixture, but although they have yielded results, still they are not sufficiently satisfactory to describe in this paper, which is the first of a series we intend publishing on alloys. This mixture was :—

Kaolin	1750 parts.
Sodium chloride	1200 "
Iron	875 "

"From this we obtained a metallic mass and a few globules which we have not yet analyzed."

(See also Benzon's process, p. 325.)

M. Chenot,[†] on the occasion of Deville's first paper on aluminium being read to the French Academy, Feb. 6, 1854, announced that in 1847, by reducing earthy oxides by means of metallic

---

[*] In the original paper it is given as 12.09 per cent. iron. The inference is unavoidable that this was a misprint, but it is not corrected in the Errata at the end of the volume.

[†] Comptes Rendue, xxxviii. 415.

sponges, he had obtained a series of alloys containing up to 40 per cent. of the earth metals. He cited from a memoir presented by him to the "Société d'Encouragement" in 1849, in which he had said, "on taking precipitates of the earths, they are all reduced by the metallic sponge (e. g., that formed by reducing iron oxide in a current of carbonic oxide gas). In this manner I have made barides, silicides, aluminides, etc., all of which are beautiful silver-white, very hard and unoxidizable in air or in contact with acid vapors. They are fusible, can be cast and work perfectly under the hammer."

Faraday and Stodart[*] made an exhaustive investigation on the preparation of iron-aluminium alloys, being started on this line by finding that Bombay "wootz" steel contained 0.0128 to 0.0695 per cent. of aluminium, while no metals of the earths were to be found in the best English steels. This led to the conclusion that the peculiar properties of the former, especially the "damasceening," were due to the small amount of aluminium. These scientists commenced by taking pure steel or sometimes good soft iron and intensely heating it for a long time imbedded in charcoal powder. Carbides were thus formed, having a very dark gray color, and highly crystalline. Average analysis of this product gave 5.64 per cent. carbon. This was broken and powdered in a mortar, mixed intimately with pure alumina and heated in a closed crucible for a long time at a high temperature. An alloy was obtained of a white color, close granular texture and very brittle, containing 3.41 per cent. of aluminium, with some carbon.

When 40 parts of this alloy were melted with 700 parts of good steel (introducing 0.184 per cent. of aluminium) a malleable button was obtained which gave a beautiful damask on treatment with acids; while 67 parts of the alloy with 500 of steel (introducing 0.4 per cent. of aluminium) gave a product which forged well, gave the damask and "had all the appreciable characters of the best Bombay wootz." This appears to be very strong synthetic evidence that alumina is reduced to a small extent even in the rude hearths in which the Indian steel is manufactured. Karsten, however, could not find weighable quantities of alu-

---

[*] Quarterly Journal, ix. 320.

minium in specimens of wootz, nor could Henry, a very expert analyst. The latter suggested that Faraday was misled by the alumina contained in intermingled slag, yet the latter obtained alumina without silica in his analyses. In the light of more recent developments we would accept Faraday's results as being very near the truth in the matter.

Ledebuhr* quotes an analysis made by Grüner in which 0.50 per cent. of aluminium was found in cast-iron containing besides 2.30 per cent. of carbon and 2.26 per cent. of silicon. This would tend to show that under certain conditions iron takes up aluminium in the blast-furnace. Karsten, however, in his many analyses of malleable iron, steel and cast-iron only found aluminium in unweighable quantities. Grüner and Laut† stated that aluminium is reduced in small quantities in the blast furnace if the temperature is high and the slag basic; a large addition of lime thus increases the reduction of alumina and hinders that of silica. Most pig irons contain very small amounts of aluminium, but some English varieties contain 0.5 to 1.0 per cent. and several Swedish pig irons 0.75 per cent. Schafhäutl‡ found as much as 1.01 per cent. of aluminium in a grey iron, and was led to consider silicide of iron and aluminide of iron as characteristic components of grey iron. Lohage§ states that adding alumina in the manufacture of cast-steel has a great influence on the grain and lustre of the steel, the effect being doubtless due to a minute quantity of aluminium taken up. Silicates of magnesium and aluminium are formed at the same time and separating out float on the surface of the molten steel. Corbin‖ reports 2.38 per cent. of aluminium in chrome steel, but Blair,¶ of Philadelphia, found no more aluminium in chrome than in other steels. This chemist has examined many irons and steels particularly for aluminium, and reports that nearly always it exists as such in steel, but never more than a few thousandths of a per cent., say 0.032

* Handbuch der Eisenhüttenkunde, p. 265.
† Berg u. Hüttenmännische Zeitung, 1862, p. 254.
‡ Erdman's Journal fr. Pr. Chemie, lxvii. 257.
§ Berg u. Hüttenmännische Zeitung, 1861, p. 160.
‖ Silliman's Journal, 1869, p. 348.
¶ H. M. Howe, E. and M. J. Oct. 29, 1887.

per cent. as a maximum. He has further been unable to connect its presence with any peculiarity in the properties of the metal or its mode of manufacture.

G. H. Billings,[*] of the Norway Iron Works, Boston, made the following experiment on reducing alumina in contact with iron :— A soft iron was used containing a trace of sulphur and phosphorus, no manganese and only 0.08 per cent. of carbon. The mixture was made of

    12 parts emery.
    18 "   alumina.
     1 "   pulverized charcoal.
    36 "   fine iron turnings.

These were mixed thoroughly, and heated to whiteness for 48 hours. The metal resulting showed a solid, homogeneous fracture with a fine crystalline structure resembling steel with 1 per cent. of carbon, and contained on analysis

    0.20 per cent. of carbon.
    0.50 "   aluminium.

It was also found that if this quantity of aluminium was added to a pot of molten iron the product obtained exhibited the same characteristics as the above.

Another attempt to produce iron-aluminium alloys directly is stated in E. Cleaver's patent specifications as follows :[†] Four parts of aluminium sulphate in solution are mixed with one part of lamp-black, the mixture dried and heated to the highest temperature attainable by using coal-gas and oxygen in a lime-lined furnace similar to those used for melting platinum. Excess of reducing gas is maintained. The charge is cooled in the furnace, removed, mixed with twenty times its weight of finely-divided cast-iron, and fused in a steel melting furnace. If copper is used, a bronze results. The alloying metal may be added in the gas furnace, but this is not recommended as economical. This inventor also claims that aluminium ferrocyanide, either alone or with carbon, can be decomposed in the above-described gas furnace, yielding a rich iron-aluminium alloy. As a higher heat

---

[*] Transactions American Inst. Mining Engineers, 1877, p. 452.
[†] English Patent, 1276, Jan. 26, 1887.

than before is needed, it is recommended that the oxygen be previously heated. The principal difficulty in this latter process would apparently be to procure the aluminium ferrocyanide to operate on.

Mr. Ostberg,* connected with the Mitis process for making wrought-iron castings, stated that the ferro-aluminium used in that process in Sweden was made by adding clays to iron in process of smelting, that it contained 7 to 8 per cent. of aluminium, and could be made very cheaply. Inquiries made for further particulars about this process have received no satisfactory reply, and there is no outside confirmation of the above statement to be found.

Brin Bros. claim that they can alloy aluminium in small quantities with iron (see p. 328). Besides the processes described as most suitable for producing bronze, they also state that if soft strap-iron is coated with the flux composed of clay and salt and heated to over 1000° C. in a muffle or a blowpipe flame, the iron absorbs aluminium and becomes tough and springy, having many of the properties of steel. They also claim that by simply charging broken lumps of cast-iron into a cupola with alternate layers of common clay and a flux, the metal run down contains as much as 1.75 per cent. of aluminium, yielding a very fluid, strong iron, which runs into the thinnest castings. The London papers state that the alloys thus produced assuredly contain aluminium, and that the contained aluminium does not cost over 25 cents per lb.

A newspaper report speaks of exactly similar processes being operated by an aluminium company in Kentucky. (See also p. 317.) It is said that they charge a cupola with scrap-iron, pig-iron, coke, clay, and a flux, and that on melting the charge down and pouring, the castings produced are similar to the best steel, the fracture of the metal being white, slightly fibrous and free from blow-holes. It it stated, further, that the castings, on analysis, contained 1.7 per cent. of aluminium. Scrap-iron is also treated in the same way as reported by Brin Bros., being simply coated with a pasty mixture of clay and a flux and heated almost white-hot, when the iron absorbs aluminium.

* .Eng. and Mining Journal, May 15, 1886.

The Aluminium Process Company, of Washington, D. C., own several patents granted to W. A. Baldwin, of Chicago, Ill. In one of these,* a bath is formed by fusing together 4 parts of ground clay, 12 parts of common salt and 1 part of charcoal powder. The metal to be alloyed, e. g., an iron bar, is thrust into the bath, which is not hot enough to melt it, and allowed to remain some time, with occasional stirring, until the alloying is complete. In the case of metals with low fusing points the metal may be melted with the mixture. It is claimed that the metal takes up a small percentage of aluminium. With a more highly aluminous material than clay, the proportions of salt and carbon are to be increased proportionately. In a modification of this process adopted for foundry practice,† a mixture of clay, salt and carbon, similar to the above, is put into a large ladle and the molten iron tapped directly from the cupola on to it. A brisk stirring up of the iron takes place, much scum rises to the surface, and the resulting iron is more fluid, can be carried further before setting, and makes sounder and stronger castings than similar iron not treated. The resulting iron does not contain enough aluminium to be detected by quantitative analysis. Old-fashioned fluxes used long ago in foundries were similar in composition to this mixture used by Baldwin, and were found efficacious in freeing dirty iron from slag and other impurities. It is hard to see how this latter process is anything but the use of a common flux, with a different explanation as to how it acts— the explanation being probably the most questionable part of the whole. The first-mentioned process, however, has the merit of novelty, and pieces of poor iron treated by it are made springy and much like steel, but whether this is due to absorption of aluminium is doubtful.

The Williams Aluminium Company, of New York City (works at Newark, N. J.), manufacture an alloy which they call aluminium-ferro-silicon and sell for foundry use. A year or more ago this alloy was represented to contain 10 per cent. of aluminium, but several analyses disproved this and recently the alloy has

---

* English Patent, 2584, Feb. 21, 1888.

† U. S. Patent, 380161, March 27, 1888.

been sold simply on the guarantee of what it will accomplish. The metal at present sold by this company does contain a small amount of aluminium, and its action on poor foundry iron is similar to that of other brands of ferro-aluminium; therefore, as long as no certain percentage of aluminium is now claimed, the company is certainly doing a legitimate business. (For method of using, etc., see Chap. XVI.) The alloy is made by melting down a mixture of iron filings, clay, salt, charcoal, and another flux whose composition is not divulged. This is put into cast-iron pots and the whole charge, crucibles and all, run down in a furnace of peculiar design constructed by Mr. Williams. The capacity of the plant is about 1000 lbs. of alloy a day, which is broken by small stamps into pieces of about an inch diameter and sold at 10 cents per lb. Mr. Williams is at present experimenting on manufacturing aluminium bronze by the same methods, and a sample piece recently forwarded the author has a very promising appearance.

Reduction by or in Presence of Zinc.

M. Bekétoff,[*] was not able to reduce vapor of aluminium chloride by vapor of zinc, although silicon chloride under the same conditions was readily reduced.

M. Dullo[†] observes that the double chloride of aluminium and sodium, which he makes directly from clay, may be reduced by zinc. He says, "the reduction by zinc presents no difficulties, but it is less easy than with sodium. An excess of zinc should be employed, which may be got rid of afterwards by distillation. The metal thus prepared possesses all the characteristics and all the properties of that obtained from beauxite with sodium."

M. N. Basset,[‡] a chemist in Paris, patented a somewhat similar process for obtaining aluminium. If the statements are correct they are of great value. The paper is as follows: "All the metalloids and the metals which form by double decomposition proto-chlorides or sesqui-chlorides more fusible or more soluble

[*] Bulletin de la Societé Chemique, 1857, p. 22.
[†] Bull. de la Soc. Chem., 1860, v. 472.
[‡] Le Genie Industriel, 1862, p. 152.

than aluminium chloride may reduce it or even aluminium-sodium chloride. Thus, arsenic, bismuth, copper, zinc, antimony, mercury, or even tin or amalgam of zinc, tin, or antimony may be employed to reduce the single or double chloride. The author employs zinc in preference to the others in consequence of its low price, the facility of its employment, its volatility, and the property which it has of metallizing easily the aluminium as it is set free. When metallic zinc is put in the presence of aluminium-sodium chloride, at 250 to 300°, zinc chloride, $ZnCl^2$, is formed and aluminium is set free. This dissolves in the zinc present in excess, the zinc chloride combines with the sodium chloride, and the mass becomes little by little pasty, then solid, while the alloy remains fluid. If the heat is now raised, the mass melts anew, the zinc reduces a new portion of the double chloride and the excess of zinc enriches itself in aluminium proportionately. These facts constitute the basis of the following general process: One equivalent of aluminium chloride is melted, two of sodium chloride added, and when the vapors of hydrochloric acid are dissipated, four equivalents of zinc, in powder or grain, are introduced. The zinc melts rapidly, and by agitation the mass of chloride thickens and solidifies. The mass is now composed of the chlorides of aluminium, zinc, and sodium, and remains in a pasty condition on top of the fluid zinc containing aluminium. This pasty mass is removed, piled up in a crucible or in a furnace, and bars of the fluid alloy of zinc and aluminium obtained from a previous operation are placed on top of it. This is gradually heated to bright redness, and kept there for an hour. The melted mass is then stirred with a rake and poured out. It is an alloy of the two metals in pretty nearly equal proportions. This alloy, melted with some chloride from the first operation, furnishes aluminium containing only a small per cent. of zinc, which disappears by a new fusion under aluminium chloride mixed with a little fluoride, providing the temperature is raised to a white heat and maintained till the cessation of the vapors of zinc, air being excluded.

"The metal is pure if the zinc employed contained no foreign materials or metals. It is melted and cast into ingots. In case the zinc contains iron, or even if the aluminium chloride contains some, the metallic product of the second operation may be treated with dilute sulphuric acid to remove it. The insoluble residue is washed and melted layer by layer with fluorspar or cryolite and a small quantity of aluminium-sodium chloride, intended solely to help the fusion."

Mr. Wedding,* makes the following remarks on this process :—

"It is some time since Mr. Basset established the possibility of replacing sodium by zinc in the manufacture of aluminium. Operating on aluminium-sodium chloride with granulated zinc, the reduction takes place towards 300°. The reduced aluminium dissolves in the excess of zinc, while the zinc chloride formed combines with the sodium chloride, forming a pasty mass if the heat is not raised. Under the action of heat the alloy enriches itself in aluminium, because the zinc volatilizes. The zinc retained by this alloy is completely eliminated by fusion with double chloride and a little fluorspar. The temperature ought to be pushed at last to a white heat, and maintained till no vapor of zinc escapes, air being excluded during the operation. These results I have confirmed, having submitted the experiments of Mr. Basset to an attentive examination, and I recommend its use. However, the process demands very much precaution because of the high temperature which it necessitates. Another chemist, Mr. Specht, even in 1860 decomposed aluminium chloride by zinc, and has the same report to make—that he thinks the process will be some time advantageously practised on a large scale."

The author made the following experiment to determine if cryolite would be reduced by zinc : One pound of finely powdered cryolite was melted in a graphite crucible and 6 ounces of granulated zinc dropped into it. No perceptible reaction took place except the volatilization of zinc when the crucible was uncovered, and the metal obtained after 15 minutes' treatment contained on analysis 0.6 per cent. of aluminium.

Mr. Fred. J. Seymour† patented the reduction of aluminium by zinc, making the following claim : An improvement in extracting aluminium from aluminous earths and ores by mixing them with

* Journal de Pharm. [4] iii. p. 155 (1866).
† U. S. Pat., No. 291631, Jan. 8, 1884.

an ore of zinc, carboniferous material and a flux, and subjecting the mixture to heat in a closed retort, whereby the zinc is liberated, is caused to assist in bringing or casting down the aluminium in a metallic state, and the alloy of aluminium and zinc is obtained.

A furnace was put up in the early part of 1884, somewhere in the vicinity of Cleveland, in a description of which by a newspaper correspondent we are told that steel retorts were charged with a mixture of zinc ore 100 parts, kaolin 50, carbon (either anthracite coal or its equivalent of some hydrocarbon) 125, pearlash 15, common salt 10 ; the heat necessary being about 1400° C. In a second patent* Mr. Seymour claimed that by heating the same mixture in a retort and introducing air he volatilized oxides of aluminium and zinc, which were caught in a condenser, mixed with carbon and reduced in a crucible. Immediately after the issue of this patent, the American Aluminium Company was organized in Detroit with a capital stock of $2,500,000, to operate the patents of a Dr. Smith, under whose name processes similar to the above had been patented in Great Britain and France. A works was then started at Findlay, Ohio, using natural gas for fuel. A gentleman who saw the plant described it as a reverberatory furnace, into which a charge of 800 lbs. of zinc ore, 900 of native alum, and 300 of charcoal was put. On heating very strongly by gas with plenty of air admitted, zinc oxide was volatilized (Mr. Seymour claimed that it carried alumina with it) and was condensed by passing the gases through large copper condensers. This fume was then collected, mixed with carbon and a metal, and run down in a crucible to an alloy. It was claimed that the plant had a capacity of 600 lbs. of pure aluminium a day, and had presumably been in operation over a year, yet there was not over 20 lbs. of alloys or $\frac{1}{2}$ lb. of pure aluminium to show for it. On closer inspection so plain indications of fraud were visible in the last part of the operation that, although veritable aluminium alloys were taken out of the crucible, the gentleman referred to refused to believe that the aluminium was produced in the process. As the sequel to this it can be stated that in the middle of 1889 the executive committee for the stockholders, being fully satisfied

* U. S. Patent, 337996, March 16, 1886.

of the worthlessness of the process, called for a meeting to wind up the company. One month later Mr. Seymour died. The story went the rounds of the daily press that the one metallurgist who commanded the secret of obtaining cheap aluminium had died taking the talisman with him, and a vivid picture was drawn of the manner in which the secret had been preserved by means of twelve-foot palisades, doubly-bolted doors, and by working at the midnight hour. Alas! the secret was out one month before he died.

A method of reducing cyanide of aluminium by means of zinc is the subject of a patent granted F. Lauterborn.* Fourteen parts of aluminium sulphate dissolved in twice its weight of water is precipitated by thirteen parts of ferro-cyanide of potassium dissolved in four times its weight of water; the precipitate of ferro-cyanide of aluminium is collected and dried. This substance is then mixed with slightly less than one-half its weight of dry, anhydrous sodium carbonate, put in a crucible and ignited with as little admission of air as possible. The ferro-cryanide is decomposed, iron carbide separates out, and, besides sodium cyanate, the double cyanide of aluminium and sodium ($Al^2Cy^6 + 3NaCy$) is obtained as a melted mass which is poured away from the heavier iron carbide at the bottom of the crucible. If two parts of this salt are then ignited with one part of zinc in a covered crucible, aluminium separates as a regulus, while double cyanide of sodium and zinc remains as slag. The slag is dissolved in water and treated with metallic iron, whereby metallic zinc is precipitated out and a solution of ferro-cyanide of soda remains and can be used over in the process.

I do not know whether Lauterborn has ever succeeded in carrying out this process; it would appear at the very beginning that the precipitation of aluminium ferro-cyanide, although appearing à priori possible, has always been found impracticable, alumina being precipitated.

J. Clark, of Birmingham, England, has taken out several patents in England and one in Germany. In the first,† hydrated

* German Patent, 39915 (1887).
† English Patent, 15946, Dec. 6, 1886.

aluminium chloride is to be mixed with lime, iron, zinc, ammonia, or any other substance which combines readily with chlorine, and finely divided coke. After drying the mixture it is introduced into the iron blast furnace or blown into the Bessemer converter, an iron-aluminium alloy being thus produced. In a second patent,[*] hydrated aluminium chloride is to be mixed with $2\frac{1}{2}$ parts of granulated zinc and 1 part of iron turnings or borings, or the alloy of the zinc and iron known as "zinc dross" might be used. The mass is let stand 24 hours and dried. The orange-colored powder resulting is mixed with borax (!) or any suitable flux and put into a crucible with 20 parts of fine granulated copper and melted down. After about an hour the zinc and iron present have probably volatilized as chlorides, while aluminium bronze remains. When the copper is to be alloyed in large quantities, it may be melted on the hearth of a reverberatory furnace and the prepared powder stirred into it.

In closing the subject of reduction by zinc, I would state that the distillation of the zinc from an aluminium-zinc alloy appears to be quite practicable. My friend, Dr. Lisle, informs me that he has taken a highly zinciferrous alloy and brought it up to 98 per cent. aluminium in this way.

REDUCTION BY LEAD.

According to the invention of Mr. A. E. Wilde,[†] of Notting Hill, lead or sulphide of lead, or a mixture of the two, is melted and in a molten state poured upon dried or burnt alum. The crucible in which the mass is contained is then placed in a furnace and heated, with suitable fluxes. The metal, when poured out of the crucible, will be found to contain aluminium. The aluminium and lead can be subsequently separated from each other by any known means, or the alloy or mixture of the two metals can be employed for the various useful purposes for which lead alone is more or less unsuited.

---

[*] English Patent, 10594, Aug. 18, 1886 ; German Patent, 40205 (1887).
[†] Sci. Am. Suppl., Aug. 11, 1887.

REDUCTION BY MANGANESE.

Walter Weldon[*] claimed to melt together cryolite with calcium chloride or some other non-metallic chloride or sulphide, and then to reduce the aluminium chloride or sulphide produced by manganese, also adding metallic sodium to promote the reaction. It is not probable that the first reaction named can be produced, but the latter part, as far as the manganese is concerned, may succeed, as is indicated by the thermo-chemical study of the reaction (see p. 191). Of course, the manganese would have to be used as ferro-manganese or spiegeleisen and an iron alloy produced. Dr. Green, of Philadelphia, mixed powdered spiegeleisen with cryolite, placed the mixture in a graphite crucible and heated it close to the ports of an open-hearth steel furnace until it softened, yet the iron contained afterwards only 0.3 per cent. of aluminium.

REDUCTION BY MAGNESIUM.

Magnesium develops more heat in forming compounds than aluminium does (see p. 186), which would indicate that it would reduce aluminium compounds easily. Only one or two statements on this point can be found. Margottet[†] states that magnesium will decompose molten cryolite, setting the aluminium at liberty. R. Gratzel[‡] patents the reduction of a double fluoride of aluminium and potassium or sodium by metallic magnesium, or by conducting magnesium vapor into the liquid compound.

Roussin[§] states that magnesium does not precipitate aluminium in a metallic state from its solutions. To test this point I placed a strip of magnesium in solution of aluminium sulphate, when magnesium sulphate went into solution and a precipitate of alumina was formed. It is apparent that the aluminium is first precipitated in the metallic state and promptly oxidized by the water as fast as set free, in a manner strictly

[*] English Patent, No. 97 (1883).
[†] Fremy's Ency. Chim.
[‡] English Patent, 14325, Nov. 25, 1885.
[§] Jrnl. de Pharm. et de Chimie, iii. 413.

analogous to the production of alumina at the negative pole when electrolyzing aqueous aluminium solutions.

The only other reference to using magnesium as a reducing agent is in a patent awarded Count R. de Montgelas, of Philadelphia,[*] in which it is stated that aluminium chloride is mixed with litharge, charcoal, and common salt; fused and crushed. It is then remelted with magnesium filings and potassium chloride. After cooling it is again crushed, mixed with more potassium chloride and nitrate of potash, fused, poured into water, and the globules of aluminium separated out.

REDUCTION BY ANTIMONY.

F. Lauterborn[†] proposes to decompose aluminium sulphide by means of antimony and carbon. One hundred parts of dried aluminium sulphate is mixed with 50 parts of charcoal and 72 parts of metallic antimony; some sodium carbonate and fluorspar is then added, and the mixture melted. It is claimed that antimony sulphide and aluminium are found in the product, the former being in the bottom of the crucible. In a modification of this process it is claimed[‡] that if a mixture of dried aluminium sulphate, sodium carbonate, and antimony sulphide (stibnite) is put into a shaft filled with incandescent coke, antimony will be first set free by the reaction

$$2Sb^2S^3 + 6Na^2CO^3 + 3C = 6Na^2S + 9CO^2 + 4Sb,$$

and that these products act further on the aluminium sulphate, setting free aluminium, by the reaction

$$2Al^2(SO^4)^3 + 6Na^2S + 4Sb + 12C = 4Na^2SbS^3 + 4Al + 12CO^2.$$

The sodium sulph-antimonide can be smelted over with soda and antimony regained.

These extraordinary formulas have little or no basis in chemical science. Dr. Fischer[§] says plainly that they are false. The author tried by direct experiment to reduce aluminium sulphide by antimony, fusing down a mixture of powdered antimony with

---

[*] English Patent, 10606, Aug. 18, 1886.
[†] German Patent, 32126 (1885).
[‡] Dingler, 256, 226, and 233.
[§] Wagner's Jahresbericht, 1885.

REDUCTION BY MISCELLANEOUS AGENTS.

aluminium sulphide, but the button of metal obtained did not contain a trace of aluminium.

REDUCTION BY TIN.

J. S. Howard and F. M. Hill, assignors to the Aluminium Product Company, of New York, make the following statements in their patent specifications :* Some aluminous material is boiled in muriatic acid, cooled, mixed with Spanish white or lime, the free acid evaporated off at a high temperature, and the heat finally increased to about 1000° F., thereby volatilizing any iron present as chloride. The product thus obtained is put into a lime-lined crucible along with lime, charcoal, fluorspar, cryolite and potassium bisulphate, the whole covered with chloride of tin, and finally by common salt. This is subjected to a smelting temperature, when on cooling it is claimed that an alloy of tin and aluminium is obtained. To separate the aluminium, the alloy is melted with lead or bismuth, which alloy with the tin, letting the aluminium float on the surface along with oxides and other impurities. This is skimmed off and purified by exposure to heat on a bed of porous material.

While we cannot altogether deny the possibility of metallic tin reducing aluminium chloride, yet it is not at all probable that this salt remains after the first ignition. It is also improbable that tin would reduce cryolite in the latter operation, but there is no direct evidence to contradict the above statement.

In an experiment made by the author, aluminium sulphide was heated with tinfoil for a short time at a red heat. The resulting metal contained 0.52 per cent. of aluminium, and from the proportions of tin and sulphide used, it was evident that a large part of the aluminium sulphide present had been reduced.

REDUCTION BY PHOSPHORUS.

†L. Grabau patented the following process, in 1883, but has since advocated altogether the use of sodium, showing that this

---

* U. S. Patent, 378136. Feb. 21, 1888.
† English Patent, 5798, Dec. 18, 1883.

method was not successful. The proposition was as follows :—
A rich alloy of aluminium and phosphorus is made by melting
the two elements together or by fusing aluminium with phosphor
salts and a reducing agent. The alloy is crushed, mixed with
alumina or clay or aluminous fluorides, covered with coal-dust
and heated to incandescence in a crucible. It was claimed that
the phosphorus combined with the oxygen or fluorine of the aluminous compound, the metal produced uniting with the aluminium
already present. To produce any given alloy, the phosphide of
the required metal is substituted for the aluminium-phosphorus
alloy. Mr. Grabau broadens out his specifications into more
probable fields by adding that "manganesic or carburetted metals
may be used instead of the phosphide alloys, as reducing agents."
It is almost needless to say that the generally small heat of combination of phosphorus compounds would show that the reactions
proposed are in a high degree improbable.

REDUCTION BY SILICON.

*M. Wanner makes the following general claims: The pro-
duction of aluminium by treating a fused bath of aluminium
fluoride or aluminous fluorides, while in a molten metallic bath
and protected from oxidizing agents, with a reagent whose elements
dissociate below the fusing point of the aluminium fluoride or the
aluminous fluorides, and having one element of such affinity that
it displaces the aluminium in the fluoride compound, but having
no element capable of combining with the reduced aluminium.
More specifically, sulphide of silicon or an equivalent reagent is
mentioned, and the reaction takes place on the hearth of a short
reverberatory furnace, the metallic bath mentioned being preferably metallic iron or copper, which are able to combine with the
resulting aluminium.

The reaction proposed is so far out of the usual run of speculations that no opinion can be hazarded as to its probability.
Further, silicon sulphide is an almost unknown compound, and

* U. S. Patent, 410568, Sept. 3, 1889.

it would be very interesting to know how it is to be prepared; it would probably be a more difficult feat to get the reagent alone than to produce aluminium by many well-known methods.

CHAPTER XIII.

WORKING IN ALUMINIUM.

MELTING ALUMINIUM.

DEVILLE: To melt aluminium it is necessary to use an ordinary earthen crucible and no flux. Fluxes are always useless and almost always harmful. The extraordinary chemical properties of the metal are the cause of this; it attacks very actively borax or glass with which one might cover it to prevent its oxidation. Fortunately this oxidation does not take place even at a high temperature. When its surface has been skimmed of all impurities it does not tarnish. Aluminium is very slow to melt, not only because its specific heat is considerable, but its latent heat appears very large. It is best to make a small fire and then wait patiently till it melts. One can very well work with an uncovered crucible.

In the fusion of impure aluminium, very different phenomena are observed according to the nature of the foreign metal which contaminates it. Ferruginous material often leaves a skeleton less fusible and pretty rich in iron; a liquation has taken place, increasing the purity of the melted material. When the aluminium contains silicon, this liquation is no longer possible, or at least it is very difficult, and I have sometimes seen some commercial aluminium so siliceous that the workmen were unable to remelt it. When it is desired to melt pieces together, they can be united by agitating the crucible or compressing the mass with a well-cleaned, cylindrical bar of iron. Clippings, filings, etc., are melted thus: Separate out first, as far as possible, foreign metals, and to avoid their combining with the aluminium heat the divided metal to as low a heat as possible, just sufficient to melt it. The oil and organic matters will burn, leaving a cinder, which hinders the reunion of the metal if one does not press firmly with the iron bar. The metal may then be cast very easily and there is found at the bottom of the crucible a little cinder, which still contains a quantity of aluminium in globules. These may be easily separated by rubbing in a mortar and then passing through a sieve, which retains the flattened globules.

Biederman gives the following directions : " The whole quantity of metal which is to be melted must not be put into the crucible at once, but little by little, so increasing the mass from time to time as the contents become fully melted. The necessary knack for attaining a good clean melt consists in dipping the pieces which are to be melted together in benzine before putting in the crucible. Mourey even pours a small quantity of benzine into the crucible after the full melting of the metal, and he recommends the employment of benzine in the melting of all the noble metals. To utilize the scrap pieces produced in working aluminium into various useful articles, one must as far as possible separate out first the pieces which have been soldered, in order that the newly melted aluminium may not be contaminated by the solder. The solder adhering to these pieces can be removed by treating them with nitric acid, by which the aluminium is not attacked."

Aluminium can be melted with perfect safety in common clay or sand crucibles if they are lined with carbon. This can be done by mixing lamp-black to a paste with molasses, plastering the inside of the crucible evenly and drying slowly for several days at a moderate temperature. A means of getting a more perfect lining is to ram the crucible full of this paste, drying slowly and then hollowing out a cavity of the required size leaving a uniform lining of sufficient thickness. In using these carbon-lined crucibles they should be kept well-covered, in order that the carbon may not burn away too quickly, taking particular care to place the cover on when the metal has been poured out and the crucible is cooling in the air. Small crucibles may be made of a single block of soapstone, and seem to last a long while, the aluminium apparently having no action on this mineral at a temperature a good deal higher than its melting point.

With extra care, heating in a furnace where the heat is under exact control, aluminium may be melted in sand crucibles or in cast-iron ones. In using sand or Hessian crucibles no flux must be added, and if the metal is heated slowly to a temperature only enough above its melting point to admit of pouring quickly, it will be found that the crucible is unattacked. If, however, the crucible is heated to a bright-red heat at any part, it will be found that at that place the aluminium has attacked the crucible, and on pouring out the metal a thin, tough skin will be left adhering to the spot attacked and generally taking with it bits of the wall when it is forcibly detached. This thin sheet is hard, tough, and rich in silicon, while the rest of the metal poured out has also absorbed a little silicon. An iron crucible acts in precisely the same way; if the heat is kept close to the melting point of aluminium the latter does not "wet" the iron, but at a bright-red heat it attacks it and adheres, as in the case of the sand crucible. I will repeat, that if the temperature is kept as low as it is possible to melt and cast the metal, and if no fluxes are used, neither the iron or sand crucibles are attacked.

I have been told that in the large European works, where 500 or 1000 lbs. of aluminium are melted at once on the bed of a reverberatory furnace, the hearth was formerly protected by pure beauxite, closely rammed in and strongly fired before using, but basic magnesia bricks have now been substituted, similar to those used in basic open-hearth furnaces but of purer materials, and they are reported as being all that can be desired in their capacity for resisting corrosion.

Casting Aluminium.

Deville: "Aluminium can be cast very easily in metallic moulds, but better in sand for complicated objects. The mould ought to be very dry, made of a porous sand, and should allow free exit to the air expelled by the metal, which is viscous when melted. The number of vents ought to be very large, and a long, perfectly round git should be provided. The aluminium, heated to redness, ought to be poured rather quickly, letting a little melted metal remain in the git till it is full, to provide for the contraction of the metal as it solidifies. In general, this precaution ought to be taken even when aluminium is cast in iron ingot moulds or moulds of any other metal. The closed ingot moulds give the best metal for rolling or hammmering. By following these precautions, castings of great beauty may be obtained, but it is not advisable to conceal the fact that to be able to succeed completely in all these various operations requires for aluminium, as for all other metals, a special familiarity with the material which practice alone is able to give."

The peculiarity of molten aluminium which a metal caster would first notice is its viscousness, that is, it runs thick. When about to pour, a thick edge or lip of metal forms which must in many cases be punctured in order to start the metal flowing. On account of this property it does not run sharply in the moulds except where a head of metal puts it under pressure. To obtain sharp castings there must be a "gate" to give pressure to the metal, and when we remember that the gravity of aluminium is only 2.6 and that, besides, it does not flow thinly, it will be seen that a much higher head of metal is necessary to ensure sharp castings than is needed for iron, brass, etc. A small slab of aluminium two inches high run in a closed iron mould with very little gate was quite sharp at the lower end but had rounded corners at the top. For casting in closed moulds, the best results as to sharp castings free from cavities are obtained if a slight artificial pressure can be applied to the still liquid metal in the mould immediately after pouring. This is accomplished by Dr. C. C. Carroll, of New York, an expert in making cast-aluminium dental plates, by closing in tightly the top of the crucible containing the molten aluminium and then by air pressure from a rubber bulb forcing the metal through a syphon-shaped tube terminating underneath the metal and connecting tightly with the pouring gate of the mould. The mould is previously heated, in order not to chill the metal too quickly, and when the metal has been forced out of the crucible by squeezing the bulb the pressure is continued for a short time in order to force the metal into every crevice of the mould and allow the casting to set under pressure.

The idea is undoubtedly correct, and the excellently sound and extremely sharp castings obtained by Dr. Carroll attest its success in practice.

PURIFICATION OF ALUMINIUM.

Freeing from slag.—Deville gives the following information on this important subject :—

"It is of great importance not to sell any aluminium except that which is entirely free from the slag with which it was produced and with which its whole mass may become impregnated. We have tried all sorts of ways of attaining this end, so as to obtain a metal which would not give any fluorides or chlorides upon boiling with water, or give a solution which would be precipitated by silver nitrate. At Glacière we granulated the metal by pouring it while in good fusion into water acidulated with sulphuric acid; this method partially succeeded. But the process which M. Paul Morin uses at present (1859), and which seems to give the best results, is yet simpler. Three or four kilos of aluminium are melted in a plumbago crucible without a lid, and kept a long time red hot in contact with the air. Almost always acid fumes exhale from the surface, indicating the decomposition by air or moisture of the saline matter impregnating the metal. The crucible being withdrawn from the fire, a skimmer is put into the metal. This skimmer is of cast iron; its surface ought not to be rough and it will not be wetted by the aluminium in the least during the skimming; it may be of advantage to oxidize its surface with nitre before using. The white and slaggy matters are then removed, carrying away also a little metal, and are put aside to be remelted. So, in this purification, there is really no loss of metal. After having thus been skimmed, the aluminium is cast into ingots. This operation is repeated three or four times until the metal is perfectly clean, which is, however, not easily told by its appearance, for, after the first fusion, the crude aluminium when cast into ingots has a brilliancy and color such as one would judge quite irreproachable, but the metal would not be clean when it was worked, and especially when polished would present a multitude of little points called technically

'piqûres,' which give to its surface, especially with time, a disagreeable look. Aluminium, pure and free from slag, improves
in color on using. It is the contrary with the impure metal or
with aluminium not freed from slag. When aluminium is submitted to a slow, corroding action, its surface will cover itself
uniformly with a white, thin coating of alumina. However, any
time that this layer is black or the aluminium tarnishes, we may
be sure that it contains a foreign metal and that the alteration is
due to this impurity."

Freeing from impurities.—Again Deville is the authority, and
we quote his advice on the subject :—

"A particular characteristic of the metallurgy of aluminium is
that it is necessary, in order to get pure metal, to obtain it so at
the first attempt. When it contains silicon, I know of no way
to eliminate it, all the experiments which I have made on the
subject have had a negative result; simple fusion of the metal in
a crucible, permitting the separation by liquation of metals more
dense, seems rather to increase the amount of silicon than to de-
c.ease it. When the aluminium contains iron or copper, each
fusion purifies it up to a certain limit, and if the operation is done
at a low heat there is found at the bottom of the crucible a metallic skeleton containing much more iron and copper than the
primitive alloy. At first I made this liquation in the muffle of
a cupel furnace, in which process the access of air permitted the
partial oxidation of these two metals. The little lead which aluminium may sometimes take up may thus be easily separated.
Unfortunately, the process does not give completely satisfactory
results. It is the same in fusing impure aluminium under a
layer of potassium sulphide, K^2S^3; there is a partial separation
of the lead, copper, and iron. That which has succeeded best
with us is the process which we have employed for a long time at
Glacière, and which consists in melting the aluminium under
nitre in an iron crucible. We have in this way improved the
quality of large quantities of aluminium. The operation is conducted as follows: Aluminium has generally been melted with
nitre in order to purify it by means of the strong disengagement
of oxygen at a red heat, no doubts being entertained as to the
certainty of the result. But it is necessary to take great care

when doing this in an earthen crucible.   The silica of the crucible is dissolved by the nitre, the glass thus formed is decomposed by the aluminium, and the siliceous aluminium thus formed is, as we know, very oxidizable, and especially in the presence of alkalies. So, the purification of aluminium by nitre ought to be done in a cast-iron crucible well oxidized itself by nitre on the inside.

"On melting aluminium containing zinc in contact with the air and at a temperature which will volatilize the zinc, the largest part of the latter burns and disappears as flaky oxide.   To obtain a complete separation of the two metals it is necessary to heat the alloy to a high temperature in a brasqued crucible.   This experiment succeeds very well, but it is here shown that the aluminium must oxidize slightly on its surface, for some carbon is reduced by the aluminium from the carbonic oxide with which the crucible is filled.   This carbon thus separated is quite amorphous."

Dr. Lisle, of Springfield, O., tested the removal of zinc from aluminium by distillation and succeeded in obtaining a product with 98 per cent. of aluminium from which the zinc had been completely removed.   If aluminium containing tin is melted with lead, the latter sinks with the tin, removing it almost completely from the aluminium, but the metal remaining retains a little lead.   In an experiment made by the author, aluminium with 10 per cent. of tin was treated in this way, but the aluminium retained nearly 7 per cent. of lead, giving it a blue color and large crystalline structure.   M. Peligot is reported by Deville to have succeeded in cupelling aluminium with lead, whereby impure, tough metal became quite malleable.   It may be that the small percentage retained by aluminium when melted with lead is removed by fusion on a cupel, but I have been unable to perform any operation with aluminium at all analogous to the cupellation of silver with lead.

G. Buchner states that commercial aluminium contains considerable quantities of silicon, which by treatment, when melted, with hydrogen evolves hydrogen silicide.   This does not result if arsenic is present.

To test this point, I took a sample of aluminium containing 4 per cent of silicon, and 94 per cent. of aluminium.   This was

23

melted in a sand crucible and hydrogen gas run in for 12 minutes. On pouring out, the crucible was unattacked, but the metal was identical in color and structure with that not treated. This was repeated several times, hydrogen gas being passed in as long as 20 minutes with the metal at red heat, but no apparent change in the purity of the metal resulted.

Similar experiments were made with a current of sulphuretted hydrogen gas, with a view of removing the iron. No improvement in the looks of the aluminium was apparent. If, however, air was blown into the melted metal, an improvement was made. The aluminium was at a bright red heat, and air blown in for 5 minutes, at the end of which time about 5 to 10 per cent. of dross composed of mixed metal and oxide was formed, but the remaining metal was whiter, of a finer grain and evidently much improved.

Prof. Mallet* made some very accurate estimations of the atomic weight of aluminium (which he found to be 27.02), and obtained the chemically pure metal required for his work by the following process, which is quite applicable when studying the properties of the pure metal, yet is, of course, altogether out of the question as an industrial operation. The purest commercial metal was bought, and on analysis was found to contain—

Aluminium	96.89 per cent.
Iron	1.84 "
Silicon	1.27 "

This was treated with liquid bromine and converted into bromide. This salt was then purified by fractional distillation, the temperature being very carefully regulated, and the operation repeated until the product was perfectly colorless, and dissolved in water without leaving any perceptible impurity. The reduction was accomplished with difficulty and much loss by treating with sodium in a crucible made of a mixture of pure alumina and sodium aluminate. The metal obtained gave on analysis no weighable quantity of impurities, and the properties mentioned in Chapters III. and IV. are those quoted from Mallet's report.

* Philosophical Magazine, 1880.

ANNEALING.

By heating to redness and cooling quickly, as by dropping into water, aluminium becomes soft. To get the best results, the metal should be heated until it just begins to glow, or the object is rubbed with a lump of fat, and the moment that the black trace left by the carbonization of the fat disappears, the metal is removed from the annealing oven. Great care is necessary in annealing thin sheets which are being beaten into leaf, to avoid melting them. Fine wire can be annealed over the chimney of an Argand burner.

HARDENING.

By hammering, rolling, or drawing, aluminium becomes sensibly harder and stiffer; also, by long, gradual cooling the same result is obtained, so that it becomes elastic enough to be used even for hair-springs for watches.

ROLLING.

Before rolling a bar of aluminium it is well to soften it, and to taper down a "lead" by hammering. The metal forges well under the hammer at a low-red heat. The metal for rolling had better be cast into plates in covered iron ingot-moulds, and the surfaces planed to remove small irregularities. The rolling is not difficult, except that a large amount of power is required (about as much for cold aluminium as for hot steel), and as the metal quickly gets hard it must be annealed often. It is recommended that the metal be brought warm under the rolls, and, if possible, elongated 10 to 20 per cent. with the first pass, in order to entirely destroy the crystalline structure of the metal. The annealing is repeated between each pass until the sheet is about 3 millimetres thick, after which it can generally be rolled with fewer annealings, sometimes without any at all. Rolls warmed to 100–150° work better than cold ones. Aluminium has been rolled down to the thinness of tissue-paper.

Thin rolled sheets may be still further extended out by beating into leaf. The gold-beaters are said to have no special difficulty in doing this, except that more frequent annealings are necessary than with gold or silver. M. Degousse was the first to make this leaf, in 1859; he states that the tempering must be done by warming only to 100° or 150°, an actual glowing heat proving very unsuitable. Aluminium leaf is made as thin as ordinary gold or silver leaf, and this property would, therefore, establish aluminium as next to these noble metals in the order of malleability.

DRAWING.

Prof. Thurston places aluminium as sixth in the order of ductility, being preceded by gold, silver, platinum, iron, and copper, but it is doubtful if it does not rank equally as high as iron. Deville states that in 1855 M. Vangeois obtained very fine wire with metal far from being pure. Bell Bros., at Newcastle-on-Tyne, recommended that the metal for drawing be run into an open mould so as to form a flat bar of about one-half inch section, the edges of which are beaten very regularly with a hammer. The diameter should be very gradually reduced at first, with frequent heating. When the threads are required very fine the heating becomes a very delicate operation, on account of the fineness of the threads and the fusibility of the metal. The gauge should be reduced by the smallest possible gradations, when wires may be obtained as fine as a hair.

Aluminium tubes are drawn, either round or square, from sheet which has been soldered together or from cast rings of the required section; the first method is said to be preferred. As the aluminium quickly loses its temper it must be annealed after each extension.

STAMPING AND SPINNING.

Aluminium can be spun on the lathe into all sorts of round and hollow forms; it may also be pressed or stamped into shape in the cold; it is of advantage in doing this to use a kind of varnish composed of 4 parts of oil of turpentine and 1 part of stearic acid. Soap-water is always to be avoided.

GRINDING, POLISHING AND BURNISHING.

When cast carefully it can be filed without fouling the file. Spun and stamped articles of aluminium can easily be ground by using olive oil and pumice. Biederman makes the following remarks on polishing: "The use of the old means of polishing and burnishing metals, such as soap, wine, vinegar, linseed-oil, decoction of marshmallow, etc., is not effective with aluminium, but, on the contrary, is even harmful; because, using them, the blood stone and the burnishing iron tear the metal as fine stone does glass. Oil of turpentine has also been used, but with no good effect. Mourey found, after many attempts, that a mixture of equal weights of olive oil and rum, which were shaken in a bottle till an emulsified mass resulted, gave a very brilliant polish. The polishing stone is dipped in this liquid, and the metal polished like silver, except that one must not press so hard in shining up. The peculiar black streaks which form under the polishing stone need cause no trouble; they do not injure the polish in the least, and can be removed from time to time by wiping with a lump of cotton. The best way to clean a soiled surface and remove grease is to dip the object in benzine, and dry it in fine sawdust."

Mr. J. Richards found that when buffing in the ordinary way the dark-colored burnishing powder cut into the metal and filled its pores with black specks. The best means of burnishing is to use a piece of soft wood soaked in olive oil, this closes the grain and gives a most brilliant finish.

ENGRAVING.

Kerl & Stohman: Aluminium resists the action of the engraving tool, which slides upon the surface of the metal as upon hard glass. But as soon as a varnish of 4 parts of oil of turpentine and 1 of stearic acid, or some olive oil mixed with rum is used, the tool cuts into it as into pure copper.

MAT.

Deville: "Aluminium, like silver, is able to take a very beautiful mat which keeps indefinitely in the air. It is obtained easily by plunging the surface for an instant in a very dilute solution of caustic soda, washing in a large quantity of water and at last dipping in strong nitric acid. Under these conditions, all the foreign materials which might contaminate it, except silicon in large proportion, dissolve and leave the metal quite white and with a very pleasing appearance." Bell Bros. recommend first washing the objects in benzole or essence of turpentine before treatment with caustic soda.

SOLDERING ALUMINIUM.

At the time Deville wrote his book, the difficulty of soldering aluminium properly was one of the greatest, if not the greatest, obstacle to the employment of the metal. His views on the question may be, therefore, very interesting; they are as follows:—

"Aluminium may be soldered, but in a very imperfect manner, either by means of zinc, or cadmium, or alloys of aluminium with these metals. But a very peculiar difficulty arises here, we know no flux to clean the aluminium which does not attack the solder, or which, protecting the solder, does not attack the aluminium. There is also an obstacle in the particular resistance of aluminium to being wetted by the more fusible metals, and on this account the solder does not run between and attach itself to the surfaces to be united. M. Christofle and M. Charrière made, in 1855, during the Exposition, solderings with zinc or tin. But this is a weak solder and does not make a firm seam. MM. Tissier, after some experiments made in my laboratory, proposed alloys of aluminium and zinc, which did not succeed any better. However, M. Denis, of Nancy, has remarked that whenever the aluminium and the solder melted on its surface are touched by a piece of zinc, the adhesion becomes manifest very rapidly, as if a particular electrical state was determined at the moment of contact. But even this produces only weak solderings insufficient in most cases.

A long time ago, M. Hulot proposed to avoid the difficulty by previously covering the piece with copper, then soldering the copper surfaces. To effect this, plunge the article, or at least the part to be soldered, into a bath of acid sulphate of copper. Put the positive pole of a battery in communication with the bath, and with the negative pole touch the places to be covered, and the copper is deposited very regularly. M. Mourey has succeeded in soldering aluminium by processes yet unknown to me ; samples which I have seen looked excellent. I hope, then, that this problem has found, thanks to his ingenuity, a solution ; a very important step in enlarging the employment of aluminium."

Mourey's first practicable solders for aluminium were of zinc and aluminium and of two kinds—hard and soft. He used a soft solder to first unite the pieces and afterwards finished the soldering with a less fusible one. These solders contained—

	I.	II.	III.	IV.	V.
Aluminium	20	15	12	8	6
Zinc	80	85	88	92	94

The alloys with the larger proportion of zinc are the easiest melting or softest ones, one such as IV being used for ordinary work, while II was used for brazing. These solders have the disadvantage that on melting they oxidize very easily, and in consequence of the film of oxide thus formed the work is so much more difficult. This difficulty can be overcome by dipping the small grains of solder in copaiva balsam and turpentine, which keep out the air and act as reducing agents during the operation. The new solders subsequently used were simpler in application, for the work was finished with one solder and the moistening with balsam was rendered unnecessary.

Mourey improved upon the zinc-aluminium solders by adding copper, using five different alloys of these three metals according to the objects to be soldered. They contained—

	I.	II.	III.	IV.	V.
Aluminium	12	9	7	6	4
Copper	8	6	5	4	2
Zinc	80	85	88	90	94

The following directions are given for preparing these alloys :—*

* Das Löthen, by Edmund Schlosser, Vienna, 1880, p. 101.

"To make the solder, first put the copper in the crucible. When it is melted, then add the aluminium in three or four portions, thereby somewhat cooling the melted mass. When both metals are melted, the mass is stirred with a small iron rod, and then the required quantity of zinc added, free from iron, and as clean as possible. It melts very rapidly. The alloy is then stirred briskly with an iron rod for a time, some fat or benzine being meanwhile put in the crucible to prevent contact of the metal with air and oxidation of the zinc. Finally the whole is poured out into an ingot mould previously rubbed with benzine. After the addition of zinc, the operation must be finished very rapidly, because the latter will volatilize and burn out. As soon as the zinc is melted, the crucible is taken out of the fire. Only zinc free from iron can be used, since even an apparently insignificant amount of this impurity injures the qualities of the solders very materially in regard to durability and fusibility.

"The separate pieces of metal to be soldered together are first well cleaned, then made somewhat rough with a file at the place of juncture, and the appropriate solder put on it in pieces about the size of millet grains. The objects are laid on some hot charcoal, and the melting of the solder effected by a blast lamp or a Rochemont turpentine-oil lamp. During the melting of the solder, it is rubbed with a little soldering iron of pure aluminium. The soldering iron of pure aluminium is essentially a necessity for the success of the operation, since an iron of any other metal will alloy with the metals composing the solder, while the melted solder does not stick to the iron made of aluminium.

"For quite small objects, as for jewelry, solder I is used; for larger objects and ordinary work IV is more suitable, and is the solder most used. These alloys work so perfectly that plates soldered together never break at the joint when bent back and forth, but always give way in other places; which is a result not always possible in the best soldering of plates of silver."

Bell Bros. used the above solders in their works at Newcastle, and in a description of the soldering operation state the following facts in addition to those already given :* "In the operation of

* Chemical News, iv. 81.

soldering, small tools of aluminium are used, which facilitate at the same time the fusion of the solder and its adhesion to the previously prepared surfaces. Tools of copper or brass must be strictly avoided, as they would form colored alloys with the aluminium and the solder. The use of the little tools of aluminium is an art which the workman must acquire by practice. At the moment of fusion the work needs the application of friction, as the solder suddenly melts very completely. In soldering it is well to have both hands free and to use only the foot for the blowing apparatus."

We also find the following alloys credited to Mourey as used by him for soldering aluminium, probably in the same manner as has just been described :—*

Aluminium	.	.	.	.	.	.	.	30	20
Copper	.	.	.	.	.	.	.	20	15
Zinc	.	.	.	.	.	.	.	50	65

Col. Wm. Frishmuth† recommends a solder containing :—

Aluminium	.	.	.	.	.	.	.	.	20
Copper	.	.	.	.	.	.	.	.	10
Zinc	.	.	.	.	.	.	.	.	30
Tin	.	.	.	.	.	.	.	.	60
Silver	.	.	.	.	.	.	.	.	10

Later, Col. Frishmuth‡ states that the solder just given is used for fine ornamental work, while for lower-grade work he uses the following :—

	I.	II.	III.
Sn	 95	97	98–99
Bi	 5	3	2–1

He recommends for a flux, in all cases, either paraffin, stearin, vaselin, copaiva balsam, or benzine. In the solder for fine work, if aluminium is used in larger quantity than recommended, the solder becomes brittle.

Schlosser§ recommends two solders containing aluminium as especially suitable for soldering dental work, on account of their resistance to chemical action. Copper cannot be allowed in alloys

* Dingler, 166, 205.   † Techniker, vi. 249.
‡ Wagner's Jahresb., 1884.   § Das Löthen, p. 103.

intended for this use, or only in very insignificant quantity, since it is so easily attacked by acid food, etc. Since these two alloys can probably be used also for aluminium dental work, we subjoin their composition—

Platinum-aluminium solder.		Gold-aluminium solder.	
Gold . . . 30		Gold . . . 50
Platinum . . 1		Silver . . . 10
Silver . . 20		Copper . . . 10
Aluminium . . 100		Aluminium . . 20

M. Bourbouze states that the difficulties met with in soldering aluminium are satisfactorily overcome by the following process :—

*" The parts to be united are subjected to the ordinary operation of tinning, except that, in place of pure tin an alloy of tin and zinc, or, better, of tin, bismuth and aluminium is used. However, preference is given to an alloy of tin and aluminium, mixed in different proportions to suit the work put on the joint. For those which are to be subsequently worked, an alloy of 45 parts tin and 10 parts aluminium should be used. This solder is malleable enough to resist hammering, drawing, or turning. Pieces not to be worked after soldering may, whatever the metal to be united to the aluminium, be solidly soldered with a tin solder containing less aluminium. Neither of these solders requires any preparation of the pieces, and the last one may be applied with a common soldering-iron. To unite other metals to aluminium it is best to coat the part with pure tin, the aluminium is coated with one of the above alloys, the joint closed and finished by heating in the usual manner.

O. M. Thowless has patented the following solder for aluminium, and method of applying it :† The alloy is composed of—

Tin	55 parts.
Zinc	23 "
Silver	5 "
Aluminium	2 "

The silver and aluminium are first melted together, the tin added, and lastly the zinc. The metallic surfaces to be united

* Comptes Rendue, 98, 1490.
† English Patent, 10237, Aug. 29, 1885.

are immersed in dilute caustic alkali or a cyanide solution, washed and dried. They are then heated over a spirit lamp, coated with the solder and clamped together, small pieces of the alloy being placed around the joints. The whole is then heated to the melting point of the solder and any excess of it removed. No flux is used.

J. S. Sellon patents the following method :* The aluminium surfaces are cleaned by scraping and covered with a layer of paraffin wax as a flux. They are then coated by fusion with a layer of an alloy of zinc, tin and lead, preferably in the proportions

Zinc	5
Tin	2
Lead	1

The metallic surfaces thus prepared are soldered together in the usual way with any good solder.

I am quite aware of the criticism which the above information about aluminium solders will meet,—that much more satisfactory alloys have been discovered and are now used ; and this is possibly the case, but it must be remembered that a metal worker who by searching patiently discovers such an improvement considers that his reward is found in keeping to himself the monopoly of its use, and that to obtain these trade secrets is in most cases impossible, especially since the more valuable they are the more carefully are they guarded.

COATING METALS WITH ALUMINIUM.

Many attempts have been made to give baser metals a thin coating of aluminium and thus impart to them superficially the resisting proporties of that metal. We may distinguish broadly two different methods of procedure used to accomplish this—the chemical (and electric) and the mechanical. Aluminium is not thrown down in a metallic state from its solutions by any other metal, and therefore it cannot be obtained as a plating by dipping in any solution of its salts. Further, it is probably not thrown

---

* English Patent, 11499, Sept. 26, 1885.

down from aqueous solution by the battery (see "Electrolytic Methods of Production") although this has been frequently asserted and patented. Aluminium can be deposited electrolytically from a fused bath of its chlorides, etc., but, although Deville says that this principle can be utilized for coating other metals with aluminium, yet the metal is never deposited in a dense, compact film but as a powder mixed with carbon and other impurities, and satisfactory results cannot be obtained. The mechanical methods alluded to are conducted in two ways, either by uniting thin sheets of aluminium to the surface of another metal (veneering) or by using aluminium powder and burning in (aluminizing).

Of the practice of veneering with aluminium, Deville says in 1859: "M. Sevrard succeeded in 1854 in plating aluminium on copper and brass with considerable perfection. The two metallic surfaces being prepared in the ordinary manner and well scoured with sand, they are placed one on the other and held tightly between two iron plates. The packet is then heated to dark redness, at which temperature it is strongly compressed. The veneer becomes very firmly attached, and sheets of it may be beaten out. I have a specimen of such work perfectly preserved. The delicate point of the operation is to heat the packet just to that point that the adherence may be produced without fusing the aluminium, for when it is not heated quite near to this fusing point the adherence is incomplete. Experiments of this kind with copper and aluminium foil did not succeed, for as soon as any adherence manifested itself the two metals combined and the foil disappeared into the copper. In an operation made at too low a temperature, the two metals, as they do not behave similarly on rolling, become detached after a few passes through the rolls. Since then, the experiments in veneering aluminium on copper, with or without the intervention of silver, have succeeded very well." Deville stated later, in 1862, that Chatel had brought this art to perfection, the veneered plates being used largely for reflectors, etc., in place of silver-plated material.

Dr. Clemens Winckler* gives his experience in this line as fol-

* Industrie Blätter, 1873.

lows: "The coating of other metals with aluminium by the so-called plating method is, according to my own experience, possible to a certain degree, but the product is entirely useless, every plating requiring an incipient fusing of both metals and their final intimate union by rolling.   The ductility of aluminium is, however, greatly injured by even a slight admixture with other metals; iron makes it brittle, and copper, in small per cent. makes it fragile as glass.   If now it were possible in any way to fuse a coating of aluminium upon another metal, there would be formed an intermediate alloy between the two metals from which all ductility would be gone and which would crumble to powder under the pressure of the rolls, thus separating the aluminium surface from the metal beneath.   But even if it were possible in this way to coat a metal with a thin plate, it is still doubtful if anything would be attained thereby.   For, while compact aluminium resists oxidizing and sulphurizing agencies, the divided metal does not.   In powder or leaves aluminium is readily oxidized, as is shown by its amalgam becoming heated in the air and quickly forming alumina.   In the form of a coating upon other metals it must necessarily be in a somewhat finely divided state, and hence would probably lose its durability."

Dr. G. Gehring* has patented a method of aluminizing by which difficulty fusible metals, stone-ware or the like, can be coated with aluminium.   A mixture is made of a fatty acid (sebacic) and acetic acid with clay, etherized oil and aluminium (or aluminium bronze) in powder.   This is spread evenly on the metal or object to be treated and then heated with a Bunsen burner using blast or in a muffle.   The coating produced is silver white, does not oxidize under ordinary conditions, stands heating in an ordinary fire and can be highly polished.   It is stated† that this process is now largely made use of in Germany.

Somewhat analogous results are obtained by Brin Bros. (p. 335) without the use of metallic aluminium (similar to those claimed also by Baldwin, p. 336), but while in Dr. Gehring's process the coating would be formed on any surface, in these other

* German Patent, 29891 (1885).
† Engineering and Mining Journal, Feb. 13, 1886.

processes the presence of the metallic base is necessary, and an alloy with a few per cent. of aluminium is formed on the surface of the object and penetrates a little way into its interior. Such a coating, then, is simply a transformation of the outer layer of metal into an alloy with a small quantity of aluminium, and could possess very few of the qualities of aluminium, but might possess, as an alloy, qualities superior to those of the original metal.

PLATING ON ALUMINIUM.

Deville says: "The gilding and silvering of aluminium by electricity is very difficult to do satisfactorily and obtain the desirable solidity. M. Paul Morin and I have often tried it by using a bath of acid sulphide of gold or of nitrate of silver with an excess of sulphurous acid. Our success has only been partial. However, M. Mourey, who has already rendered great services in galvano-plasty, gilds and silvers the aluminium of commerce with a surprising perfection considering the little time he has had to study the question. I also know that Mr. Christofle has gilded it, but I am entirely ignorant of the methods employed by these gentlemen.* The coppering of aluminium by the battery is easily effected by M. Hulot by using an acid bath of sulphate of copper."

Tissier Bros. state that aluminium can be gilded without using a battery by preparing a solution as follows: " Eight grammes of gold are dissolved in aqua regia, the solution diluted with water and left to digest twenty-four hours with an excess of lime. The precipitate, with the lime, is well washed, and then treated with a solution of twenty grammes of hyposulphite of soda. The liquid resulting serves for the gilding of aluminium without the aid of heat or electricity, the metal being simply immersed in it after being previously well cleaned by the successive use of caustic potash, nitric acid, and pure water."

Aluminium can be veneered with other metals in a manner

---

* M. Mourey has stated that his means are galvanic, and that he has no trouble in depositing silver and gold in six different colors—shining, matt, or dull—but does not describe his methods.

strictly analogous to the reverse process described by Deville. For example, Morin describes the veneering with silver as follows: "Sheet silver is laid on the clean aluminium surface, a steel plate placed over the silver and the whole bound into a packet with fine copper wire. Two large cast-iron blocks are heated to a dark red heat, the packet placed between them and a pressure of 1 ton to the square centimetre (10 tons per square inch) applied gradually and sustained for 15 minutes. When removed from the hydraulic press they can be rolled like silvered copper when brought to the proper heat. The plating with gold succeeds best if a thin leaf of silver is slipped between the two sheets of metal, the operation proceeding then exactly as above. Platinum may be plated on aluminium just as easily as silver.

Uses of Aluminium.

Deville wrote in 1862: "Aluminium is the intermediate metal between the noble and the base metals." This was true then of its price as well as of its properties; it does not withstand chemical agents in general as strongly as the noble metals, but it withstands air, water, sulphuric acid, nitric acid and sulphuretted hydrogen—which is not the case with iron, copper or even silver. We have then a semi-noble metal; but, while silver, gold and platinum have extremely small prospect of becoming noticeably cheaper, yet the time is probably not far distant when we shall have our semi-noble metal at the price of the base ones. This affords the immense future for aluminium. Whatever its price, it can only replace gold or platinum because of its lightness; it already replaces silver especially because of its resistance to sulphur, as well as for its lightness, besides being cheaper; it can only replace the common metals, at its present price, for uses where its lightness is an extraordinary advantage. But, when its price is down to that of these baser metals, it will begin to replace them by virtue of its other superior qualities, chemical and physical; aside from its lightness it will win a large field simply in comparison with them on its merits as a metal. Thus, there are wide applications now almost unthought of, because the high price has been a blank wall to stop its use, but as it cheapens more and more we hear every day of new uses brought to light. Thus its sphere will widen until, since its ores are as cheap as those of iron, it will approximate in utility to that universal metal. For the above reasons, aluminium has really not yet won a very large field, and perhaps not a little disappointment is felt on finding out exactly the few uses it has been put to.

Chronologically, the first article made of aluminium was a baby-rattle intended for the infant Prince Imperial of France, in 1856. For this purpose it no doubt answered excellently, from its brightness, lightness, ring and cleanliness, but only a prince could afford to possess one in those days. In the next few years it was used for all sorts of articles of ornament and luxury. It was found well-suited for fine jewelry by reason of its adaptability to being cast and carved, the beautiful reflections from a chased surface, its color, matching well with gold, and the absence of odor. But it did not keep its polish as well as gold, and, perhaps more to the point, it did not stay in fashion long, so that the rage for aluminium jewelry subsided almost as fast as it had arisen, and it is only quite lately that it is being used again in this way to any extent. Then the French, with their ability for producing artistic furniture, used it for inlaid work on carved mouldings, cabinets, table tops, etc., but this application never exceeded a limited extent.

It is said that the Emperor's interest in aluminium, in 1854, was aroused partly by the idea that if it could be had cheaply it would wonderfully lighten the weight of military equipments, such as spurs, buttons, sword-handles, sabre-sheaths, helmets, and the imperial eagles. A helmet was made for the Emperor's cousin, the King of Denmark, which when gilded, ornamented, and fitted up complete weighed only 1½ lbs. The weight of the imperial eagles was lessened from 8 lbs. to nearly 3 lbs., it being remarked that "since they were gilded, only the bearer perceived the difference." When Garopon, in Paris, was furnished with very fine wire by Vaugeois, he was immediately successful in working it into embroidery, lace, and passementerre. This use of aluminium has also a military bearing, since aluminium wire can be used instead of silver in embroidering banners, and especially in working figures and epaulets on soldiers' uniforms.

On account of its resistance to sulphur, aluminium was early proposed in place of silver for many uses. M. Morin made an aluminium plate for use in place of a silver one for cooking eggs, and found that it answered perfectly, not being blackened in the least. This would also suggest its use for egg-cups in place of silver ones, and, again, for the spoons with which the egg is eaten. For these uses it is much superior to silver. Whole services of plate have been made of aluminium instead of silver, and present a brilliant appearance. A service made recently by Tiffany & Co., of New York, and placed in their window attracted general admiration. We would recall, however, an expression of Otto's on this subject, which runs as follows: "Were aluminium spoons as beautiful and durable as silver spoons they would not find place at the tables of the rich, because they are cheaper. It is more agreeable to use a light spoon than a heavy one, yet silver spoons are made as heavy as possible, and as large as small ladles, simply to show the wealth of the owner. The heavier the spoons the more is the man worth." When aluminium becomes cheaper it will without doubt be used for culinary articles of many kinds, replacing copper and tin vessels, for it is attacked to a less degree by the acids and salts ordinarily found in food than either of those metals, and possesses the great superiority that if dissolved its salts are not poisonous like those of copper or tin, being, on the contrary, perfectly harmless. The sulphurous acid of the air or of the products of combustion likewise leave aluminium untouched, while they quickly blacken silver. This caused its early use for reflectors, for, while not taking at the start as high a polish as silver, yet it keeps its lustre indefinitely; it has also the added superiority that its slight blue tint partly neutralizes the yellow color of artificial light, thus reflecting a very soft, white light. Even the unconsumed gas itself, containing sulphuretted hydrogen, does not blacken the aluminium reflector in the least. It would follow that as a material for candelabra, chandeliers, or, in general, for any objects exposed to the air in dwellings, aluminium keeps its color in a manner far superior to silver. This explains the superiority of aluminium leaf to silver leaf for almost any use, either for picture-frames or mural decorations indoors, or for outside decorations, especially in large cities where

24

the air contains much sulphurous acid gas. Since aluminium leaf can be purchased in books at as low a price as silver leaf, its use by gilders is becoming quite general.

Aluminium has often been proposed as a material for coinage, but the only recommendation it ever possessed for this purpose was its high price. Again, the primary property of a metal for coining is that its value should be as nearly fixed as possible. Aluminium, of all metals, is the one whose price has been most uncertain, and therefore its use, for this reason, has been out of the question. For instance, suppose that when, for twenty years, aluminium sold at $11 per lb., the Government had made the experiment of coining several million aluminium dollars. They would have weighed one-third more than a silver dollar, and would have been five times as large. This would have been the immediate disadvantage, but it would have been as nothing compared to the result of a single invention which reduced the price of aluminium one-half. At one single stroke the value of the metal in the aluminium dollar would have been reduced to fifty cents, with no reasonable probability that it would stay there for any length of time. Indeed, in another five years a further cut of 50 per cent. would have come along. So then, to begin with, the use of aluminium for coinage is economically impossible. It is said that the United States Government made experiments, in 1865, in making aluminium coins, but that the results were not sufficiently successful to induce its adoption. What the difficulties were I cannot find out, but they were—aside from the uncertain value—probably the fact of the great power required to stamp the coins, which ·is stated to be several times that needed for silver unless the metal is of exceptional purity. The problem of hardening it by adding a little silver or nickel did not probably stand in the way of its adoption. However, as an alloy in ordinary silver coins to replace copper, aluminium can be successfully used, since 5 per cent. of aluminium added to silver makes an alloy as durable as ordinary coin silver with 10 per cent. of copper, without giving it the yellow color of coin silver.

The harmlessness (innocuousness) of aluminium gives it exceptional advantages for use in surgery. M. Charrière made, in 1857, a small aluminium tube for a patient on whom tracheotomy had been practised. The tube was very light and therefore of little inconvenience to carry, and after wearing for some time the metal was very little attacked. After a long time a very thin, almost invisible coat of alumina formed, which was absolutely without harmful effect on the patient. Under the same circumstances a silver tube would have been blackened and corroded by the purulent matter. Aluminium has been used very advantageously for suture wire, and we can also deduce from this the great advantage there would be in making various surgical instruments of this material, not only from their not being corroded but also because of the decrease in weight of the instrument case which the physician has to carry, often for long distances. We would also notice the comfort to be derived from this large decrease of weight in any sort of surgical appliances, braces, trusses, etc., which have to be worn and carried about continually on the person.

The cause for the use of aluminium in the great majority of cases is its low specific gravity. We can see further that this property will be of the maximum utility where an object is of a certain fixed size, which is so in very many cases. For instance, for mountings of opera glasses, marine and field glasses, sextants, surveyors' instruments, portable electric instruments, portable astronomical instruments. We have long been familiar with the appearance and advantages of the aluminium opera and field glasses, but the difficulties met in working the metal and more especially the monopoly of their manufacture by a few firms have kept the price of these desirable instruments at unreasonable figures. Since there are but a few ounces of aluminium in the frames of these glasses, there is no reason at all why purchasers should have to pay double price for aluminium mountings over those of other metals ; and with the present wide development of the employment of aluminium I hope it will not be long before some enterprising American firm will make these instruments and sell them more nearly at their proper cost. Long before 1860, Loiseau, of Paris, made for Captain Gordon a beautiful sextant, which only weighed one-third as much as those ordinarily made of brass. For an instrument which one is obliged to hold to the eye by one hand for several minutes, making observations from the rolling deck of a vessel, this property is of the greatest convenience, as any one will attest who has had his wrist ache after making the noon observation. Similarly, a difference of a pound in the weight of a tourist's glass may hardly seem much on lifting the glass, but we have trustworthy witnesses who say that it makes twenty pounds difference towards the end of a long walk. I suppose that sextants have not been more generally made of aluminium because of its high price; this is now much less of an obstacle than formerly, and it is to be hoped that the instrument-makers will take up this subject again with fresh vigor. I have heard that engineering instruments, as transits, levels, etc., have been made in France with aluminium frames, but it is certain that they have not come as yet into anything like common use. With cheaper aluminium it is to be hoped that the makers of these instruments will lighten the burdens of our surveyors by bringing about their general adoption.

It has been well said that if the problem of aerial flight is ever to be solved, aluminium will be the chief agent in its solution. We are not going beyond the bounds of legitimate speculation when we predict that the cheapening of aluminium will result in such a revival of the numerous projects to attain this end that it is not impossible that success will be achieved. However, we would here point out the fact that magnesium is almost as strong as aluminium and it is only seven-tenths as heavy (1.75 to 2.6), and since it is unalterable enough, if properly protected, to stand the weather, it would be of still greater promise in this line. Again, in all positions where the dead weight of a moving object must be diminished in order that greater speed can be attained, aluminium will come into play, but not until it is less expensive. For instance, in torpedo boats every ounce of weight is considered and cost is almost of no moment compared with speed; also on fast express trains a like principle is involved, and the large decrease of the dead weight which could be made by substituting aluminium for brass in carriage fittings, water tanks, etc., would result in a noticeable increase of speed. Most of these speculations will only be realized when aluminium approximates more nearly in price to the common metals; but I would say a word or two about the popular fallacy of aluminium replacing steel as a constructive material for bridges, railway cars, steamships, wagons, or in any position where its strength is of importance. It is urged that aluminium is one-third the weight of steel, but it is forgotten that it is only one-third as strong; therefore the tension member of a bridge, if made of aluminium, would have to be of three times the section in order to have the same strength, and we would be simply substituting a large rod for a small one without any decrease of weight. The added disadvantage would then be met of a much larger surface to oppose resistance to the wind; the only advantage we see would be that the aluminium bar would not rust. It would be the same in using it for any constructive purpose where strain is to be met. Our steel steamships are made as thin as safety will allow; to substitute aluminium therefore would be simply to put a plate three inches thick in place of a one inch plate, without any decrease of weight.

However, for all parts where stress is not considered or where lightness and beauty are desired, aluminium may be substituted. It has been used for the handles and fixed parts of bicycles and tricycles; for similar parts of sulkies for racing; for carriage trimmings; for the metallic parts of travelling bags and trunks. It has been suggested that many of our heavy keys so burdensome to carry in the pocket could be made strongly enough of aluminium. The largest bells in the world are very seldom rung, because, being in towers, the motion of such heavy weights endangers the safety of the tower; this difficulty would be in great measure removed if aluminium could be made to answer the place of bronze in their composition.

Dental plates have been cast of aluminium, and, when complete, their weight is only a fraction of that of gold plates. But two difficulties are met in this application;. aluminium contracts very much in solidifying, and it is found almost impossible to cast it solidly on to the teeth; also, pure aluminium is slightly corroded by the acids of the food and the saliva. To overcome these difficulties, Dr. Carroll (see also p. 407) adds a little copper, which he says decreases the contraction so much that the teeth remain solidly imbedded in the plate; while the addition of some platinum and gold renders it unalterable in the mouth. The aluminium plates possess the added advantage that on contact with metallic substances no disagreeable electric current is set up. It is a matter of common experience that if a bit of iron, e. g., a carpet tack, is held in the mouth and touches a gold plate, a disagreeable bitter sensation is at once felt, due to electro-magnetic action. For this reason, some persons even refuse to wear the gold plates; but it is stated by those who have worn aluminium plates that no such effect occurs with this metal. Further, broken teeth, etc., can be again attached by means of rubber cement, the sulphur of the rubber having no action on the aluminium.

Dr. Fowler* obtained a patent for using aluminium in dentistry in combination with vulcanite, which consisted in mixing granulated aluminium with a vulcanizable compound and then vulcanizing in the usual manner. The patent also claimed the inlaying of vulcanite articles with aluminium, the joining of articles made of vulcanite or rubber with clasps or rivets of aluminium, and the use of aluminium tacks, nails, etc., in making rubber shoes.

Wheatstone determined that in a solution of caustic alkali, aluminium was electro-negative towards zinc but positive towards cadmium, tin, lead, iron, copper or platinum. In hydrochloric acid it is negative towards zinc and cadmium, but in dilute nitric or sulphuric acid negative to all the above metals except platinum and copper. E. St. Edme determined that in caustic alkali aluminium was positive to zinc and lead, in hydrochloric acid negative to both those metals, in dilute nitric or sulphuric acids strongly negative to zinc or iron and positive towards gold and platinum. It results from these properties that aluminium can be used in the battery. In a solution of caustic alkali it is said to form a very strong couple with copper, but this arrangement would entail the destruction of the aluminium. Since aluminium is so inert in presence of nitric acid it forms a good substitute for platinum in the Grove battery. Hulot used a couple composed of an aluminium plate and a zinc plate amalgamated for some time previous to use. The exciting fluid may be either dilute nitric or sulphuric acid. With water, charged with $\frac{1}{20}$ part of sulphuric acid at 66°, the cell gave for some hours a current at

* U. S. Patent, 46230, Feb. 7, 1865.

least equal to that afforded by platinum in the same conditions. After six hours its original force was diminished one-fifth, and at the end of 24 hours the cell was not entirely polarized but still gave one-fourth its original current. To restore the electronegative character of the aluminium it was only necessary to immerse it an instant in nitric acid and wash well. Col. Frishmuth, of Philadelphia, has used aluminium-zinc batteries for several years in electrolytic experiments as well as for ordinary house use, and has stated that it answers as well as platinum and in some cases gives greater power.

Aluminium has been used for the beams of fine chemical balances as well as for the very small weights used with them. The aluminium weights for this purpose are in general use; they are quite rigid, unattacked by the air, and the smallest weights are of such size as to be quite manageable. The 50 milligramme weight can still be formed into a cylinder and terminated with a button, while the tenth of a milligramme is sufficiently large to be easily handled. The only other metal used for these small weights is platinum, and when the same weights in each metal are placed side by side the difference in size is very striking. Aluminium balances are not yet so frequently seen. Collot Bros., of Paris, made a balance which, with the exception of the aqua-marine bearings, was composed entirely of aluminium. Pure aluminium, however, is hardly rigid enough for the beams, and Sartorius, of Göttingen, was the first to stiffen these by adding 4 per cent. of silver. With this improvement an aluminium balance has no equal. They are now made by almost all the fine scale-makers. Troemner, of Philadelphia, places aluminium beams on all his assayers'.button balances, while his analytical balance, entirely of aluminium except the bearings, is pronounced the chef d'œuvre of the scale-makers' art. Dr. A. A. Blair, the noted analytical chemist, after using one for several years, states that for sensitiveness and quickness it is unsurpassed, while the gases of the laboratory have not had the slightest effect on it.

Besides the uses already enumerated we may refer to the following: Aluminium has been used with great advantage in replacing other metals in delicate physical instruments where it is necessary to avoid the inertia of heavy masses; it has been used with advantage in making portable barometers, galvanometers, and electrical instruments which have to be carried about. For any articles which are usually carried around in the pocket, such as watches, compasses, knife-handles, match-cases, spectacle-cases, etc., the decrease of weight by making them of aluminium is conducive to comfort in carrying them. Aluminium has also been used to a small extent for statuettes and small works of art. It has been suggested, if it ever becomes cheap enough, as a material for telegraph wires, for which its high conductivity would fit it, but it is no stronger than copper and not nearly so good a conductor, so that it is not likely that this use will ever be made of it. It is said that in Germany experiments have been made to coat it on iron as a substitute for tin plate, but no definite results have been reported.

CHAPTER XIV.

ALLOYS OF ALUMINIUM.

ALUMINIUM unites easily with most of the metals, the combination being usually accompanied by a disengagement of heat, which is particularly active in the case of copper. (I do not know that any attempt has been made to measure this heat quantitatively.) This circumstance is thought to be an indication that these alloys are chemical combinations of the metals rather than mere mechanical mixtures. Lead, antimony and mercury appear to be the only metals not alloying with it easily. The practical production of these alloys from the metals is in general a very easy operation. The aluminium may be melted in a clean crucible without a flux and the other metal simply thrown in; it falls to the bottom, melts, and is absorbed by the aluminium. In some few cases the alloying metal must be mixed in powder with finely-divided aluminium and heated together in a closed crucible, but this is only exceptionally the case. Again, a bar of aluminium may be taken in the tongs and held under the surface of another metal already melted. This is the best method of introducing small percentages of aluminium into other metals, unless we may except the adding of a small quantity of a rich alloy to pure metal, thus diluting the percentage of aluminium to the desired quantity. Most of the alloys thus produced are improved by careful remelting, the aluminium seeming to become more intimately combined. The alloy made in the first operation is often not entirely homogeneous, but becomes more uniform, and finally perfectly so, by repeated fusions. Very few of the alloys will liquate; in general the alloy acts as a single metal. However, in some cases where the alloy is not of a very definite or certain composition, a liquation may take place, leaving as a residue an alloy with different proportions from the fluid metal running off. In the case of volatile metals, they can usually be driven out of the aluminium by keeping the alloy melted and exposed to a heat sufficient to drive off the volatile metal.

The useful alloys of aluminium seem to fall naturally into two groups: 1. Aluminium containing not over 10 to 15 per cent. of other metals. 2. Other metals containing not over 10 to 15 per cent. of aluminium. In almost every case, alloys between these limits possess no useful properties, and are mere chemical curiosities.

1. Aluminium is too soft to stand much wear or to keep a high polish, and too weak to support much stress. In order, then, to make it harder, stronger and better wearing, and at the same time to keep its valuable lightness and beautiful color, it is alloyed with a small percentage of some suitable metal. Silver, nickel, copper, or tin is frequently used for this purpose, as well as some other metals, as will be explained at length in the succeeding consideration of the alloys. It might here be remarked that the color of aluminium is not radically altered except by very large proportions of the foreign metal, by reason doubtless of our proportions being expressed by weights, while the influence of a metal in changing the color of another depends more on its volume. For instance, an alloy of 50 per cent. aluminium and 50 per cent. copper has the color of aluminium, an alloy with 70 per cent. of copper still has the white color of aluminium, but with 85–95 per cent. of copper the alloy is yellow. It has experi-

mentally been observed that the color appears to change from white to yellow at about 82 per cent. of copper. If we combine equal volumes of copper and aluminium, our alloy would contain about 77.5 per cent. of copper. If, then, we acknowledge the principle that the metals affect each other's color according to the proportions by volume in which they combine, we see the explanation of both facts—the very small influence of foreign metals in changing the color of aluminium, and the great influence aluminium has in whitening or changing the color of other metals. Again, precisely the same principle holds when we consider the specific gravity of these alloys, except that in this case our fundamental proposition — that the metals affect each other's specific gravity according to the proportions by volume in which they combine—is capable of mathematical demonstration. But we have also to consider in the case of the specific gravity a most curious phenomenon, which is, that aluminium seems to be able to absorb several per cent. of certain metals without increasing in volume, and, in some cases, it even decreases in volume. The basis for this statement is easily recognized. For instance, some aluminium was cast in a mould which gave a piece of a certain size weighing 480 grains. The aluminium was melted and 5 per cent. of silver added to it; a piece was then cast in the same mould. Now, if the aluminium had absorbed the silver without increasing in volume, the second test piece should have weighed 504 grains; it weighed 502 grains, showing only the merest dilatation of the aluminium in absorbing 5 per cent. of silver. So, if we calculate from the analyses of commercial aluminium given in Chapter III., the specific gravity of the alloy (for we can so consider it), on the supposition that all the foreign elements are absorbed by the aluminium without change of volume, it will be found that in almost every case this calculated specific gravity is very close to or even below the observed gravity, showing in the latter event that even a condensation beyond the volume of the aluminium had taken place. This condensation seems to offer a natural explanation of the hardening and strengthening effect produced by the addition of a small quantity of the metals named.

2. At the other extreme of the scale of alloys we have those containing a few per cent. of aluminium. In general, the effect

of a small quantity seems to be principally a notable increase in strength and a striking change in color of the highly colored metals. A very small quantity has little effect in reducing the specific gravity, but as the quantity increases, the effect is what we would infer from the previous remarks on specific gravity. The reason of this is that the condensation in alloying taking place with these alloys is so great that the metal absorbs the aluminium without any noticeable increase of volume, and its specific gravity may be even increased slightly at first. This great condensation offers a partial explanation of the strengthening of the original metal, since its texture is finer and its hardness increased. After a certain small limit in the percentage of aluminium is passed, the beneficial effects alluded to are overpowered by the influence of crystalline chemical combinations between the alloying metals, and the alloy quickly loses strength and malleability. With the exception of copper and tin, 5 per cent. of aluminium is the limit of the useful alloys at this end of the scale.

The alloys of aluminium with copper and iron have become so important that it seems proper to devote separate chapters to their consideration, the remainder of this one will therefore treat of the alloys with metals other than copper and iron. I wish to remark, that as the tertiary alloys cannot be rigorously classified, we will have to place them under the alloys of that metal which, besides aluminium, seems to be their characteristic ingredient. Some of the combinations of aluminium with metalloidal elements, which might possibly be looked for in this chapter, are described in Chapter V. under the compounds of aluminium.

ALUMINIUM AND NICKEL.

Tissier: "An alloy with 50 per cent. of nickel was made by melting together the metals in equal proportions under sodium chloride; the heat evolved was sufficient to raise the mass to incandescence. This alloy remains pasty at the temperature of melting copper. It is so brittle that it pulverizes under the hammer. By melting proper proportions of this alloy with more aluminium, an alloy with 25 per cent. nickel was produced. This is less fusible than aluminium, and as brittle as the 50 per cent.

alloy. By melting some 25 per cent. nickel alloy with aluminium, a 5 per cent. nickel alloy was obtained. This is much less brittle than the preceding, but is still very far from being easy to work. From the 5 per cent. alloy one with 3 per cent. was made. With this amount of nickel the aluminium acquired much hardness and rigidity, and was easy to work. A curious fact with this alloy is that it may be melted on a plate of aluminium, showing its fusion point to be less than that of pure aluminium, the reverse effect to what iron produces, which if present in the same proportion would diminish the fusibility of the aluminium. To sum up, the action of nickel on aluminium is much analogous to that of iron, for nickel, like iron, produces crystalline alloys with aluminium, but if employed with care it gives to it certain desirable qualities such as hardness, elasticity, etc."

Michel[*] melted together aluminium with nickel chloride, and obtained an alloy with nearly 25 per cent. of nickel, which was tin-white, crystalline, specific gravity 3.65, but too brittle to be of any practical use.

ALUMINIUM-NICKEL-COPPER ALLOYS.

The following alloy has a beautiful white color and takes a high polish. It resembles some of the finer grades of German silver :—

Copper	.	.	.	.	.	.	.	70 parts.
Nickel	.	.	.	.	.	.	.	23 "
Aluminium	.	.	.	.	.	.	.	7 "

A similar alloy, but somewhat harder, is called Minargent. It contains

Copper	.	.	.	.	100 parts	= 56.5 per cent.
Nickel	.	.	.	.	70 "	= 39.5 "
Antimony	.	.	.	.	5 "	= 2.8 "
Aluminium	.	.	.	.	2 "	= 1.2 "

To make this alloy, the directions are first to melt together the copper, nickel and antimony, and then granulate the resulting alloy in water. The dried granules are mixed with the alumin-

[*] Am. der Chem. und Pharm., 115, 102.

ium and with 1.5 per cent. of a flux consisting of 2 parts borax and 1 part fluorspar, and then remelted.

F. H. Sauvage states that the following alloy resembles pure silver. He gives it the name Neogen :—

Copper	58 parts.
Zinc	27 "
Nickel	12 "
Tin	2 "
Bismuth	$\frac{1}{2}$ "
Aluminium	$\frac{1}{2}$ "

P. Baudrin claims that the following alloy resembles silver very closely in color, malleability, ring and even specific gravity(!) :—

Copper	75 parts.
Nickel	16 "
Zinc	$2\frac{1}{4}$ "
Tin	$2\frac{3}{4}$ "
Cobalt	2 "
Iron	$1\frac{1}{2}$ "
Aluminium	$\frac{1}{2}$ "

Mr. Jas. Webster has patented the composition of several bronzes containing nickel. Prof. Kirkaldy's tests on these alloys, made by the "Webster Crown Metal Company," now the "Aluminium Company, Limited," gave results from 82,000 to over 100,000 lbs. per square inch with 20 to 30 per cent. elongation.

a)* Copper is melted and aluminium added to it until a ten per cent. bronze is made. There is then added to it 1 to 6 per cent. of an alloy, ready prepared, containing

Copper	20 parts.
Nickel	20 "
Tin	30 "
Aluminium	7 "

The alloy thus prepared would contain, as represented by the two extremes,

	I.	II.
Copper	89.3	86.4 per cent.
Nickel	0.3	1.4 "
Tin	0.4	2.0 "
Aluminium	10.0	10.2 "

* German Patent, 11577.

*b)** The two following alloys are prepared in the usual way, under a flux consisting of equal parts of potassium and sodium chlorides, and are cast into bars :—

I.		II.	
Aluminium . . 15 parts.		Nickel . . 17 parts.	
Tin . . . 85 "		Copper . . 17 "	
	100 "	Tin . . . 66 "	
			100 "

To make the bronzes, equal parts of these two alloys are melted with copper, the more of the alloys used the harder and better the bronze.   The best mixture is of

Copper .   .   .   . 84 parts.
Alloy I .   .   .   .   8 "
  " II .   .   .   .   8 "
                                                       100 "

The copper is first melted, then the alloys put in together and stirred well.   As iron is harmful to this bronze, the stirrer must be of wood or clay.   This alloy is suitable for art castings, kitchen utensils, etc., or anywhere where durability, hardness, malleability, polish and very slight oxidizability are required.   A cheaper and more common alloy may be made of

Copper .   .   .   . 91 parts.
Alloy I .   .   .   .   4 "
  " II .   .   .   .   5 "

These two bronzes would contain centesimally

	Rich alloy.	Poorer alloy.
Copper	85.36	94.58
Tin	12.08	6.70
Nickel	1.36	3.85
Aluminium	1.20	0.60

c)† The following alloy is said to withstand oxidation well, to have great tenacity, durability, capability to bear vibrations, and to take a high polish.   A preliminary alloy is made of

* German Patent, 28117.
† English Patent, 8320, June 23, 1886.

Copper	.	.	.	.	.	.	.	.	.	200 parts.
Tin	.	.	.	.	.	.	.	.	.	80 "
Bismuth	.	.	.	.	.	.	.	.	.	10 "
Aluminium	.	.	.	.	.	.	.	.	.	10 "

The alloy proper is formed by melting together

Preliminary alloy	.	.	.	.	.	.	.	.	4½ parts.
Copper	.	.	.	.	.	.	.	.	.	164 "
Nickel	.	.	.	.	.	.	.	.	.	70 "
Zinc	.	.	.	.	.	.	.	.	.	61½ "

The final composition would be by calculation

Copper	.	.	.	.	.	.	.	.	55.67
Nickel	.	.	.	.	.	.	.	.	23.33
Zinc	.	.	.	.	.	.	.	.	20.50
Tin	.	.	.	.	.	.	.	.	0.40
Bismuth	.	.	.	.	.	.	.	.	0.05
Aluminium	.	.	.	.	.	.	.	.	0.05
								100.00

It will be noticed that this alloy contains the same ingredients as Baudrin's alloy, though not in exactly the same proportions, the principal change being that it contains more nickel and less aluminium.

d)* Another alloy patented by Mr. Webster contains

Copper	.	.	.	.	.	53 parts	= 51.0 per cent.
Nickel	.	.	.	.	.	22½ "	= 21.6 "
Zinc	.	.	.	.	.	22 "	= 21.2 "
Tin	.	.	.	.	.	5 "	= 4.8 "
Bismuth	.	.	.	.	.	¾ "	= 0.7 "
Aluminium	.	.	.	.	.	¾ "	= 0.7 "
							100.0

"Lechesne" is an alloy not very different from some already mentioned, said to be invented by M. Thirion, but the English patent being taken out by the Societe Anonyme La Ferro-Nickel, of Paris. The patent mentions two alloys, containing

					I.	II.
Copper	.	.	.	.	900 parts.	600 parts.
Nickel	.	.	.	.	100 "	400 "
Aluminium	.	.	.	.	1¾ "	½ "

* United States Patent, 377918, Feb. 14, 1888.

Which would give in per cents.—

	I.	II.
Copper	89.84	59.97
Nickel	9.98	39.98
Aluminium	0.18	0.05
	100.00	100.00

The first of these alloys is the one to which the name "Lechesne" appears to be given. In a description of the manufacture of this alloy a French magazine states that the nickel is first put into a crucible and melted, the copper stirred in gradually, then the heat raised and the aluminium added. The alloy is heated almost to boiling and cast very hot. It is claimed that this alloy is equal to the finest German silver, being very malleable, homogeneous, strong and ductile, and stands hammering, chasing, punching, etc., perfectly.

The proportion of aluminium in these bronzes last described is so small that it appears probable that its chief function must be as a deoxidizing (perhaps also desiliconizing) agent, since a very small amount of oxygen dissolved in the bath would quickly combine with all the aluminium added. In any event, very little can be left over, apparently too little to be able to account for all the improvement made in the alloy.

Cowles Bros. have manufactured some of these bronzes, and have appropriated the name "Aluminium Silver" to the alloy made by adding aluminium to German silver, or to their alloy of aluminium, nickel, and copper called "Hercules Metal." Two of their alloys were made of—

	I.	II.
5 per cent. aluminium bronze	1	2
Nickel	2	1

containing respectively—

	I.	II.
Copper	31.67	63.33
Nickel	66.67	33.33
Aluminium	1.67	3.33

These alloys, on being tested, gave tensile strengths of 79,163 and 118,000 lbs. per square inch respectively, the first showing 33 per cent. elongation. They claim that an alloy of the proportions of II. will show an average strength of over 90,000 lbs. per square inch, but with very little elongation, and will take an edge in a manner that makes it quite suitable for table-knives.

ALUMINIUM AND SILVER.

These alloys are easily made by direct fusion together of the two metals. All the alloys containing up to 50 per cent. of silver are more fusible than pure aluminium. In general, the introduction of a few per cent. of silver into aluminium benefits it considerably, increasing its hardness, capability of polish, making it whiter, denser and stronger. It has already been remarked that aluminium will absorb almost 5 per cent. of silver without increasing in volume. A great advantage gained by this small amount of silver is also the increased facility of casting, the metal filling the moulds better and shrinking less; this alloy also rolls, draws, and works under the hammer like pure aluminium, requiring, however, more power to work it. Deville states that the alloy containing 3 per cent. of silver is unattacked by sulphuretted hydrogen, but Mierzinski states that every alloy of aluminium and silver is blackened more quickly than pure silver. I think the latter remark untenable, since the alloys with 3 to 5 per cent. of silver are noted for keeping their color like pure aluminium in places where silver would be immediately tarnished. With over 10 per cent. of silver, Mierzinski's remark may be true. M. Christophle made statuettes of the alloy with 3 per cent. of silver, which kept their beautiful white color permanently.

With 5 per cent. of silver aluminium becomes elastic and as hard as coin silver with 10 per cent. of copper, but it is still as malleable as pure aluminium. It has a specific gravity of 2.8, casts better than aluminium, shrinks less, and can be rolled or drawn perfectly, but requires more power than pure aluminium. It has been used for dessert-spoons, knife-blades, and even for watch-springs. This is the alloy which has often been proposed as a substitute for coin silver, since it contains no such poisonous metal as copper, and the color is not appreciably different from

25

that of pure silver. The beams of fine balances when made of aluminium contain from 3 to 5 per cent. of silver.

The alloy containing 10 per cent. of silver is much harder than the foregoing. It casts well, and can be rolled at a particular heat, but it does not work well under the hammer. Dr. Carroll, manufacturer of dental plates, uses an alloy for casting these articles composed of—*

Aluminium	90 to 93 parts.
Silver	5 to 9 "
Copper	1

This alloy, when cast under slight pressure, gives perfect castings, is very white and easy to work. The addition of copper is said to decrease to a minimum the shrinkage of the alloy, also giving it a closer grain.

The alloys containing from 10 to 50 per cent. of silver are all brittle and cannot be worked under the hammer. Debray states that the 50 per cent. alloy is as hard as bronze. " Tiers Argent" is an alloy of two-thirds aluminium and one-third silver. It is chiefly made in Paris; its advantages over silver are that it is cheaper, harder, and can be stamped and engraved with greater ease than the alloys of silver and copper. Some difficulty was met, at first, in getting the alloy homogeneous, but this has been overcome, and spoons, forks, salvers and articles generally made of silver are now made of this metal with an appearance equal to that of any other silver alloy. Tissier Bros. stated that the alloy with 33 per cent. of silver (the same as " Tiers Argent") is fusible enough to be used as a solder for aluminium, but they found difficulty in running it out, and also found that it made a brittle joint.

†Hirzel made alloys of aluminium and silver in atomic proportions, containing from 6 to 20 per cent. of aluminium. He found the alloy AlAg, containing 20 per cent. of aluminium, to be silver-white, very porous, tarnishing in the air, with a specific gravity of 6.73; the alloy $AlAg^2$, containing 11.11 per cent. of aluminium to be also silver-white, less porous, also tarnishing in

---

* U. S. Patent, 373221, Nov. 15, 1887.

† Bayerisches-Kunst und Gewerb-Blatt, 1858, p. 451.

the air, specific gravity 8.744; and the alloy AlAg⁴, containing 5.9 per cent. of aluminium, to be pure silver-white, very malleable and forgeable, tarnishing in the air and with a specific gravity of 9.376.

ALUMINIUM AND GOLD.

Tissier Bros. state that aluminium can contain as much as 10 per cent. of gold without its malleability or ductility being impaired. The alloy with 10 per cent. of gold can be forged at a red heat as well as aluminium, is a little harder than aluminium but polishes scarcely any better. The color of the alloy is a peculiar brownish tint. The alloy with 15 per cent. of gold can no longer be forged.

The addition of a small amount of aluminium to gold quickly takes away all its malleability. Ten per cent. of aluminium makes a white, crystalline, brittle alloy; five per cent. is said to be extremely brittle, as much so as glass; one per cent. gives an alloy similar to the gold-silver alloy called by the jewellers "green gold." It is very hard but still malleable. Professor W. Chandler Roberts-Austin, in a lecture on the influence of other metals on gold, stated that while a sample of pure gold had a tensile strength of 7 tons per square inch with 25 per cent. elongation, the addition of 0.186 per cent. of aluminium increased its strength to 8.87 tons (26 per cent.), its elongation remaining practically the same—25.5 per cent.[*]

"Nürnberg gold" is an alloy used to make cheap imitation-gold ware, resembling gold in color and not tarnishing in the air. Its composition is said to be

Copper	90 per cent.
Gold	2½ "
Aluminium	7½ "

ALUMINIUM AND PLATINUM.

Aluminium alloys readily with platinum, forming alloys more or less fusible according to the proportion of aluminium. Tissier

[*] Journal of the Society of Arts, vol. 36, p. 1125.

Bros. state that an alloy with 5 per cent. of platinum approached in color gold containing 5 per cent. of silver, but was not malleable enough to be worked. Debray says that a small quantity of platinum (no definite amount named) can be tolerated in aluminium without the malleability being destroyed.

ALUMINIUM AND TIN.

These two metals unite readily, small amounts of either metal changing the properties of the other quite materially.

A small amount of tin renders aluminium brittle. Tissier Bros. made an alloy with 3 per cent. of tin, melting the metals under sodium chloride, then remelting once without any flux. This alloy was a little more fusible than aluminium, but very brittle; the grain was very fine and crossed, but the bar broke at the first blow. Deville stated that these alloys, with a small proportion of tin, may be used as solders for aluminium, but they answer only imperfectly.

*M. Bourbouze has recommended the use of an aluminium-tin alloy for the interior parts, especially of optical instruments, in place of brass. The alloy formed of 100 aluminium to 10 of tin, or 9 per cent. of tin, is recommended as being the best for this purpose. It is white and has a specific gravity of 2.85, only slightly above that of aluminium itself. It may therefore be used in place of aluminium where great lightness is desired, and it is further superior to aluminium itself in resisting alteration better, being more easy to work; and, finally, it can be soldered without any special apparatus as easily as brass, particularly if the solder recommended by M. Bourbouze (p. 362) is used. If the alloy does work as well as represented by Bourbouze, there must be a very sudden change in the properties of aluminium alloys between the 3 per cent. tin alloy described by Tissier Bros. and this 9 per cent. alloy; but from analogy with other aluminium alloys we can admit that this is not impossible. Having so many advantages over aluminium it should replace it for many pur-

* Comptes Rendue, c. 11, p. 1317.

poses, especially in instruments of a portable character; it would, besides, be somewhat cheaper than pure aluminium.

At the other end of the scale we have alloys containing only a few per cent. of aluminium. Aluminium gives to tin greater hardness and tenacity, if it is not present in too large an amount. The alloy with 3 per cent. of aluminium is harder than tin and less acted on by acids; 5 per cent. of aluminium gives a much stronger and more elastic metal. The alloy with 7 per cent. of aluminium is especially recommended as being easy to work, being malleable at a red heat, and capable of a good polish, but possessing the drawback that it cannot be melted without a part of the tin separating from the aluminium. Tissier Bros. state that tin will not combine with more than 7 per cent. of aluminium; for they state that the alloy with 10 per cent. is not homogeneous, and on cooling in a mould arranges itself in two layers, an upper brittle one, a little more fusible than aluminium, and a lower one containing nearly all the tin, but rendered harder and less fusible than pure tin by a small quantity of aluminium. I have not been able to notice this liquation. Mr. Joseph Richards prepared an alloy with 10 per cent. of aluminium, which was whiter and much stronger than tin, and kept its color perfectly in the air. It had a specific gravity of 6.45 (calculated value 6.28) and melted only imperfectly at a temperature slightly above that of tin. It was quite malleable, but became hard by rolling. On heating a piece of this, in the form of sheet, at a gradually increasing temperature, small globules sprouted out in all directions, but they were identical in composition with the bulk of the alloy, so that no separation had taken place. The surface of the sheet appeared harder than the interior, and did not melt so easily, but this property seemed to result from other causes than difference in composition. After standing several months, a peculiar internal change took place in this metal by which it lost all its malleability, and became as rotten as baked clay, and annealing could not restore its strength.

The alloy with 19 per cent. of aluminium is said to be malleable and workable at a red heat, though not so much so as the 7 per cent. alloy. The alloys with over 30 per cent. of aluminium are described as silver-white, porous and brittle.

ALUMINIUM AND ZINC.

Aluminium unites readily with zinc, the alloys being in general harder and more fusible than aluminium. Some of the first attempts to solder aluminium were with these alloys, containing 6, 8, 12, 15, and 20 per cent. of aluminium. They answered better than any other solders which had been tried, but, unfortunately, when melted they are thick and cast with difficulty, so much so that it is necessary to spread them over the joint as a plumber does when he wipes the joints of lead pipes. Joints thus made stand hammer blows or rough usage very poorly.

The alloy containing 10 per cent. of aluminium is brittle, has the appearance of zinc, is more fusible than aluminium and less so than zinc. The alloy with 25 per cent. of aluminium has a fine, even grain, and is of about the same fusibility as the preceding. The alloy AlZn, containing 29.5 per cent. of aluminium, formed a silver-white, very brittle, crystalline alloy, with a specific gravity of 4.53. It was noticed that in the preparation, when the two metals were fused together in this proportion under a layer of sodium and potassium chlorides, they united with incandescence. The 50 per cent. alloy is white, crystalline, brittle, and does not appear to be homogeneous, for the Tissiers report that when heated on an aluminium plate it separated into a fusible portion, which ran off, and a less fusible part which did not melt until the plate did.

When zinc is in small proportion in aluminium it makes it brittle, unless it is below a few per cent. in quantity. The alloy containing 3 per cent. of zinc is described by Debray as harder than aluminium, very brilliant, but still quite malleable. Some of the aluminium first made by Deville contained zinc, the presence of which he accounted for as follows: "The retorts used for making the aluminium were made at the Vieille Montagne Zinc Works, and having in their mixture some ground-up old zinc retorts, the new retorts contained zinc, which passed into the aluminium and altered its properties in a very evident manner. Some analyses of this metal having been made in England, some asserted that French aluminium was only an alloy to which zinc gave a fusibility which might be wanting in pure aluminium."

I have been told that when an alloy of zinc and aluminium is highly heated and air passed over it, considerable alumina is found in the zinc oxide produced. If the alloy is distilled, pure zinc passes over and almost pure aluminium remains; the last traces of zinc are only expelled at a very high temperature.

Aluminium-Zinc-Copper Alloys.

These alloys are generally known as "aluminium brasses," and are about as much superior to ordinary brass as aluminium bronze is to ordinary bronze. They are made in two general ways; either by introducing metallic aluminium into melted brass, or by introducing zinc into melted aluminium bronze. The latter method is pursued by the Cowles Smelting Company, because they produce the bronze directly, while the makers of pure aluminium claim that the first method is superior, because, by adding aluminium to melted brass the dissolved cuprous oxide and zinc oxide are removed, producing a dense metal, casting without pores, while the aluminium already combined with copper does not have this effect. It appears, however, that the Cowles brasses are equal in strength, elongation, and casting qualities to those made with pure aluminium. Repeated remeltings of aluminium brass are not advisable, since, like all brasses, it changes in composition on melting, though not to so large a degree. After mixing, it need be remelted only once in a clean crucible. Aluminium brasses flow well, give sharp, sound castings, are more ductile, malleable, and have greatly increased strength and power to resist corrosion. The working qualities are said to be governed largely by the percentage of zinc present, an increase of which makes the brass harder, but does not injure its malleability.

Numerous tests have been made of the strength and elongation under stress of aluminium brasses. Even 1 per cent. of aluminium is found of benefit, while the strength increases with the amount of aluminium.

As early as 1863, Julius Baur, of New York, obtained a patent* for alloys containing—

* U. S. Patent, 40388, Oct. 1863.

Copper . . . . 14–16 parts.
Zinc . . . . 10 "
Aluminium . . . . 0.1–3 "

which he stated were of great hardness, toughness and durability. This alloy differs from the aluminium brasses now made, only in containing about half as much again of zinc. The next year, M. G. Farmer, of Salem, Mass., patented* an alloy containing—

Copper . . . . 65–80 parts.
White metals . . . . . 35–10 "
Aluminium . . . . 0.3–10 "

Cowles Bros. report the following series of tests made, in 1886, at their works in Lockport, their alloys all being made by adding zinc to aluminium bronze :—

COMPOSITION.			Tensile strength per sq. inch (castings).	Elongation, per cent.
Aluminium.	Copper.	Zinc.		
5.8	67.4	26.8	95,712	1.0
3.3	63.3	33.3	85,867	7.6
3.0	67.0	30.0	67,341	12.5
1.5	77.5	21.0	32,356	41.7
1.5	71.0	27.5	41,952	27.0
1.25	70.0	28.0	35,059	25.0
2.5	70.0	27.5	40,982	28.0
1.0	57.0	42.0	68,218	2.0
1.15	55.8	43.0	69,520	4.0

When it is remembered that ordinary brass rarely has a tensile strength of over 30,000 lbs., with an elongation of about 10 per cent., the benefit of the aluminium can be easily realized. Government tests of this company's brasses, to determine their suitability for steamship propellers, were made on the alloy composed of 1 part 10 per cent. bronze, 1 part copper and 1 part zinc, containing therefore—

Aluminium . . . . 3.3 per cent.
Copper . . . . 63.3 "
Zinc . . . . 33.3 "

The test pieces were 22 inches long, 1⅛ inches diameter, and 10 inches between elongation marks. The results showed a tensile strength of 70,000 to 82,500 lbs., elastic limit 55,000 to

* U. S. Patent, 44086, Aug. 1864.

65,000 lbs., and elongation 1.6 to 2.5 per cent.; having, therefore, a tensile strength three times and elastic limit four times as great as the best government bronzes.

The Aluminium und Magnesium Fabrik, of Bremen, state the strength of aluminium brass according to their tests to be—

2 aluminium, 23 zinc, 75 copper	.	.	.	41,000 lbs.
2 " 30 " 68 "	.	.	.	49,530 "
2½ " 30 " 67½ "	.	.	.	65,400 "

Prof. Tetmayer, of Zurich, has tested the strength of aluminium brasses with the following results:—

	STRENGTH.		Elongation,
Content of aluminium.	Kilos per sq. mm.	Lbs. per sq. in.	per cent.
4 per cent.	69	98,100	6½
3 "	60	85,300	7½
2½ "	52	73,900	20
2 "	48	68,250	30
1½ "	45	64,000	39
1 "	40	56,900	50

The amount of zinc in these brasses is not stated, but was probably 25 to 30 per cent.

In general, then, we can conclude that even a fraction of 1 per cent. of aluminium added to brass will increase very remarkably its strength and ductility, while about 5 per cent. will make an alloy, having brought its tensile strength close up to 100,000 lbs. per inch. This indicates that as far as strength alone is concerned the cost of aluminium brass castings is less than aluminium bronze castings of the same strength, for they contain less aluminium and a quantity of the cheap metal zinc. The principal disadvantage of the brass compared with the bronze is, that it cannot be remelted without changing its quality, by reason of its containing zinc.

ALUMINIUM AND CADMIUM.

Cadmium is said to unite easily with aluminium, producing alloys which are malleable and easily fusible, and which were used for soldering aluminium, but only answered that purpose imperfectly.

ALUMINIUM AND MERCURY.

It is not easy to understand why Deville said in his Treatise, " Mercury is not able to unite with aluminium. Experiments of this nature, which I have made myself, and which Mr. Wollaston has confirmed, prove it most clearly." Several years ago, before reading anything about the failure of mercury to unite with aluminium, I remember that on taking a clean, bright piece of aluminium foil, putting on it a small globule of mercury and rubbing in hard with the finger, I felt the foil become hot and a white powder appeared immediately. On rubbing still more a hole was eaten through the foil. I concluded at once that the mercury amalgamated the aluminium and that the latter when distributed through the amalgam was in such a fine state of division that the air readily oxidized it, forming the white powder. In looking up the literature on the subject the following information has been collected :—

*Caillet stated that aluminium could be amalgamated by the action of ammonium or sodium amalgam in the presence of water ; also when the aluminium is connected with the negative pole of a voltaic battery and dipped into mercury overlaid with acidulated water, or into a solution of mercuric nitrate. Tissier confirmed the latter method, adding that if the aluminium foil used is not very thick it becomes amalgamated throughout and very brittle. Tissier also found that aluminium may be made to unite with mercury merely by the intervention of a solution of caustic potash or soda, without the intervention of the battery. If the surface of the metal be well cleaned, or moistened with the alkaline solution, it is immediately melted by the mercury, and a shining amalgam forms on its surface.

†Joule states that if a solution of an aluminium salt is electrolyzed, using mercury as the negative pole, it will form an amalgam with the aluminium set free. Since the amalgam decomposes water, setting free hydrogen, it is probable that all the aluminium deposited would be promptly oxidized.

* Comptes Rendue, 49, p. 56.
† Chemical Gazette, 1850, p. 339.

*Gmelin states that potassium amalgam introduced into a hole bored in a crystal of alum immediately acquires a rotary motion, which lasts sometimes half an hour. At the same time, it takes up a considerable quantity of aluminium and becomes more viscid.

It is first stated in Watts' Dictionary (vol. viii.), that "aluminium oxidizes when its surface is simply rubbed with a piece of soft leather impregnated with mercury; the rubbed surface becomes warm and in a few seconds whitish excrescences appear consisting of pure alumina." In Watts' Supplement I., the best method of preparing the amalgam for use is stated to be by heating the two metals together in a gas which does not act on either of them. The operation is performed by placing a piece of aluminium foil at the bottom of a thick-walled test-tube, and pouring well-dried mercury on it, the tube having been previously drawn out at the middle to prevent the foil rising to the surface. The air is then expelled by a stream of carbonic acid gas and the tube is heated, without interrupting the current of gas, till the metal is all dissolved.

†J. B. Baille and C. Féry made a study of the production of aluminium amalgam, using the method just described; their results were as follows: "If aluminium foil is placed in a tube with mercury, it oxidizes very rapidly, becoming heated, while the mercury loses quickly its ordinary fluidity and becomes covered with a layer of alumina. By constructing thin glass tubes and filling them with known weights of aluminium and mercury, in an atmosphere of carbonic acid gas, heating the tubes on a sand bath, we verified these facts:—

a. The amalgamation proceeds more rapidly the higher the temperature; at the boiling point of mercury the solution is very active.

b. The vapor of mercury does not attack aluminium; only liquid mercury attacks it.

c. The weight of aluminium dissolved is proportional to the weight of mercury used, and reaches a certain maximum after a given time.

* Gmelin's Hand Book, vi. 3.
† Ann. de Chim. et de Phys., 1889, p. 246.

On cooling the bath of mercury obtained, it became evident that the amalgam consisted of a definite compound of aluminium and mercury dissolved in excess of mercury, for, on cooling, a crystalline paste separates out, floating on the bath. This paste was strained out, put in a covered crucible and heated in a current of hydrogen gas. The mercury distilled, leaving arborescent crystals of aluminium. We thus determined the composition of this compound to be, in 3.181 grammes,

Mercury . . . 2 902 grammes = 91.26 per cent.
Aluminium . . . 0.279 " = 8.74 "

Its formula is probably Al^2Hg^3, which would require 8.26 per cent. of aluminium."

These investigators further noticed that if a leaf of aluminium has been once attacked by mercury and afterwards exposed to the air, it cannot be again attacked, since the layer of alumina produced adheres so closely as to protect it perfectly. In order to attack it again it is necessary to drive off all the mercury by heat and remove the alumina by acid. A leaf may be completely dissolved in a current of mercury.

Properties of aluminium amalgam.—The aluminium in its amalgam is very easily acted upon, indeed, it behaves like a metal of the alkaline earths. If let stand exposed to the air it covers itself immediately with gelatinous, opalescent excrescences of pure hydrated alumina, exhibiting both in their form and growth considerable resemblance to the so-called Pharoah's serpents. This hydrated alumina is perfectly soluble in acids and alkalies. However, the coating protects the portion underneath to some extent, so that it takes a long time, say 24 hours, for the mercury to free itself completely of aluminium. If the mercury containing aluminium is heated and agitated in the air, more or less anhydrous alumina is formed, colored reddish from a little mercuric oxide, but all the aluminium is speedily oxidized. Water is decomposed by it at ordinary temperatures, hydrogen being liberated.

Acids attack the amalgam, dissolving the aluminium, even nitric acid (which does not attack aluminium *en masse*) dissolving out the aluminium completely. Caustic potash acts similarly, forming potassium aluminate.

Baille and Féry determined also that if antimony amalgam is mixed with aluminium amalgam, small crystals of antimony form immediately on the surface; later, the aluminium oxidizes and a bath of mercury remains, free from both metals. Lead amalgam produces a similar effect, except that some lead remains with the mercury. It is interesting to note that this is quite similar to the action of lead on a molten alloy of aluminium and tin, the aluminium being driven out of combination.

*Krauchkoll states that if aluminium and iron together are connected with the negative pole of a battery, and dipped into mercury covered with acidulated water, an amalgam of both iron and aluminium is obtained, which oxidizes more slowly in the air than aluminium amalgam.

ALUMINIUM AND LEAD.

Deville remarked that these two metals had so little tendency to combine that there may be recovered intact at the bottom of an ingot of aluminium any small pieces of lead which may accidentally have dropped into the metal. Later, however, Deville remarked that M. Peligot was able to cupel buttons of impure aluminium with lead, thereby purifying the metal, and thought that an alloy may exist in certain proportions at the temperature necessary for cupellation. It is well known that if aluminium and lead are melted down together and cooled they separate, the aluminium chilling first and floating on the fluid lead. Mierzinski remarks that this property would render it possible to use aluminium for de-silverizing bullion, if its price allowed.

I do not think that the separation is quite as absolute as is indicated above. On melting the two metals together, they separated, with a sharp line of demarkation, so that there appeared to be no combination; but the lead was hardly as blue as at first, and contained about $\frac{1}{2}$ per cent. of aluminium, while the alumin-

---

* Journal de Physique, iii. 139.

ium had visibly deteriorated, was darker, more crystalline, heavier, and contained at least 5 per cent. of lead, a test for which could easily be had before the blowpipe on charcoal. A small percentage of lead appears to be very harmful to aluminium.

ALUMINIUM AND ANTIMONY.

Aluminium appears to have as little tendency to unite with antimony as with lead. Tissier Bros. state that they were unable to get a homogeneous alloy of these two metals.

ALUMINIUM AND BISMUTH.

These two metals combine easily, the alloys being very fusible; unchanged in the air at ordinary temperatures but oxidizing rapidly when melted. As small a quantity of bismuth as 0.1 per cent. in aluminium makes it so brittle that it will crack under the hammer in spite of repeated annealings. Tissier Bros. tried 0.5, 2.5, 3 and 5 per cent. of bismuth, but with similar results; with 10 per cent. of bismuth, however, the alloy was not so brittle and could be worked under the hammer to a certain extent but could not be rolled or drawn. It takes a fine polish and is not attacked by nitric acid or blackened by sulphuretted hydrogen. The same chemists found that on melting 1 part of aluminium with 2 parts of bismuth they obtained in the crucible an alloy of the two metals floating on top of pure bismuth. The alloy contained approximately 75 per cent. of aluminium, showing that aluminium does not appear to be able to take up over 25 per cent. of bismuth. This alloy was not so brittle as pure bismuth, and was so distinct from it in the crucible that the two layers could be separated by a blow of the hammer.

ALUMINIUM AND SILICON.

Deville: "Any siliceous material whatever, put in contact with aluminium at a high temperature, is always decomposed; and if the metal is in excess there is formed an alloy or a combination of silicon and aluminium in which the two bodies may be united in almost any proportions. Glass, clay, and the earth of crucibles act in this way. However, aluminium may be melted in glassware or earthen crucibles without the least contamination of the metal if there is no contact between the metal and the material; the aluminium will not wet the crucible if put into it alone. But the moment that any flux whatever facilitates immediate contact (even sodium chloride does this), the reaction begins to take place, and the metal obtained is always more or less siliceous. It is for this reason that I have prescribed in melting aluminium not to add any kind of flux, even when the flux would not be attacked by the metal. Among the fusible materials which facilitate the melting of aluminium, it is necessary to remark of the fluorides that they attack the siliceous materials of the crucible, dissolving them with great energy, and then the siliceous materials thus brought into solution are decomposed by the aluminium with quite remarkable facility. Aluminium charged with silicon presents quite different qualities according to the proportion of the alloy. When the aluminium is in large excess, there is obtained what I have called the 'cast' state of aluminium, by means of which I discovered crystallized silicon in 1854. This 'cast' aluminium, gray and brittle, contains according to my analysis, 10.3 per cent. of silicon and traces of iron. When siliceous aluminium is attacked by hydrochloric acid, the hydrogen which it disengages has an infected odor, which I formerly attributed to the presence of a hydrocarbon, but which we now know is due to hydrogen silicide, SiH^4, thanks to the experiments of MM. Wöhler and Buff. It is by the production of this gas that may be explained the iron smell which is given out by aluminium more or less contaminated with silicon. But aluminium may absorb much larger proportions of silicon, for, on treating fluo-silicate of potash with aluminium, M. Wöhler obtained a material still metallic containing about 70 per cent. of silicon, sometimes occurring as easily separable crystals. Since I had the occasion in a work which I published on silicon to examine a large number of these combinations, I found that they were much more alterable than pure aluminium or silicon, without doubt because of the affinity which exists between silica and alumina. I have, therefore, dwelt on and tried to explain the importance of this point in obtaining perfectly pure aluminium."

A small amount of silicon does not appear to be very injurious to the malleability of aluminium, which bears it much as iron and copper do, but over 1 or 2 per cent. commences to change its color, make it harder and, especially, crystalline, so that its malleability is rapidly impaired. The silicon, however, may be present up to even 5 per cent. without preventing the use of the metal for castings and articles not to be worked. Silicon plays a role with aluminium quite analogous to carbon in iron, occurring both free and combined, as has been noted by Prof. Rammelsberg (see p. 55).

ALUMINIUM AND MAGNESIUM.

Wöhler[*] fused these two metals together in the proportions represented by Al^2Mg, forming an alloy with 69.2 per cent. of aluminium. The product was a tin-white mass, very brittle, igniting at a red heat and burning with a white flame similar to magnesium alone. The alloy, in the proportions Mg^2Al, containing 36 per cent. of aluminium, was malleable but completely destroyed by leaving in water for a day, without any evolution of hydrogen. Both these mixtures appeared to contain a compound of definite composition, for when treated with a solution of sal-ammoniac they disengaged hydrogen abundantly and deposited a brilliant, tin-white metallic powder. The solution contained magnesium chloride, while the residue was rich in aluminium, and appeared to produce some aluminate of magnesium, which clouded the solution. The metallic residue was washed with water, again treated with sal-ammoniac solution and then with solution of caustic soda until it no longer evolved hydrogen. This residue was about one-third the total alloy, and was insoluble in both the above reagents. It burnt with brilliant sparks, when thrown into a flame.

[*] Ann. der Chemie und Pharm. 138, p. 253.

ALUMINIUM AND CHROMIUM.

Wöhler* heated violet chromium chloride with aluminium wire, obtaining an alloy of the two metals, while aluminium chloride volatilized. When 1 part of aluminium was used to 2 of the salt, a gray, crystalline mass resulted, from which excess of aluminium was removed by caustic soda; leaving lustrous, tin-white crystals. When these were heated in the air they became steel-gray, but did not oxidize further. They were unattacked by caustic soda or concentrated nitric acid, but dissolved in hydrochloric acid, with separation of a little silica. Concentrated sulphuric acid oxidized them to a green mass. They were fusible only at a very high temperature. On analysis, this alloy was found to contain both iron and silicon, coming from the aluminium used. Taking these out, the composition, as regards aluminium and chromium, was—

								Calculated for AlCr.
Aluminium	.	.	.	.	.	.	31.6	33.92
Chromium	.	.	.	.	.	.	68.4	66.08
							100.0	100.00

ALUMINIUM AND MANGANESE.

Michel† (a pupil of Wöhler) obtained an alloy of these metals by melting together—

Anhydrous manganous chloride	.	.	.	.	2 parts.
Potassium and sodium chlorides	.	.	.	.	6 "
Aluminium	.	.	.	.	.	3 "

On treating the regulus with hydrochloric acid the excess of aluminium was removed, leaving a dark-gray, crystalline powder. This was unattacked by concentrated sulphuric acid in the cold, but dissolved on warming. Dilute caustic soda dissolved out the aluminium, leaving a residue of manganese. Its specific gravity was 3.4, and analysis showed it to correspond to the formula $MnAl^3$, containing 60 per cent. of aluminium.

* Ann. der Chem. und Pharm., 106, 118.
† Ann. der Chem. und Pharm., 115, 102.

ALUMINIUM AND TITANIUM.

Wöhler* obtained an alloy by melting together—

Titanic oxide	2 parts.
Cryolite	6 "
Potassium and sodium chlorides	6 "
Aluminium	1 "

The excess of aluminium being dissolved out of the mass by caustic soda, bright, steel-colored crystals remained, containing aluminium, manganese, and a little iron and silicon from the aluminium used. The specific gravity was 3.3; the alloy was infusible before the blowpipe, but on ignition in chlorine gas burnt, forming chlorides of all the metals present. Its composition seemed to vary, for another experiment at a lower temperature gave an alloy richer in silicon, with a specific gravity of 2.7. On heating these alloys in the air they first become yellow, then steel-blue, and after that oxidize no further.

Michel, proceeding in a similar way, obtained an alloy whose analysis denoted the formula Al^3Ti, containing about 62 per cent. of aluminium.

L. Levy† describes an alloy which he obtained by similar processes, as being in crystalline plates, insoluble in water, alcohol, or ether, steel-gray in color, brittle, and conducting heat and electricity. Specific gravity 3.11; composition on analysis—

Aluminium	70.92
Titanium	26.80
Silicon	2.17

ALUMINIUM AND TUNGSTEN.

Michel fused together—

Tungstic acid	3 parts.
Cryolite	6 "
Potassium and sodium chlorides	6 "
Aluminium	3 "

at a strong red heat. The fusion was afterwards treated with hydrochloric acid, leaving an iron-gray, crystalline, brittle pow-

* Ann. der Chem. und Pharm., 113, 248.
† Comptes Rendue, 106, 66 (1888).

der, single crystals being several millimetres long. Hot caustic soda extracted from them all their aluminium, leaving pure tungsten. Their specific gravity was 5.58, and the composition corresponded to the formula Al^4W, containing 37 per cent. of aluminium.

ALUMINIUM AND MOLYBDENUM.

Molybdic acetate was dissolved in hydrofluoric acid, the solution evaporated to dryness, and the residue mixed with cryolite, flux and aluminium, in the same proportions as given for tungsten. Excess of aluminium was dissolved from the product with caustic soda, and there remained a black, crystalline powder consisting of iron-gray rhombic prisms, soluble in hot nitric or hydrochloric acid, and containing only aluminium and molybdenum in proportions corresponding to the formula Al^4Mo.

ALUMINIUM AND GALLIUM.

Lecoq de Boisbaudran has stated that alloys can be formed by melting these metals together at dull redness. The alloys thus obtained remain brilliant, and do not sensibly absorb the oxygen of the air in their preparation. After cooling they are solid but brittle, even when the excess of aluminium has raised the melting point to incipient redness. They decompose water in the cold, but better at 40°, with rise of temperature, evolution of hydrogen, and formation of a chocolate-brown powder, which is ultimately resolved into white flakes of alumina.

ALUMINIUM AND CALCIUM.

Wöhler states that an alloy of these metals was obtained by fusing together equal parts of aluminium and sodium with a large excess of calcium chloride. The alloy produced had a lead color, easy cleavage, specific gravity 2.57, and was unalterable in air or water. Analysis showed it to be evidently a mixture, as it contained—

Aluminium	.	.	.	.	.	.	88.0 per cent.
Calcium .	.	.	.	.	.	.	8.6 "
Iron	.	.	.	.	.	.	2.0 "

Prof. Mabery describes a peculiar product formed in the electric furnace, consisting principally of aluminium, copper, and up to 3 per cent. of calcium (p. 306).

ALUMINIUM AND SODIUM.

Deville states that aluminium unites easily with small proportions of sodium; with 1 to 2 per cent. it decomposes water in the cold. It follows from this that the properties of the metal made carelessly by using sodium are completely altered. The last traces of sodium can be removed only with great trouble, especially when the aluminium has been produced in presence of fluoride, because of the marked affinity of aluminium for fluorine at the temperature at which aluminium fluoride, Al^2F^6, commences to volatilize.

ALUMINIUM AND BORON.

Deville obtained an alloy rich in boron by melting aluminium with borax, boracic acid or fluo-borate of potassium. The alloy is very white, only able to bear slight bending and splits in the rolls. It exhales a strong odor of hydrogen silicide, due to its having absorbed silicon from the vessel in which it was prepared. Metallic boron may be easily extracted from the alloy as both graphitic and diamantine boron. (See also Chapter V.)

ALUMINIUM AND ARSENIC.

Wöhler[*] stated that when these two metals were mixed intimately and heated together they combine with a flame, forming a dark-gray metallic powder which smells a little of arsenuretted hydrogen. It slowly evolves that gas in cold water, rapidly in hot.

[*] Pogg. Ann. 1827, ii. 160.

ALUMINIUM AND SELENIUM.

Wöhler* stated that if selenium were used instead of arsenic, in the previous case, the elements united with a flame, leaving a black, pulverulent, metallic powder, which smells strongly of silicuretted hydrogen and evolves that gas violently when dropped into water. The liquor is colored red from precipitated selenium.

ALUMINIUM AND TELLURIUM.

Wöhler* states that when heated together in powder, these elements combine very violently, leaving a black, brittle, metallic mass, smelling strongly of telluretted hydrogen and evolving that gas actively when put into water. The water becomes first red, then brown, and finally opaque, from precipitated tellurium. Put on paper, a piece of this alloy forms a brown, metallic ring around it. Wöhler noted that it decomposed water even more energetically than aluminium sulphide.

ALUMINIUM AND PHOSPHORUS.

Wöhler* found that finely-divided aluminium, heated in phosphorus vapors, burns and forms a dark-gray metallic mass, smelling strongly of phosphuretted hydrogen; it also evolves this gas copiously when placed in water.

Shaw's phosphor-aluminium bronze is described in Chap. XV.

ALUMINIUM AND CARBON.

Deville stated that he was unable to combine carbon with aluminium. On decomposing carbon tetrachloride by aluminium, ordinary carbon was formed while the aluminium remaining was unchanged. The Cowles Company obtain in their electric furnace, when reducing a mixture of alumina and carbon alone, a yellow, crystalline substance which was exhibited by Dr. T. Sterry

---

* Ante, cit.

Hunt,* as an alloy of aluminium and carbon, but that this is the case is not yet accepted as certain.

On dissolving impure aluminium in solution of caustic potash, a black residue was obtained which behaved, when filtered out and dried, exactly like amorphous carbon. I have heard it stated that molten aluminium does dissolve appreciable quantities of carbon, and that its properties are affected considerably thereby; but I am not able to give any further light on the subject than the above experiment, which seemed to show considerable carbon in an impure metal. This is a subject which needs thorough investigation.

The Newport Steel and Aluminium Company, of Kentucky, sell a substance which they call "Aluminium Plumbago," claiming it to be a combination of equal parts of aluminium and plumbago. It is probably make by stirring powdered graphite into melted aluminium, and can be nothing more than a mechanical mixture, the small difference in their specific gravities probably not being sufficient to cause a separation of the two ingredients after a thorough mixing.

CHAPTER XV.

ALUMINIUM-COPPER ALLOYS.

THESE two metals unite readily in any proportions, the union being attended with evolution of heat, which in some cases is very large in amount. As has been noticed with regard to the alloys of other metals with aluminium, so here we note that the useful alloys of these two metals are in two groups, 1st—those in which a small percentage of copper imparts certain advantageous properties to aluminium; 2nd—those in which a limited quantity of aluminium enhances the useful properties of copper. The latter is by far the class of most importance industrially.

* Halifax Meeting, Am. Ins. Mining Engineers, Sept. 16, 1885.

ALLOYS OF THE FIRST CLASS.

A small percentage of copper hardens aluminium, but does not take away its malleability. I have observed that a very small amount, less than 0.1 per cent., closes the grain of commercial aluminium, making it look more compact, and for that reason whiter, on a fractured surface. This small amount makes the aluminium perceptibly harder. It has been observed that by adding a small amount of copper to aluminium much better castings are obtained and with greater ease than with pure aluminium. The reason is probably that the copper is mostly absorbed into the aluminium so that it shrinks less in cooling and casts more solidly. Christofle, of Paris, exhibited at the London Exhibition in 1862, statuettes of a beautiful silver-white color made of aluminium with 1 per cent. of copper. Deville states that the addition of 2 or 3 per cent. of copper was found useful in making large art-castings, and that the alloy produced worked very well under the chisel and burin. The metal reduced by Deville in copper boats (see p. 205) contained from 5 to over 6 per cent. of copper, yet it could be worked easily. This malleability is retained until the copper exceeds 10 per cent., above which quantity brittleness sets in. Cowles Bros. report the alloy with 16.8 per cent. of copper as having a specific gravity of 3.23, tensile strength 29,370 lbs. per square inch, but elongation almost nothing, the alloy being too brittle for any practical use.

Alloys containing from 30 to 40 per cent. of copper are very brittle, as hard as glass and beautifully crystalline. The 50 per cent. alloy is said to be quite soft, but as the percentage of copper increases to 70 the hardness and brittleness return. The alloys remain white or, according to some writers, bluish or grayish-white if the proportion of copper does not pass 80 per cent. With this quantity the alloy is white, brittle and resembles speculum metal. The alloy with 85 per cent. of copper is yet brittle but has a yellow color. Debray concluded that " it is probable that the copper loses its color when it falls below 82 per cent., a proportion corresponding to the alloy Cu^2Al."

As is seen in the foregoing paragraph, these alloys of the first class are of little practical application, simply because there are several other metals which improve the qualities of aluminium much more than copper does. I find no reference to any supposed combination in atomic proportions, except the suggestion of Debray's in connection with the change of color.

ALLOYS OF THE SECOND CLASS.

Aluminium is more efficient than any other metal in improving the qualities of copper. Under this second head we will consider these alloys made by adding to copper any quantity of aluminium up to the limit within which the alloys are of practical value. This limit has been definitely established at about 11 per cent.; and the alloys here included are generally known as "the aluminium bronzes," being particularized as 1 per cent. bronze, 5 per cent. bronze, etc., but on account of its general superiority over all the rest, the alloy with about 10 per cent. of aluminium has received the title of "aluminium bronze"—without any qualifications. This distinction has come into general use, and it will be well to keep it in mind in going over the succeeding pages, for, whenever the expression "the aluminium bronze" or "aluminium-bronze" occurs without the percentage of aluminium being specified, the 10 per cent. alloy is signified.

The late Dr. Percy seems to have been the first to call attention to these beautiful alloys, but I am unable to find any account given by him beyond the statement that "a small proportion of aluminium increases the hardness of copper, does not injure its malleability, makes it susceptible of a beautiful polish, and varies its color from red-gold to pale-yellow." This statement must have been made prior to 1856. In that year, Tissier Bros. brought aluminium bronze to the notice of the French Academy,[*] and a week later a paper by Debray,[†] hurriedly put together, made known the results obtained up to that time by Messrs. Rousseau, Morin and himself at La Glaciére.

Aluminium bronze went through the same experience that aluminium itself and all its other alloys underwent during the

---

[*] Comptes Rendues, 43, 885 (Nov. 3, 1856).

[†] Comptes Rendues, 43, 925 (Nov. 10, 1856).

first decade after its discovery. It was unduly praised, too much claimed for it, and so, while its wonderful properties did sustain for a season all the exaggerations heaped upon it, yet some un- prejudiced observers soon made known the true state of the case, and determined its proper place among the alloys; but, as even the high place finally accorded it was far below the first expecta- tions, we have seen the alloy become the subject of unmerited fault-finding. If some people expect too much of the alloy and are therefore disappointed, let them blame themselves or, perhaps, those who led them to expect too much, and not the metal.

The history of the application of aluminium bronze is summed up in the off-repeated expression, " it would come into extensive use in the arts if its price would permit." Since, until recently, it was made by melting directly together the copper and alu- minium, its price was naturally dependent upon that of the latter metal, and by reference to the price of that metal in different years, it is an easy matter to figure out what one-tenth of a kilo or pound of aluminium would cost to make a kilo or pound of bronze. In 1864, Morin, of Paris, quoted the aluminium bronzes at the following prices :—

10 per cent. aluminium	.	.	. 15 francs per kilo	($1.36 per lb.)
7½ "	.	"	. . . 12½ "	" ($1.14 ")
5 "		"	. . . 10 "	" ($0.91 ")

During the years from 1860 to 1883 the price of aluminium re- mained almost constant, its use was not extended, and its bronze shared the same apathy. In 1879, the Société Anonyme de l'Aluminium quoted—

10 per cent. aluminium . . . 18 francs per kilo ($1.64 per lb.)

which does not show much improvement on the price of fifteen years before. And so the bronze was kept out of almost every possible industrial application until 1885, when the application by Cowles Bros. of the electric furnace to the reduction of alu- mina in presence of copper brought the question of aluminium bronze nearer to a practical solution than it had ever been before. They sold the alloy at the start on the basis of $3.50 per lb. for the contained aluminium, which brought the price down to—

10 per cent. aluminium . . . . $0.50 per lb.

This was a reduction of 70 per cent. at one stroke, and the company owning the process experienced little trouble in finding a market for their product even at that yet comparatively high price. But, the alloy thus sold was slightly inferior to that made by mixing the metals, being noticeably harder. This defect has been almost completely overcome by the manufacturers, so that their bronze, sold at present on a basis of $2.50 per lb. for the contained aluminium, is reaching a large sale. Their present prices are, I understand, 30 to 40 cents per lb. according to quantity. The lower limit would represent contained aluminium at about $1.50 per lb. The Heroult process claims to produce aluminium bronze at a still lower figure; we gather from the statements made that the aluminium in their bronzes costs them about 50 cents per lb., which would make the bronze cost only about 5 cents more than the copper from which it is made. This company's alloys are not yet on the market in America, but we understand that they are in Europe, and are giving general satisfaction. The output of the new plant in process of erection by this company at the Rhine-Falls, Neuhausen, will be about 10 tons of bronze per day. To be able to obtain such large quantities at a price comparable with ordinary tin bronze, will indeed mark the beginning of a new era in the application of aluminium alloys. Although the price of commercial aluminium has at present dropped to $2.00 per lb. yet this still means 20 cents as the cost of the aluminium alone in 1 lb. of bronze, so that the processes producing bronze directly seem to still have the advantage and will probably retain that advantage for some time to come.

Composition and nature of the bronzes.—The question whether the aluminium bronzes are chemical combinations of the two metals composing them has long been argued *pro* and *con*. It is acknowledged that the bronzes containing about $2\frac{1}{2}$, 5, $7\frac{1}{2}$ and a little less than 10 per cent. of aluminium behave most like true alloys, or chemical combinations in which the identity of the constituents is sunk completely in that of the compound. There is a coincidence to notice here, which would be remarkable if it had no significance. The alloys represented by the following formulas would contain respectively—

Cu^4Al . . . . . . 9.61 per cent. aluminium.

Cu^8Al . . . . . . 5.05 " "

$Cu^{16}Al$ . . . . . 2.59 " "

Cu^5Al . . . . . . 7.84 " "

With regard to these bronzes, Morin advances the following arguments to prove that they are true chemical combinations according to the formulas given :—*

1) The alloy made by melting 10 parts of aluminium with 90 parts of copper is a very brittle mixture, which only takes on its best properties after two or three repeated fusions, during which the excess of aluminium above that called for by the formula is oxidized and separated out. When this point has been reached, further meltings do not alter the properties any more.

2) The addition of 5, $7\frac{1}{2}$ and 10 per cent. of aluminium give perfectly homogeneous alloys, but if 6.7 or 8 per cent. is added, only metallic mixtures result in which can be distinguished uncombined aluminium. A point worthy of remark is that the color of the 5 and 10 per cent. bronzes is similar to gold, the latter giving a brighter shade, but the $7\frac{1}{2}$ per cent. bronze is of a greenish cast, and has an entirely different appearance from the other two.

3) When 10 per cent. of aluminium is added to molten copper, the large amount of heat absorbed by the aluminium cools the copper so much that almost all of it becomes solid. However, as the whole becomes warmer, and the chilled part is stirred in that remaining melted, the mass gradually warms up as the copper combines with the aluminium, until towards the end the crucible is raised to a white heat by the heat set free within it. This phenomenon can only be explained on the basis of chemical action and combination between the two metals.

4) If a piece of aluminium bronze is heated nearly to its melting point and hammered at that temperature, it splits into fragments showing peculiar cleavage planes. These particles are evidently crystalline, and are smaller the nearer the temperature has been brought to the melting point. If they are analyzed it will be found that they are all identical in composition with the

* Genie Industriel, 1864, p. 167.

mass, and that therefore no liquation or separation of any kind has taken place. If the bronze were a mere mixture, the temperature to which these pieces were heated would have caused the more easily fusible aluminium to sweat out.

Almost all the arguments advanced to prove that the aluminium bronzes are true alloys are along these four lines discussed by Morin. A German pamphlet describes the $2\frac{1}{2}$, 5 and 10 per cent. bronzes as not being mixtures, like the alloys of copper and zinc, but perfect combinations of absolute homogeneity and uniform density and not altered by remelting. When, however, it is known that 1 per cent. or less of aluminium has a considerable influence on copper, it is almost begging the question to claim all the improvement to be due to the formation of some alloy in atomic proportion something like $Cu^{40}Al$. It is almost certain that the first effect of adding the aluminium, and in the case of very small proportions the whole effect, is that the aluminium acts as a deoxidizing agent, somewhat similar to the action of phosphorus in phosphorizing bronze. It probably reduces all dissolved oxides, combines with any occluded gases and protects the copper from oxidation during cooling. Thus a fraction of a per cent. of aluminium is of considerable benefit to brass and bronze, adding just before pouring. I have in mind the statement of an English metallurgist who reported that on adding 1 per cent. of aluminium to a not very pure copper he was unable to find any aluminium on analyzing the resulting metal, but its properties were a great improvement on those of the original metal. It is probable that in this case the aluminium was entirely slagged off. On adding the same quantity to pure copper it is certain that a larger part of the aluminium remains in the copper, as is shown by its color, but part of the improvement is undoubtedly due to the elimination of gases and traces of impurities.

It is on the strength of these facts that some makers of pure aluminium have advanced the argument (directed against those makers producing bronzes directly) that the aluminium forms valuable bronzes *principally* by its action on the foreign substances in the copper, and that therefore a really valuable aluminium bronze can only be made by introducing pure aluminium into copper; that it is wholly wrong practice to reduce an aluminium bronze of high percentage to one of low percentage by adding fresh copper, since, although the bronze used may be of first quality, yet the already combined aluminium cannot produce any effect on the copper added, and the resulting bronze will be of inferior quality. The weak point in this argument is the last statement, that "the already combined aluminium cannot produce any effect on the copper added." We know that the aluminium and copper had previously combined with evolution of much heat, but there is reason to suppose that if subsequently a substance were brought into contact with this bronze which could be decomposed by aluminium with the evolution of much more heat than was set free in the formation of the bronze—the second reaction would take place. For instance, the reaction

$$Al^2 + 3Cu^2O = Al^2O^3 + 3Cu^2$$

is very strongly exothermic, to the amount of something like 270,000 calories, and it is to a high degree probable that the heat of combination of this amount of aluminium with 9 times its weight of copper is not more than a small fraction of this amount. Therefore, the reaction

$$2Cu^4Al + 3Cu^2O^3 = Al^2O^3 + 7Cu^2$$

which differs from the former one in involving the decomposition of the bronze, Cu^4Al, is still largely exothermic and therefore liable to occur. To sum up, even considering that the heat developed in the formation of the copper-aluminium alloy is very large, say even 3 or 4 times that of mercury combining with potassium, the largest so far measured experimentally, yet it would be so small in comparison with the heat developed by the reaction of the aluminium on the impurities present as to theoretically have very little effect in retarding that reaction. I would therefore conclude that the already combined aluminium would be very slightly less powerful than free aluminium in reducing and eliminating impurities in the copper, and that the practical effects would be identical.

Further, it is clear that when a considerable percentage of aluminium occurs in the bronze, only a part of its beneficial effect is due to the removal of impurities, a considerable effect being necessarily produced by the presence of the aluminium in the bronze. Therefore, even if the argument advanced as to the combined aluminium not being able to remove impurities were true, yet this can only be part of the function of the aluminium, and if the aluminium is really present, the results will be identical in the two cases as far as the influence of the continued presence of the aluminium is concerned.

Therefore, it may be safely concluded, that if the bronzes are made with copper of the same purity, the one made by diluting a good quality high-priced bronze with copper will be practically as good as the one made directly from the metals. If it is true that the bronze placed on the market by those following the former method is not as good as that made from the metals directly, the reason is that the high per cent. bronze is not pure, due to the manner of its production. This is a defect inherent in the first production of the alloy, and not chargeable to the subsequent inability of the aluminium to perform its function during dilution.

The company producing bronzes by direct reduction have been understood to claim that since their alloy is made by copper or copper vapor absorbing aluminium vapor, the result is a bronze far more homogeneous than is produced by melting the two metals together. This is partially true, since the metals in the latter case may not be kept molten long enough for complete combination to take place, and therefore would lack homogeneity to a certain degree if not carefully made. This opportunity for carelessness to enter into the question and affect the result does not occur in the reduction process, and the latter bronze would for this reason be at an advantage. But if the alloy is made carefully and with the precautions dictated by experience in melting the metals together, there is no reason why the bronzes should differ. Two specimens containing nothing but aluminium and copper in like proportions will be identical no matter how they are produced.

Looking at the industrial side of the question, it is safe to say that users of bronze generally prefer to buy the constituent metals and make their own mixtures in whatever proportions their experience shows to be best for the kind of work they execute. Other things being equal, I think a bronze worker would prefer to purchase the metals, rather than buy the bronze ready made, as in the former case he can always be sure of the composition of his alloy, and can profit by all those little variations in the manipulations and in the proportions of metals used which experience suggests, and which form so valuable a part of the successful metal-worker's art.

Preparing the bronzes.—In making aluminium-copper alloys great attention must be paid to the quality of the copper used. Ordinary commercial copper may contain small amounts of antimony, arsenic, or iron, which the aluminium can in no way remove, and which affect very injuriously the quality of the bronze. The aluminium bronzes seem to be extremely sensitive to the above metals, particularly to iron. This necessitates the employment of the very purest copper; electrolytic is sometimes used when not too high priced, but Lake Superior is generally found satisfactory enough. Even the purest copper may contain dissolved cuprous oxide or occluded gases, and it is one of the functions of the aluminium to reduce these oxides and gases, forming slag which rises to the surface and leaving the bronze free from their influences. If tin occurs in the copper, it lowers very greatly the ductility and strength of the bronzes, but zinc is not so harmful.

Care should also be taken as to the purity of the aluminium used, though its impurities are not as harmful as they would be if occurring in similar percentage in the copper, since so much more copper than aluminium is used in these alloys. Yet, the bronzes are so sensitive to the presence of iron that an aluminium with as small a percentage of this metal as possible should be used. The silicon in commercial aluminium is not so harmful as the iron, but it does harden the bronze considerably and increases its tensile strength. The purest aluminium alloyed with the purest copper always produces the highest quality of bronze.

The "Magnesium und Aluminium Fabrik," of Hemelingen, give the following directions for preparing the bronzes: "Melt the copper in a plumbago crucible and heat it somewhat hotter than its melting point. When quite fluid and the surface clean, sticks of aluminium of a suitable size are taken in tongs and pushed down under the surface, thus protecting the aluminium from oxidizing. The first effect is necessarily to chill the copper more or less in contact with the aluminium, but if the copper was at a good heat to start with, the chilled part is speedily dissolved and the aluminium attacked. The chemical action of the aluminium is then shown by a rise of temperature which may even reach a white heat, considerable commotion may take place at first, but this gradually subsides. When the required amount of aluminium has been introduced, the bronze is let alone for a few minutes and then well stirred, taking care not to rub or scrape the sides of the crucible. By the stirring, the slag, which commenced to rise even during the alloying, is brought almost entirely to the surface. The crucible is then taken out of the furnace, the slag removed from the surface with a skimmer, the melt again stirred to bring up what little slag may still remain in it, and is then ready for casting. It is very injurious to leave it longer in the fire than is absolutely necessary; also, any flux is unnecessary, the bronze needing only to be covered with charcoal powder. The particular point to be attended to in melting these bronzes is to handle as quickly as possible when once melted."

As with ordinary brass and bronze, two or three remeltings are needed before the combination of the metals appears to be perfect and the bronze takes on its best qualities. When the alloy is thus made perfect, the bronze is not altered by remelting, and the aluminium, which in the first instance removed the dissolved oxides and occluded gases from the copper, now prevents the copper from taking them up again and so keeps the bronze up to quality. If, however, the bronze is kept melted a long time, and subject to oxidizing influences, the tendency of the copper to absorb oxygen will cause some loss of aluminium by the action of the latter in removing the oxygen taken up, and a slag consisting principally of alumina will result; but if the remelting of the bronze is done quickly and the surface covered with charcoal or coke, the loss from this cause will be very trifling, and the percentage of aluminium will remain practically constant.

We have already discussed at length the dilution of a high per cent. bronze to a lower one. This operation is practised on a large scale by the companies which produce aluminium bronze.

directly in their reduction furnaces. I understand that they simply melt the high-percent. bronze in a crucible and stir into it pure copper in the required proportions, or else melt the two down together on the hearth of a reverberatory furnace. The combined aluminium thus cleanses the added copper and produces a lower bronze of right quality if the high-percent. bronze used is pure and the copper added of the proper quality. It is, of course, quite certain that no difficulty can occur in adding aluminium to a low-percent. bronze, to increase its percentage, other than that of imperfect combination, which may be overcome by one or two remeltings.

Fusibility.—Aluminium lowers the fusing point of copper, so that the bronzes melt quite readily. No accurate observations have been made as to their exact melting points, but they lie somewhere between that of copper (1050°) and that of aluminium (600°), with probably a nearer approach to the latter than would be inferred simply from the proportion of aluminium present. The Cowles Electric Smelting Company state in one of their pamphlets that their A grade bronze (containing about $9\frac{1}{2}$ per cent. of aluminium) melts at about 1700° F. (925° C.), but I do not know with what degree of accuracy this figure was determined. These alloys pass quickly from the solid to a quite fluid condition, with a very small intermediate period of softening.

Casting.—Aluminium bronze is not an easy metal to cast perfectly until the moulder is familiar with its peculiarities. The great enemies of steel castings, dissolved oxides and gases, forming blowholes, are here absent. As we have seen, the aluminium removes these impurities from the original copper and by its presence afterwards keeps the bronze free from them. This difficulty, therefore, is not met in casting aluminium bronze, but the obstacles which afford most trouble are the shrinkage in setting and contraction in cooling. These two factors are extraordinarily large, and must be met by provisions made in moulding, as shown later.

A plumbago crucible is the best to use for melting the bronze, the melt being kept covered with powdered charcoal. I would recommend that the stirrers and skimmers used be coated with a wash made of plumbago and a little fire-clay, as the contact of

27

bronze with bare iron tools cannot but injure its quality. The crucible should not be kept in the fire any longer than is absolutely required to bring the bronze to proper heat for casting. In casting, it is of considerable advantage to use a casting ladle, into which the bronze is poured, which is arranged so as to tap from the bottom. This effectually keeps any slag or scum from being entangled in the casting. The same result is also obtained by arranging a large basin on top of the pouring gate, which is temporarily closed by an iron or clay stopper. Enough bronze is then poured into this basin to fill the mould, and after the dirt is all well up to the surface the plug is withdrawn and the mould fills with clean metal. For very small work the ordinary skim-gate will answer the above purpose; for large castings the tapping ladle is preferred. Plain castings, such as pump-rods, shafting, etc., and especially billets for rolling and drawing, are cast advantageously in iron moulds, which should be provided with a large sinking-head on top to feed the casting as it cools. Rubbing with a mixture of plumbago, kaolin and oil is said to protect the iron moulds from sticking. The chilling makes the bronze soft, and the slabs and cylinders thus cast for rolling and drawing are in good condition to be worked at once. For ordinary foundry castings, sand moulds are used. The slower cooling makes the castings more or less hard ; if soft castings are wanted they can be subsequently annealed.

Thomas D. West, the author of " American Foundry Practice," read a paper on " Casting Aluminium Bronze and other strong metals" before the American Society of Mechanical Engineers, November, 1886, from which we make the following extracts :—

" The difficulties which beset the casting of aluminium bronze are in some respects similar to those which were encountered in perfecting methods for casting steel. There is much small work which can be successfully cast by methods used in the ordinary moulding of cast-iron, but in peculiarly proportioned and in large bronze castings other means and extra display of skill and judgment will be generally required. In strong metals there appears to be a ' red shortness,' or degree of temperature after it becomes solidified at which it may be torn apart if it meets a very little resistance to its contraction, and the separation may be such as cannot be detected by the eye, but will be made known only when pressure is put upon the casting. To overcome this evil and to make allowances for sufficient freedom in contraction much judgment will often be required, and different modes must be adopted to suit varying conditions. One factor often met with is that of the incompressibility of cores or parts forming the interior portion of castings, while another is the resistance which flanges, etc., upon an exterior surface oppose to freedom of contraction of the mass. The core must generally be 'rotten' and of a yielding character. This is obtained by using rosin in coarse sand and filling the core as full of cinders and large vent-holes as possible, and by not using any core rods of iron. The rosin would cause the core when heated to become soft, and would make it very nearly as compressible as a 'green-sand' core when the pressure of the contraction of the metal would come upon it.

"By means of dried rosin or green-sand cores we were able to meet almost any difficulties which might arise in ordinary work from the evils of contraction, so far as cores were concerned. For large cylinders or castings which might require large round cores which could be 'swept,' a hay rope wound around a core barrel would often prove an excellent yielding backing, and allow freedom for contraction sufficient to insure no rents or invisible strain in the body of the casting. To provide means for freedom in the contraction of exterior portions of castings, which may be supposed to offer resistance sufficient to cause an injury, different methods will have to be employed in almost every new form of such patterns. It may be that conditions will permit the mould to be of a sufficiently yielding character, and again it may be necessary to dig away portions of the mould or loosen bolts, etc., as soon as the liquid metal is thought to have solidified. In any metal there may be invisible rents or strains left in a casting through tension when cooling sufficient to make it fragile or crack of its own accord, and it is an element which from its very deceptive nature should command the closest attention of all interested in the manufacture of castings.

"Like contraction, the element of shrinkage is often found seriously to impede the attaining of perfect castings from strong metals. In steel castings much labor has to be expended in providing risers sufficient to 'feed solid' or prevent 'draw-holes' from being formed, and in casting aluminium bronze a similar necessity is found. The only way to insure against the evils of shrinkage in this metal was to have the 'risers' larger than the body or part of the castings which they were intended to 'feed.' The feeder or riser being the largest body, it will, of course, remain fluid longer than the casting, and, as in cast-iron, that part which solidifies first will draw from the nearest uppermost fluid body, and thus leave holes in the part which remains longest fluid. The above principle will be seen to be effective in obtaining the end sought. It is to be remembered that it is not practical to 'churn' this bronze, as is done with cast-iron. A long cast-iron roll, 1 foot in diameter, can by means of a feeder 5 inches in diameter and a $\frac{1}{2}$ inch wrought-iron rod be made perfectly sound for its full length. To cast such a solid in bronze, the feeding head should be at least as large as the diameter of the roll, and the casting moulded about one-quarter longer than the length of roll desired. The extra length would contain the shrinkage hole, and when cut off a solid casting would be left. This is a plan often practised in the making of guns, etc., in cast-iron, and is done partly to insure against the inability of many moulders to feed solid, and to save that labor. A method which the writer found to work well in assisting to avoid shrinkage in ordinary castings in aluminium bronze was to 'gate' a mould so that it could be filled or poured as quickly as possible, and to have the metal as dull as it would flow to warrant a full run casting. By this plan very disproportionate castings were made without feeders on the heavier parts, and upon which draw or shrinkage holes would surely have appeared had the metal been poured hot.

"The metal works well in our ordinary moulding sands and 'peels' extra well. As a general thing, disproportionate castings weighing over 100 pounds are best made in 'dry' instead of 'green'-sand moulds, as such will permit of cleaner work and a duller pouring of the metal, for in this method there is not that dampness which is given off from a green-sand mould and which is so liable to cause 'cold shots.' When the position of the casting work will permit, many forms which are proportionate in thickness can be well made in green-sand by coating the surface of the moulds and gates with silver lead or plumbago.

"From 'blow-holes,' which are another characteristic element likely to exist in strong metals, it can be said that aluminium bronze is free. Should any exist it is the fault of the moulder or his mould, as the metal itself runs in iron moulds as sound and close as gold. Sand moulds to procure good work must be well vented, and, if of 'dry-sand,' thoroughly open sand mixture should be used and well dried. The sand for 'green-sand' work is best fine, similar to what will work well for brass castings. For 'dry-sand' work the mixture should be as open in nature as possible, and, for blacking the mould, use the same mixtures as are found to work well with cast-iron."

In view of the above-recommended precautions, the reader recognizes at once the falsity of the statement in the pamphlet of a German manufacturer that "aluminium bronze shrinks almost none at all." I have, in fact, measured the contraction of a 5 per cent. bronze as $\frac{1}{4}$ inch to the foot, just twice that of cast-iron. Such wild statements, at a time when aluminium bronze is coming into such extended use, can only bring discredit on the firm making them.

Color.—One per cent. of aluminium changes the color of copper considerably, making it like red brass. The influence of the aluminium is properly seen when two ingots, one of pure copper and the other of the bronze, are chilled from a high temperature, in which case the yellower color of the latter is plainly perceptible. If, however, the two ingots are let cool slowly in the air, the effect of the aluminium is more striking, since the copper oxidizes so as to turn quite black while the bronze keeps as untarnished as if it had been chilled in water. Two and one-half per cent. of aluminium makes a bronze resembling in color gold of low carat alloyed with copper. The five per cent. bronze is pure yellow and the nearest approach to the color of pure gold of any known metal or alloy. The seven and one-half per cent. bronze has the color of the jeweller's green gold. The ten per cent. bronze, or aluminium bronze proper, is a bright light-yellow, similar to the jeweller's pale gold. The eleven per cent. bronze is still paler.

Fifteen per cent. makes a yellowish-white alloy which is too brittle to be of any practical use.

Specific gravity.—The aluminium bronzes are not as much lighter than copper as the percentage of aluminium would indicate. If the specific gravity of a mixture of copper and aluminium in the given proportions is calculated, it will be found uniformly lower than the observed specific gravities, showing the amount of contraction in alloying to be large. This will account for the denseness and very close grain of these bronzes. The following table will set forth these data more plainly :—

| | SPECIFIC GRAVITY. | | Contraction in alloying |
Per cent. of aluminium.	Observed.	Calculated.	(per cent.)
2½	8.60*	8.40	2.3
3	8.69	8.33	4.1
4	8.62	8.13	5.7
5	{ 8.37	8.0	4.4 }
	{ 8.20*	"	2.4 }
7½	8.00†	7.60	5.0
10	{ 7.69	7.25	5.5 }
	{ 7.56†	"	4.1 }
11	7.23†	7.10	1.8

Hardness.—Accurate observations of the hardness of the aluminium bronzes are wanting, with perhaps a single exception. It is known that the annealed metal, which has been chilled from a red heat, is much softer than that which has been allowed to cool very slowly. The metal which has been worked some time becomes almost as hard as steel. I think that the pure aluminium bronzes, when softened, are yet harder than all ordinary bronzes.

The Cowles Company's bronzes almost invariably contain a small amount of silicon, which slightly increases the tensile strength, but is principally active in increasing the hardness of the bronze. For this reason, the following determinations, made on their alloys at the Washington Navy Yard, graded according to the government standard of hardness, must be considered as maximum figures :—

* According to Saarburger.
† Cowles Bros. alloys. The rest were given by Bell Bros.

Metal.	Hardness.	Remarks.
Average of gun-steel forgings, oil-tempered and annealed . .	21.4	Elongation 20 per cent.
Same, not oil-tempered and annealed	14.9	" 18.7 "
11 per cent. aluminium bronze, cast in sand	20.0	Elongation 4.5 per cent.
Same, forged at low redness . .	18.0	" 5.2 "
Same, rolled at red heat . . .	21.2	" 6.5 "
7½ per cent. aluminium bronze, rolled hot	16.9	Elongation 30.0 per cent.
Same, cast in chill moulds . .	{ 13.4 { 11.8	" 32.1 " } " 26.1 " }
Government bronzes . . .	{ 3.3 { 6.6	Elongation 12.5 per cent. " 33.6 "

Transverse strength.—The transverse strength or rigidity of aluminium bronze is one of its most noticeable qualities. Strange measured the amount of deflection in bars of aluminium bronze, gun-bronze, and ordinary yellow brass, laid on horizontal supports with the weight in the centre. From these experiments he concluded that the aluminium alloy was three times as rigid as gun-bronze and forty-four times as much so as ordinary brass.

Compressive strength.—Mr. Anderson made a test of aluminium bronze for compressive strength in the Royal Gun Foundry at Woolwich. The piece taken had a height and diameter of 15 millimetres ($\frac{9}{16}$ inch). The results were as follows :—

STRAIN APPLIED.		Shortening,	Permanent set,
kilos per sq. mm.	lbs. per sq. in.	per cent.	per cent.
14.84	21,100	1.01	0.17
96.42	137,140	(Specimen crushed.)	

It is seen from this test that the elastic limit for compression is comparatively low, but that the metal gives very slowly, as is shown by the large interval between this point and the ultimate crushing strength.

According to two tests made at the Watertown Arsenal, on Cowles' bronzes, their compressive strength was as follows :—

11 per cent. aluminium . . . 160,400 lbs. per sq. inch.
10   "   aluminium . . . 153,600   "   "

The test pieces being 2 inches long and $\frac{1}{2}$ inch diameter, and cast specimens. The shortening up to the crushing point was 15 and 23.7 per cent. respectively.

Tensile strength.—One of the first properties of the aluminium bronzes to draw attention to them was their great tensile strength. Since this property is attended by a large extensibility under strain, and a high elastic limit, we see that they are very valuable metals for engineering uses. For such uses, however, a metal costing $1.50 to $2 per lb. is almost entirely out of question, and it is only recently that the lower price has permitted placing the metal in situations where its great strength is of most use.

Taking the various determinations chronologically, we have, first, those made by Lechatelier, in 1858. The alloy was cast in cylinders 10 millimetres (0.37 inch) in diameter, length of test-pieces not given. The results were as follows :—

Percentage of aluminium.	STRENGTH.	
	kilos per sq. mm.	lbs. per sq. in.
10	58.36	83,000
10	55.35	78,720
8	33.18	47,190
5	32.20	45,800
5	31.43	44,700
French wrought-iron . . .	35.00	49,780

Deville determined the strength of aluminium bronze drawn into wire, compared with that of iron and steel wire of the same size (1 mm. diameter = No. 19 B. W. G.), as—

	Kilos per sq. mm.	lbs. per sq. in.
Aluminium bronze . . .	85	120,900
Best iron	60	85,340
Steel	{ 90 / 100	{ 128,000 / 142,000

Experiments made in 1861,* on the relative strengths of aluminium bronze and the common metals gave—

* Chemical News, v. p. 318.

Aluminium bronze 19
Gun metal (copper 89, tin 11) 10
Drawn brass wire 8
, Drawn copper wire 7
Tin bronze (copper 96, tin 4) 4
Same, with 1 per cent. aluminium 10
Same, with 2 " " 16

In 1862, Anderson tested the strength of aluminium bronze at the Woolwich Arsenal; testing pieces $3\frac{1}{4}$ inches long and 0.6 inch diameter, with the following results :—

	Lbs. per sq. in.
Aluminium bronze	73,185
Gun metal	35,000
Hardest steel	118,000
Medium cast-steel	82,850
Steel from a Krupp cannon	74,670

The Cowles bronzes have been tested officially at the Watertown Arsenal and at the Washington Navy Yard, with especial reference to their comparison with the government bronzes. The Cowles alloys contain small variable quantities of silicon and the following content of aluminium :—

Grade.	
Special " A "	11 per cent.
A	10 "
B	$7\frac{1}{2}$ "
C	5–$5\frac{1}{2}$ "
D	$2\frac{1}{2}$ "
E	$1\frac{1}{4}$ "

Three tests of the "special A" grade, made on the Watertown testing machine, gave the following results :—

1. Cast in sand.   Test piece 2 inches long, 0.2 sq. inch area.
2. Forged hot (some flaws).   "   10   "   0.5   "   "
3. Rolled hot   "   2   "   0.2   "   "

	1.	2.	3.
Tensile strength (lbs. per sq. in.)	109,800	87,600	111,400
Elastic limit " "	79,900	41,000	84,000
Total elongation (per cent.)	0.	2.	6.5
Modulus of elasticity	—	17,240,000	15,625,000

The curve of number 3 is given on the diagram (Fig. 29) as A.

Two tests of the "A" grade, on the same machine, gave

5. Cast in sand.   Test piece 1 inch   long, 0.08 sq. inch area.
6. Forged hot   "   10 inches "   0.25   "   "

	5.	**6.**
Tensile strength (lbs. per sq. in.)	87,510	89,680
Elastic limit " "	—	50,000
Total elongation (per cent.)	17.	29.7
Modulus of elasticity	—	15,741,000

The curve of number 6 is given on the diagram (Fig. 29) as B.

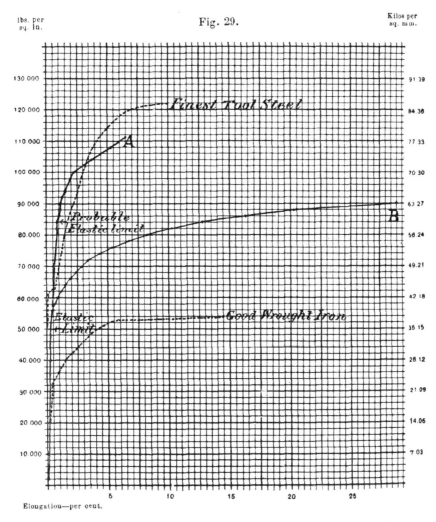

Fig. 29.

The test of a specimen of Cowles' bronzes of "A 3" grade, containing 8½ per cent. of aluminium made by Professor Unwin, F. R. S., gave

Tensile strength . . . . . . 82,389 lbs. per sq. in.
Elastic limit . . . . . 39,738 " "
Elongation . . . . . . 33.26 per cent.
Test piece 10 inches long and 0.2 sq. inch area.

The B, C, D, and E grades decrease in tensile, transverse, torsional and compressive strength and in elastic limit in the order in which they are named, but the extensibility increases as the other properties decrease.

Samples of Cowles' B and C grades, tested by Mr. Edw. D. Self, at the Stevens Institute, Hoboken, N. J., gave

	B.	C.
Tensile strength (lbs. per sq. in.) . .	51,680	40,845
Elongation (per cent.)	4.1	11.2

A sample of the D grade, tested on the manufacturer's machine, gave a strength of 42,770 lbs., with 53 per cent. elongation.

The bronzes made at Neuhausen by the Heroult process have been tested in Zurich with the following results :—

Grade.	Per cent. aluminium	TENSILE STRENGTH.		Elongation, per cent.
		kilos per sq. mm.	lbs. per sq. in.	
A	7	35.9	49,200	25.4
		38.7	55,060	27.3
B	7½	38.4	54,600	27.4
		40.7	58,320	25.5
C	8	36.4	51,760	34.3
		45.0	64,000	45.7
		46.3	65,950	48.4
D	8½	48.0	68,270	37.5
E	9	50.6	72,250	32.9
		51.6	73,380	39.2
F	9½	52.2	74,240	23.5
		56.0	79,650	16.1
G	10	55.3	78,620	18.5
		62.1	88,325	10.5
H	10½	59.0	83,915	12.0
		64.0	91,000	6.3

Professor Tetmayer, under whose supervision the above tests were made, made a series of bronzes from the pure metals, for the express purpose of testing, and obtained the following results :—

Per cent. of aluminium.	TENSILE STRENGTH.		Elongation, per cent.
	kilos per sq. mm.	lbs. per sq. in.	
5½	44	62,580	64.0
8½	50	71,115	52.5
9	57.5	81,780	32.
9½	62.	88,180	19.
10	64.	91,000	11.
11	68.	96,720	1.
11½	80.	113,780	0.5

The annexed diagram (Fig. 30) shows by its curves the variation of tensile strength and elongation of the aluminium bronzes with the increasing percentage of aluminium, the curves C and C' being taken from Cowles' advertised guarantee (1889), the elongation being the minimum and the strength the average values guaranteed in castings; H and H' represent the average values given by Professor Tetmayer for bronzes made by the Heroult process; T and T' represent Tetmayer's determinations with bronzes made from the pure metals.

Cowles Bros. tested the effect of temperature on the strength of aluminium bronze. A bar was tested, and showed a tensile strength of 109,120 lbs. per square inch with 5 per cent. elongation. A duplicate bar was then put in the machine and 100,000 lbs. per square inch put on it. It was then heated (still under stress) by a blowpipe flame to about 400° F., and the strain increased to 107,000 lbs. The bar was then cooled down to the temperature of the room, and afterwards stood 110,160 lbs. per square inch without breaking. It thus appears that aluminium bronze does not seem to lose any strength up to a heat of about 400° F.

Annealing and hardening.—Aluminium bronze acts like ordinary brass in these respects. It is softened by chilling; the best procedure is said to be to heat the articles to bright redness for some time, to destroy all crystalline structure, then cool in still air to full redness and plunge into cold water. The metal becomes very hard and stiff when worked for some time without annealing. To get the bronze to its maximum elasticity and hardness it must be cooled very slowly. Articles of bronze can be heated red-hot in charcoal powder and allowed to cool embedded in it. It is said that the bronze can be thus made elastic enough for the hair-springs of watches.

Tensile strength,
lbs. per sq. in.

Fig. 30.

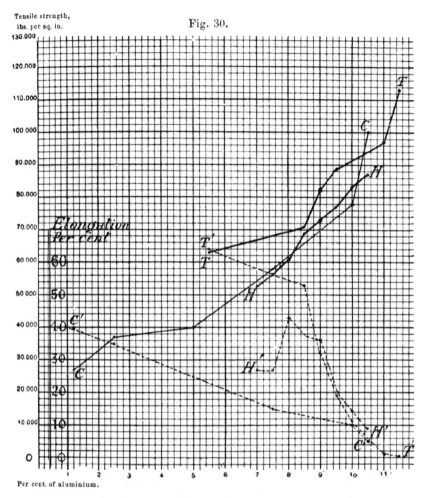

Per cent. of aluminium.

Working.—Despite the assertions that aluminium bronze can be heated to bright redness and then hammered out until quite cold, yet it is not so easily worked. Persons who have had a large experience with the alloy say that it possesses peculiarities in working which must be learned and strictly attended to. As with many other alloys, there is a certain temperature at which it works perfectly, but this is contained within narrow limits. This heat is rather indefinitely stated as full-redness. At a little above this heat (bright-red) or at a little below (low-red) it works with much less ease. If it is rolled at this temperature, it does not become brittle by working. When the bronzes are worked cold they will quickly stiffen up, and will crush or split unless annealed frequently. Working increases the strength and rigidity, but diminishes the extensibility. As aluminium. bronze is such a strong metal, it is natural to expect that it would require a large amount of power to work it. This is the case, and accords with the fact that when Cowles Bros. first introduced their bronze the workers of copper and ordinary bronze came very nearly breaking their rolls in endeavoring to work it; the steel workers, however, had no difficulty from this source. The Aluminium Brass and Bronze Company, who work up the Cowles bronzes, have put in rolls which are even more powerful than those used for working steel billets of corresponding size. It is recommended to subject the billet at the first pass to the heaviest pressure the rolls will give, in order to destroy at once any crystalline structure. When rolling cold, the lower grade bronzes can be worked for a longer time between annealings, but when rolling hot the higher bronzes give the least trouble.

When drawing the bronzes, they are first cast into rods of small diameter, annealed, and then drawn. Very hard dies are required, or the ordinary bronzes, especially the higher grades, are apt to become so hard between annealings as to cut them. The speed of drawing must be slow, and the reduction effected very gradually.

In regard to forging aluminium bronze, the statement that it can be forged perfectly at all temperatures from bright red to cold does not coincide with the experience of many workers. At a cherry red, the suitable temperature for rolling, it hardly forges at all. A much lower temperature must be used, a low redness, and at that heat it forges perfectly. Metal hammered from this heat until it is cold has its strength much increased.

Aluminium bronze can be spun, stamped, or pressed like ordinary brass. In these operations it must be annealed after each successive treatment since it hardens up so quickly under the great pressure.

In working with the file it is noticeable that the aluminium bronzes do not clog the file. Before the chisel it gives long, clean chips. On the lathe and planing machine the tool takes off long elastic threads and leaves a fine surface. It is especially in these cutting operations that the great advantages of a metal as strong as steel and yet as easy to work as brass become apparent. Simms states that aluminium bronze can be easily engraved, taking the sharp lines and deep cuts of the tool as well as any cast metal. Since it can be soldered, it forms a very suitable material for drawing into tubes. It is stated that it can even be hammered out into leaf, but this statement needs further proof.

Anti-friction qualities.—Along with the great toughness and malleability of aluminium bronze, as also its fine grain, we note a peculiar unctuousness or smoothness which seems to resist abrasion and to render it one of the best anti-friction metals known. Morin held that to bring out the metal's best qualities it should not be used in castings but as forged or rolled forms. He cites an instance where it was applied to the bearings of a large lathe, the axle having a diameter of 60 millimetres and making up to 1800 revolutions per minute. A plate of aluminium bronze was soldered in place on a common bearing metal back and then carefully bored true. This bearing had been in use four years at the time of Morin's statement, and after $2\frac{1}{2}$ years' working the wear was only 0.4 millimetre. Cowles Bros. recommend using the higher grades for bearings working under very heavy pressures, as their great strength prevents crushing, while the lower grades are more suitable for high speeds and low pressures. They recommend casting the bearings, bringing the working surface up to a high polish and using them only under steel axles.

An incident occurring under the author's immediate notice may be more to the point than quotations from other sources. A smooth metallic ball was required for a rotary steam engine, to work between steel guides in such a position that almost all the work of the engine passed through it. Steel balls were impracticable because they cut the guides. All sorts of bronzes and anti-friction metals were tried, but a few hours' work would cut them out of shape. Mr. Joseph Richards suggested to the maker that an aluminium bronze ball be tried. A ball was cast of ten per cent. bronze, but the casting not being solid it was warmed up and hammered until all the holes were closed. The ball was then turned smooth and polished. On putting into place it was worked several days at high speed without giving way, and on taking it out for examination the only sign of wear was that the polish was a little higher. I think that the hammering increased its capability of resistance, and would recommend that if the cast aluminium-bronze bearings do not meet any extra requirements put on them, the hammered or rolled metal might be successfully substituted.

Conductivity.—The aluminium bronzes are said to conduct heat and electricity better the lower the proportion of aluminium. Benoit measured the electrical conductivity of 10 per cent. bronze as 13, silver being 100 and copper 90. Edw. D. Self states that he found the heat conductivity of the 6 per cent. bronze almost the same as for pure copper and that of the 10 per cent. bronze very little less.

Resistance to corrosion.—Deville said, "in chemical properties aluminium bronze cannot differ much from the other alloys of copper," yet from numerous experiments we have noticed that it does resist most chemical agents better, particularly sea-water and sulphuretted hydrogen.

Bernard S. Procter,* after describing thirty-one experiments comparing aluminium bronze and brass, sums up the conclusions as follows :—

"From the above experiments it appears that aluminium bronze has a little advantage over ordinary brass in power to withstand corrosion, and its surface, when tarnished, is more easily cleaned. This should give it general preference where cost of material is not an important consideration, especially if strength, lightness and durability are at the same time desirable. It is out of my power to say anything about its fitness for delicate machinery, except that its chemical examination has revealed nothing which can detract from the preference its mechanical superiority should give it. Being so much less acted on by ammonia and coal-gas suggests its suitability for chemical scales, weights, scoops, etc. Its resistance to the action of the weather and the ease with which tarnish is removed render it especially applicable for door-plates, bell-handles, etc. Its mechanical strength and chemical inactivity together recommend it for hinges exposed to the weather. In experiments 18, 22, etc., the tendency of brass to corrode on the edges and at any roughness on its surface will

* Chem. News, 1861, vol. iv. p. 59.

be observed, while the bronze is free from this defect. In several cases the bronze seemed to be more quickly covered with a slight tarnish which did not increase perceptibly, probably the tarnish acting as a protection to the metal; but the brass, though less rapidly discolored, continued to be corroded and apparently with increased speed as the action was continued. The bronze is more easily cleaned. For culinary vessels its superiority to metals now in use appears questionable."; The author states that he wrote the article with a home-made pen of aluminium bronze, and suggests that it is well worthy of the attention of pen-makers.

The Cowles Co. state in a pamphlet that salt-water, soap-water and urine have no effect on aluminium bronze, which would make it a very valuable metal for ship builders and sanitary engineers. It is well known that the fatty acids, as in tallow, corrode brass and bronze, as is seen in old-fashioned brass candle-sticks, but aluminium bronze is said to be untouched by these agents. It is said that for preserving pans, in which fruit juices are boiled, aluminium bronze is superior in resisting power to the coppers now used. Sulphuretted hydrogen and coal-gas have hardly any effect on it. Hydrochloric acid dissolves out the aluminium, nitric acid dissolves out the copper. Tissier Bros. remarked that the color of the bronzes containing from 5 to 10 per cent. of aluminium could thus be altered at will by leaving in dilute nitric or hydrochloric acid. Concentrated sulphuric acid attacks it at first, but a coating is speedily formed which seems to protect the metal against further injury; dilute sulphuric acid attacks the alloy more deeply. Hot soda solution slowly removes the aluminium near the surface, leaving a velvety coating of copper oxide. The acid of perspiration is quite active in attacking aluminium bronze superficially, as any one can test by noticing how quickly a polished surface is tarnished. I do not think, however, that the corrosion is any more marked than other bronzes would show, and the tarnish is very easily removed and the polish restored by a little rubbing with a dry woolen cloth.

When heated in the air, aluminium bronze remains unoxidized even if kept at a red heat for a long time. This is a remarkable property when contrasted with the behavior of brass wire, which would quickly turn to oxide. It is stated that it has been kept

28

at a bright red heat for several months without showing any oxidation. As a chemist, who has in common with all other chemists been annoyed by the oxidation of brass wire when heated over a Bunsen burner, I hope that aluminium bronze wire gauze will soon be made for this purpose.

When exposed to the weather, the tarnish formed is very superficial and protects the metal underneath from continued oxidation. The aluminium bronzes take on a darker hue, after some exposure, but they never turn black, like ordinary bronzes, or green like brass. Since the color of out-door bronze statues is only preserved by a coat of lacquer, which hides many of the most artistic strokes, the value of a bronze which will stand the weather unprotected should be utilized by the art casters. It has been proposed to cast cannon of aluminium bronze, and this incorrodibility gives it an advantage over all other bronzes for this purpose. A commission appointed by the Austrian Government in 1888, to report on various types of rifle barrels, tested, among other materials, aluminium bronze. Two rifles, one with a steel barrel the other of aluminium bronze, were left in the gutter on a roof for several weeks. On examining them, the steel gun was rusted so far as to be completely useless, while the bronze barrel was only slightly tarnished and was loaded and fired without cleaning and without accident.

Uses of the aluminium bronzes.—The previous remarks have necessarily contained allusions to the many uses to which these bronzes may be advantageously applied. Their resemblance to gold early caused their use for jewelry, watch-chains, etc., but the ease with which their fine polish is tarnished by perspiration makes them unsuitable in this direction. Their wearing qualities make them valuable bearing metals, being also especially applicable to parts of machinery subject to heavy strain and rapid work, such as weavers' shuttles, ball-bearings, pin pivots, etc. It seems to be the great strength united with considerable malleability which give these wonderful wearing qualities.

The Engineering and Mining Journal* suggested that aluminium bronze should make excellent battery, sizing and jig-screens

* August, 1887.

for mining and ore working machinery. Being stronger and as hard as the mild steel and iron now used for perforated plates and mesh, and many times more durable than brass, with no tendency to rust or corrode and wonderful capacity for wear, it would appear to be particularly desirable for these purposes.

The considerable resistance of aluminium bronze to chemical agents has suggested its use as a material for pumps and machinery subjected to the action of acid mine-water. It is said that Worthington & Company have used it in their high pressure mine pumps.

Mr. Strange,* an English engineer, made a thorough discussion of the suitability of aluminium bronze for the construction of astronomical and philosophical instruments, and concluded that it was superior not only in some, but in every respect to any metal hitherto used for that purpose.

Mr. A. H. Cowles read a paper before the U. S. Naval Institute, October 27, 1887, urging the merits of aluminium bronze as a metal for casting heavy guns. Reference was made to the casting of a mountain howitzer of this bronze in 1860, which stood every test put on it by the French artillery officers, and which would have caused a revolution in cannon material but for the fact that its cost was prohibitory. The principal points made by Mr. Cowles are—1. The great strength and high elastic limit of aluminium bronze. 2. Its ductility. 3. The sound castings it produces. 4. The fact that no liquation takes place during cooling, as in ordinary gun-bronze. 4. No tendency to crystallization, as is the case with steel. 5. No rusting or corrosion. 6. Seventy per cent. of the cost of the gun is represented by the metal, which may be melted over when the gun is worn out. 7. Even with this high cost of metal, the total cost of the gun would be only four-fifths that of a built-up steel gun. While the discussion following the reading of this paper was not altogether in favor of the solid gun *versus* the built-up gun, yet no doubt was expressed that for casting solid guns, aluminium bronze was undoubtedly the best metal that could be used, and as regards its comparison with built-up steel guns, the sum total of the discussion seemed to

* Chemical News, vii. p. 220.

point to the expectancy that the bronze gun would be as good, if not better. But, as Deville was wont to say, " It is not necessary to theorize if you can make the experiment," and we hope soon to see a large aluminium-bronze cannon successfully cast and its exact worth demonstrated.

Aluminium bronze has been used very advantageously as a material for propeller blades. Its freedom from galvanic action, its non-corrosion by sea water and its great strength make it particularly well adapted for this use. Quite a long discussion on this subject was printed in London *Engineering* during April and May, 1888, raised principally by the makers of manganese bronze, but it was shown conclusively that in every point in which the latter was claimed to be of special advantage, aluminium bronze was still more advantageous. The Webster Aluminium Co. cast a propeller for a vessel whose bottom was subjected to the destructive influence of tropical waters, and it was used with the most satisfactory results as regards freedom from corrosion. I understand that one of the new U. S. gunboats will have an aluminium bronze propeller, since as speed is a great requisite, the lightening of the blades made possible by the use of a much stronger metal will be of considerable advantage.

Besides the items mentioned, we might conclude by saying that if aluminium bronze were as cheap as brass or ordinary bronze, there is hardly a single use that can be mentioned for which it would not be preferred ; since it costs more than these, it will be used for all those purposes for which it is of particular excellence, and its use will extend in a largely increasing proportion as its cost is lowered. Its specific gravity is not so much lower than ordinary brass or bronze to cause its use on this account, but the fact that its superior strength allows a large decrease in weight without any less strength than with the other metals, will aid greatly in increasing its sphere of usefulness. The fact that it will not rust gives it always the advantage over steel in out-door objects, and the fact that it can be cast perfectly will cause it to replace complicated steel forgings.

With aluminium bronze at the prices promised by the Heroult process, we will have to wait but a very few years to see it in as extensive use as ordinary tin bronze or, perchance, as brass itself.

Brazing.—Aluminium bronze can be brazed easily at a red heat, using ordinary brazing solder (zinc 1, copper 1) and 3 parts of borax.

Soldering.—Deville stated that aluminium bronze could be soldered with ordinary hard solder at a low red heat. It resists the common soft solders at ordinary temperatures, but if some zinc amalgam is added it can be soldered cold. Schlosser* gives the following directions for preparing this solder: White solder is alloyed with zinc amalgam in the proportions

White solder	.	.	.	.	.	2	4	8
Zing amalgam	.	.	.	.	.	1	1	1

The white solder may be composed as follows :—

Brass	.	.	.	.	.	.	40	22	18
Zinc	.	.	.	.	.	.	2	2	12
Tin	.	.	.	.	.	.	8	4	30

The zinc amalgam is made by melting 2 parts of zinc, adding 1 part of mercury, stirring briskly and cooling the amalgam quickly. It forms a silver-white, very brittle alloy. The white solder is first melted, the finely-powdered zinc amalgam added and the alloy stirred until uniform and poured into bars.

The Cowles Co. recommend the following solders as effective and convenient for aluminium bronze jewelry :—

Hard solder for 10 per cent. bronze—

Gold	.	.	.	.	.	.	.	88.88
Silver	.	.	.	.	.	.	.	4.68
Copper	.	.	.	.	.	.	.	6.44

Middling hard solder for 10 per cent. bronze—

Gold	.	.	.	.	.	.	.	54.40
Silver	.	.	.	.	.	.	.	27.00
Copper	.	.	.	.	.	.	.	18.00

Soft solder for Al bronze—

Copper 70 per cent. } Bronze	.	.	.	.	14.30
Tin 30 "					
Gold	.	.	.	.	14.30
Silver	.	.	.	.	57.10
Copper	.	.	.	.	14.30

* Das Löthen, p. 180.

Silicon-aluminium bronze.—Cowles Bros. have, by reducing fire-clay in presence of copper, obtained alloys of aluminium, silicon and copper. This alloy is white and brittle if it contains over 10 per cent. of aluminium and silicon together. With from 2 to 6 per cent. of these in equal proportions, the alloy is stronger than gun metal, is very tough, does not oxidize when heated in the air, and has a fine color. With 10 per cent. of aluminium and 2 or 3 per cent. of silicon, Cowles Bros. claim to have produced one of the strongest metals known.

Phosphor-aluminium bronze.—Thos. Shaw, of Newark, N. J.,[*] patents a phosphor-aluminium bronze, making the following claims: First, an alloy of copper, aluminium and phosphorus, containing 0.33 to 5 per cent. of aluminium, 0.05 to 1 per cent. of phosphorus, and the remainder copper. Second, its manufacture by melting a bath of copper, adding to it aluminium in the proportion stated, the bath being covered with a layer of palm oil to prevent oxidation, and then adding a small proportion of phosphorus.

It has been stated that this alloy has a high conductivity (presumably for electricity), but I am unable to find any determinations or evidence of any kind to substantiate this statement.

CHAPTER XVI.

ALUMINIUM-IRON ALLOYS.

ALTHOUGH aluminium does not appear to combine as energetically with iron as with copper, yet the affinity between these two metals is sufficient to cause their combination in all proportions. The useful alloys, however, are confined to those containing a small amount of aluminium, the addition of small quantities of iron to aluminium producing no useful result.

Iron is one of the most obstinate impurities in commercial aluminium. About the only way to obtain aluminium free from

---

[*] U. S. Patent, 303236, Aug. 1884.

iron is to keep the materials from which it is made scrupulously free from that metal, since if it once gets into the aluminium, it is almost impossible to remove it. As to its exact influence on the properties of aluminium, when in small quantities, we may say that it renders the color more of a gray, the aluminium becomes harder, less malleable, and appears to crystallize more readily. The most noticeable effect, however, is on the fusibility. Tissier Bros. observed, somewhere about 1857, that aluminium free from iron could be melted on a plate of aluminium containing 4 to 5 per cent. of that metal. Deville also observed that if a large amount of iron was present (10 per cent.), the aluminium could be liquated by careful heating, a ferruginous skeleton remaining, while aluminium containing less iron flowed away. This process, however, could not be used for the ultimate purification of commercial aluminium, since when the percentage of iron is low no liquation takes place. The exact rise in the melting point due to the presence of iron has been recently determined by Prof. Carnelly. He found that a specimen of aluminium containing 0.5 per cent. of iron melted very close to 700°, whereas a specimen containing 5 per cent. of iron did not even soften at that temperature, but commenced to fuse at about 730°. The effect of the iron is particularly seen in rendering the fusion pasty. Since silicon acts in an almost similar manner, it is important to observe that commercial aluminium can contain a certain small quantity of iron with very little detriment only on condition that the amount of silicon present is small.

Tissier Bros. took pure aluminium in small pieces, mixed it with bits of pure iron wire, and melted in a crucible under common salt. The alloy with 5 per cent. of iron thus made was harder, more brittle, and less fusible than pure aluminium. The alloy with 7 per cent. of iron differed from the preceding, principally in showing a stronger tendency to crystallize. With 8 per cent. of iron the alloy crystallized in long needles. Deville states that the alloy containing 10 per cent. of iron has the color and brittleness of native antimony sulphide (stibnite).

By melting together 10 parts of aluminium, 5 parts of ferric chloride, and 10 parts each of potassium and sodium chlorides, Michel obtained a crystalline mass, which by careful treatment with very dilute hydrochloric acid left some six-sided crystals, having the color of iron, and agreeing nearly to the formula Al^2Fe, containing 51 per cent. of iron.* These crystals dissolved easily in hydrochloric acid ; caustic soda dissolved out the aluminium. Calvert and Johnson obtained the alloy Al^2F^3 (see p. 330), containing on analysis 24.55 per cent. of aluminium—the formula calling for 24.34 per cent. Globules of this alloy were white, did not rust in moist air, and when treated with weak sulphuric acid gave up the iron, while the aluminium remained as a skeleton, having the shape of the original button. These experimenters also obtained an alloy containing, in two different experiments, 12.00 and 12.09 per cent. of aluminium. The formula AlF^4 requires 10.76 per cent. This alloy was extremely hard, and rusted on exposure to the air ; but could be forged and welded.

The iron-aluminium alloy which is being largely used at present for introducing aluminium into iron and steel is generally made with 5 to 15 per cent. of aluminium and has received the trade name of ferro-aluminium. Several different makes of this alloy are on the market, some made directly from alumina, others made by adding aluminium to iron. An analysis of the Cowles Company's ferro-aluminium has already been given ; the grade mostly supplied by this company averages 6 to 9 per cent. of aluminium, with $2\frac{1}{2}$ to $3\frac{1}{2}$ per cent. of silicon, and about 3 per cent. of carbon. When ferro-aluminium is made by alloying aluminium with iron, a good quality of pig-iron is chosen, and when melted the aluminium, in bars, is seized in tongs and dipped under the surface. A rise of temperature occurs, and a noticeable separation of graphitic carbon, causing "kish" to collect on the surface. It is said that the pig-iron thus alloyed has its combined carbon almost entirely converted into free carbon, losing sometimes as much as $2\frac{1}{2}$ per cent. in weight thereby. When all the aluminium required has been added, the melt is stirred, the crucible remaining in the furnace, then it is let stand for a few minutes, taken out of the fire, skimmed clean and cast into slabs or bars.

These alloys (ferro-aluminiums) are very hard, brittle, easily

* Ann. der Chemie und Pharm., 115, 102.

broken, and yellowish-white in color. Ledebuhr states that as the percentage of aluminium increases the iron becomes less magnetic, and at 17 per cent. the alloy is non-magnetic. It is said that in the works where the alloy is being made a workman can, after a few practices, test the alloys roughly by a simple magnetic test, even some degree of accuracy being finally attainable. Since wrought-iron and steel are very sensitive to small amounts of impurities, it is important that the ferro-aluminium added to them should be as pure as possible, being made from the purest pig-iron; this condition is of less importance when operating on cast-iron. This has led to the manufacture of two grades of ferro-aluminium, one for ordinary foundry use the other for steel and mitis castings.

Pig-irons and commercial iron and steel do not take up any appreciable quantity of aluminium in the process of their manufacture. Blair states that he finds aluminium nearly always present as such in steels, but only in quantities of a few thousandths of a per cent., such as, from actual analyses, 0.026, 0.029, 0.034 per cent. Corbin reported 2.38 per cent. in chrome steel, but Blair finds chrome steel to contain no more aluminium than other steels. Aluminium is very seldom reported even in the smallest amount in wrought-iron, since the puddling process may be reasonably expected to eliminate almost entirely any that might be in the pig-iron. Conflicting views have been expressed as to the presence of aluminium in pig-iron. As much as 1 per cent. has been reported in German gray-iron; Grüner states that some English pig-irons contain 0.5 to 1 per cent., and some Swedish irons 0.75 per cent. Percy quotes an analysis of pig-iron showing 0.97 per cent. of aluminium, but questions its correctness, supposing that much of this might have come from aluminium contained in included slag.

Faraday and Stodart obtained an alloy containing 3.41 per cent. of aluminium by melting an iron carbide with alumina at a very high heat. This alloy is described as being white, close-grained and very brittle. G. H. Billings (see p. 334) made an alloy containing 0.52 per cent. of aluminium and 0.2 per cent. of carbon. Its fracture showed solid, homogeneous and finely crystalline, like steel with 1 per cent. of carbon. It forged very well at cherry redness but crumbled to fragments at a yellow heat; it would not harden.

When we come to the consideration of the iron-aluminium alloys with a small content of aluminium, say up to 3 per cent., we soon find that the other substances present in the iron affect the result so materially that it is necessary to bring them into the discussion. In other words, in indeavoring to describe the effects of small quantities of aluminium on iron it is necessary to particularize the kind of iron—cast-iron, wrought-iron, or steel—to which the aluminium is added. The result of the addition of the aluminium is, moreover, as it was with copper, not entirely the effect due to the formation of an alloy, but also a chemical effect on various impurities present. Indeed, in some cases the latter may be the whole function of the aluminium added, determining its whole effect. We will, therefore, divide the remainder of this chapter into three parts; namely, the effect of small quantities of aluminium on (1) steel, (2) wrought-iron, (3) cast-iron.

EFFECT OF ALUMINIUM ON STEEL.

Although A. A. Blair found aluminium almost always present in steel as a few thousanths of a per cent., yet he was not able to determine that so small a quantity had any appreciable effect on the properties of the metal.

Faraday, in seeking for the distinguishing ingredient of the famous Bombay Wootz-steel, found that it always contained aluminium in quantities varying from 0.0128 to 0.0695 per cent. In order, then, to prove the case synthetically, Faraday & Stodart took an alloy of iron with 3.41 per cent. of aluminium and a little carbon, and melted it in various proportions with steel. On melting 40 parts of the alloy with 700 of good steel (introducing 0.18 per cent. of aluminium) a malleable button was obtained, which on treatment with acid on a polished surface gave the beautiful damask peculiar to Wootz. On melting 67 parts of alloy with 500 of steel (introducing 0.4 per cent. of aluminium) the resulting button forged well, gave the damask, and "had all the appreciable characteristics of the best Bombay Wootz." Karsten could not find any aluminium in specimens of Wootz which he examined, and suggested that that found by Faraday was due to intermingled slag; but the latter found aluminium in the steel without silica, which seems to prove that his results are beyond question.

Rogers* corroborated the above results obtained by Faraday. He melted an iron-aluminium alloy with steel in quantity sufficient to introduce 0.8 per cent. of aluminium. The product had great hardness, a bright silver-like polish, and when treated with dilute acid gave the undulating markings peculiar to Damascus steel and Wootz.

With aluminium costing $12 to $16 per lb. it can be readily seen that this use could not be practised commercially; but, with aluminium at less than half that price, and especially with even more economical ferro-aluminium to be had, these old references were looked up and many steelmakers began trying the virtues of aluminium. Since 1885 hardly a maker of crucible steel and steel castings but has made some experiments in this line. It had long been known that if alumina is added during the melting of steel in a crucible, the grain and lustre are improved. It was found that ferro-aluminium, added just before pouring, had the same effect, and since the latter operation is under exact control it is preferred to the former practice, providing that the ferro-aluminium can be obtained pure enough for this use.

From experiments made at Faustman & Ostberg's Mitis Foundry at Carlsvick, Sweden, in 1885, it was proved that ferro-aluminium is of great use in making steel castings. Wrought-iron scrap was melted in crucibles, carbonized to hard steel by adding pure pig-iron, and, before pouring, ferro-aluminium added to supply 0.1 per cent. of aluminium. The steel was then cast into the shape of ordinary edge-tools, which needed only to be hardened and ground in order to be ready for use. The surface of these tools was very clean, and took a high polish. It was found that manganese, which is so often purposely introduced into steel, was deleterious in its action on steel containing aluminium, and that mild steel almost free from manganese gave by far the best results when aluminium was added. If true, this is a curious

* Moniteur Industriel, 1859, p. 2379.

fact which it is not easy to see the cause of. In Bessemer practice it is evidently of no use to add ferro-aluminium before blowing, but it has been thrown into the converter just before tipping into the ladle, and Mr. Ostberg states that Bessemer ingots containing only 0.06 per cent. of carbon have been thus made which did not rise in the mould at all, and were solid and of good quality throughout. Similar advantages should also result on treating Siemens-Martin steel in this way, and perhaps even greater advantages, since mild steels are so much more difficult to cast solid than high-carbon steel.

The Cowles Co. claim that "the addition of 0.1 per cent. of aluminium to molten steel just before pouring renders it more fluid and insures the production of sound castings of increased strength and free from blow holes." As substantiating this statement, the Phoenix Iron Co. of Germany report that 0.2 per cent. of aluminium added (as ferro-aluminium) to their basic Siemens-Martin steel gave them metal with a tensile strength of 112,000 lbs. per sq. in., with 12.5 per cent. elongation, whereas the best results previously attained without aluminium were from 96,000 to 98,500 lbs. per sq. in. The Cleveland Rolling Mill Co. have reported that experiments made at their works by Mr. Cole show that the addition of 0.05 to 0.1 per cent of aluminium to Siemens-Martin steel increases its fluidity, thereby producing sharper castings, decreases the number of blow-holes, and increases the strength of the metal without affecting the elongation. Another German firm report that 2 per cent. of aluminium added to their Siemens-Martin steel increased its strength 20 per cent. without decreasing the extensibility.

If the above reports can be relied on, they seem to show that 0.1 per cent. of aluminium has as much strengthening effect on mild steel as 2 per cent. This would point to the following explanation : The effect of the aluminium is primarily to combine with dissolved gases and to reduce dissolved oxides. It takes a very small amount to do this work ; when this is done the steel has received almost all the strengthening which the aluminium can give it. Any larger amount of aluminium acts by alloying with the steel, thereby increasing its fluidity correspondingly, by reducing its melting point.

An interesting effect of the addition of aluminium to soft steel is the increased ease of welding. Specimens have been shown by the Cowles Syndicate Co., in England, of iron welded to Siemens-Martin steel with and without aluminium. With the ordinary steel the line of weld was clearly visible, but with steel containing 0.2 per cent. of aluminium no such line could be seen, the crystalline structure of the iron appearing to merge gradually into the fine grain of the steel, even under the microscope.

We shall see later that the addition of aluminium to cast-iron tends to separate combined carbon as graphite. This probably accounts for the poor results obtained by adding ferro-aluminium to high-carbon steels; for these, melting more easily and fluidly than mild steels, would be made less fusible by the decrease in combined carbon and possibly also made pasty by graphite being entangled in the metal as it thickens. As illustrative of this point we will quote the experiments made by R. W. Davenport.* A large charge of carbonless ingot iron holding about 0.08 per cent. carbon, and boiling strongly was tapped into two similar ladles and ferro-manganese added in order to convert it into a low-carbon steel. Into one ladle was put, in addition, ferro-aluminium sufficient to introduce 0.064 per cent. of aluminium. Both ladles were then teemed into sand castings and ingot moulds. The steel treated with ferro-aluminium lay perfectly dead and piped in the ingot moulds, and yielded practically solid sand castings; the other rose in the moulds, had to be stoppered and gave very porous sand castings. On another occasion, ferro-aluminium sufficient to introduce 0.04 per cent. of aluminium and 0.10 per cent. of silicon was added to molten crucible steel, which owing to the presence of carbon and manganese evolved no important quantity of gas. This steel contained 0.25 per cent. of carbon. The result of adding the aluminium was to stiffen this steel, make it hard to pour and difficult to get solid castings.

†Mr. J. W. Spencer, of the Newbern Steel Works, Newcastle-on-Tyne, made a series of tests on this subject of the effect of aluminium on crucible steel, and reached a similar conclusion. With a low carbon steel, the effect on the tensile strength in-

---

* Howe's Metallurgy of Steel.   † Iron Age, Dec. 22, 1887.

creased with the increase of aluminium, but it was found on an alysis that the amount of silicon in the metal was also increased by the treatment, which may partly account for the difference in strength. The following table shows these results :—

Carbon in steel. (per cent.)	Aluminium added. (per cent.)	Silicon present. (per cent.)	Elastic limit. (tons per sq. in.)	Tensile strength. (tons per sq. in.)
0.10	0.12	0.06	9.8	·20.8
0.15	0.22	0.08	10.2	21.8
0.28	0.43	0.22	12.0	25.5

These three steels were described as "fluid, sound and tough," excepting the last which was brittle before annealing. They were all stronger before annealing. With high carbon steels most of these properties were reversed, as is seen by the following results:

Carbon in steel. (per cent.)	Aluminium added. (per cent.)	Silicon present. (per cent.)	Elastic limit. (tons per sq. in.)	Tensile strength. (tons per sq. in.)
0.53	0.12	0.28	14.38	29.60
0.65	0.22	0.28	14.38	26.28
0.85	0.43	0.40	15.80	21.87

In these cases, as in the previous ones, the increase in carbon and particularly silicon would cause a corresponding increase in tensile strength, but it is very noticeable that with carbon over 0.5 per cent. the strength decreases with the increase of aluminium in spite of a simultaneous increase in both carbon and silicon. These steels are also described as fluid and running into sound castings, but they were brittle and hard before annealing, particularly the one containing most aluminium. Mr. Spencer sums up his experience as follows : "The result is satisfactory in every instance so far as soundness and the usual attributes of good castings are concerned, running fluid and without ebullition into sharp, clear castings ; the milder mixtures, under the hammer, breaking very strong, though unannealed. The general conclusion from the mechanical tests is that though aluminium may increase the elastic limit and tensile strength slightly, yet this is done at the expense of ductility, while in presence of high carbon it is disadvantageous in all these respects. It is also probable that the increase of elastic limit and tensile strength, when it does occur, is not more than can be accounted for by the carbon and silicon present. The chemical reactions of the aluminium in the crucible may be various, but the prevention of blow-holes and increased fluidity are the chief advantages."

In order to avoid introducing impurities into steel by using ferro-aluminium made from ordinary pig-iron, a German firm has recently put on the market a steel aluminium "containing 10 per cent. of aluminium and 90 per cent. of pure cast-steel." This can of course be used instead of ferro-aluminium for any purpose, but is particularly preferable in treating steel, which is so extremely sensitive to minute quantities of certain impurities—sulphur, phosphorus, etc.

The *rationale* of the action of aluminium in preventing blow-holes and increasing the fluidity of the metal will be discussed more at length in considering the action of aluminium in the mitis process, a little further on. It has been found that the best time to add the aluminium is just before pouring. In large mills, the ferro-aluminium is heated red hot and shovelled into the casting ladle as the liquid steel runs into it from the converter or open-hearth furnace. The force of the molten stream of steel carries the pieces almost down to the bottom of the ladle, thus diffusing it uniformly through the bath. Stirring with an iron rod coated with clay helps the mixture. When the steel thus treated is cast into ingots it lies still in the mould and makes castings as free from blow-holes as ordinary cast-iron castings.

EFFECT OF ALUMINIUM ON WROUGHT-IRON.

When wrought-iron is heated to a high temperature, it does not pass quickly into the fluid state, but for a large increase of temperature above the point at which it first softens it will remain thick or mushy. At a very high temperature it can be made sufficiently fluid to pour into moulds, but the castings thus made are notably unsound and weak. It was discovered by Mr. Wittenstroem, of Stockholm, working with the co-operation of Mr. L. Nobel, of St. Petersburg, that if a small amount of aluminium is added to a charge of wrought-iron which has been heated until pasty, the iron immediately liquefies and can be poured into castings having all the properties of wrought-iron except fibre, and as sound as if of cast-iron. This idea was investigated thoroughly at Nordenfelt's malleable-iron foundry, in Carlsvick, Sweden, by Messrs. Wittenstroem, Nordenfelt, Faustman and Ostberg. The result of two years' experimenting during 1883 and 1884 was so successful that the malleable-iron plant was pulled down and a new foundry, operated by Faustman and Ostberg, supplied their former trade with wrought-iron castings, which were called "Mitis" castings by Mr. Nordenfelt because of their softness in contrast with cast-iron castings. This plant began operations in January, 1885, and its product soon reached a larger sale than that of the malleable castings which it has supplanted. Mr. Ludwig Nobel also installed the process in his foundry at St. Petersburg at about the same time. The process was represented by a fine display at the International Inventions Exhibition in London, in 1885, and received a gold medal. In 1885, Mr. Ostberg visited the United States for the purpose of establishing the process here,* and a plant was erected and put in operation at Worcester, Mass. In February, 1886, Mr. Ostberg spoke before the Institute of Mining Engineers at their Pittsburgh meeting, showing specimens of the castings; an experienced iron worker said on that occasion that he would not have believed the statements if they had not been proved by the sight of the castings. In the same month, the "United States Mitis Company" was incorporated in New Jersey, W. F. Durfee, M. E., of New York, being general manager, Mr. Robt. H. Sayre, of Bethlehem, Pa., president, and the list of directors including Mr. John Fritz, manager of the Bethlehem Iron Works, and several other well-known gentlemen. The object of this company, which owns the "Mitis" patents for the United States, is to regulate the use and sell rights to work under these patents; and it is said that five plants are now in operation in the United States. Abroad, plants are working in England, France, Germany, Austria, Sweden and Russia. The experimental plant started at Worcester, Mass., has been abandoned for some time, it being said that the success achieved there was anything but brilliant, but since the process

* U. S. Patent, 333373.

does succeed in other places, the plant in question was probably closed for reasons satisfactory to those concerned.

Such, in brief, has been the rise of the mitis process. It seems to have found its sphere in replacing malleable iron castings, because principally of the superior toughness of mitis metal, although the castings are not so uniformly sound and trustworthy, or hardly so cheap as those of malleable iron.

The following details of the production of mitis castings are from descriptions by Nordenfelt, Ostberg, and E. A. Cowper, of London :—

Raw material.—As the raw material to operate on, wrought-iron scrap or mild steel are equally suitable. It was found that some of the best results are to be obtained by using Swedish scrap-iron or English hematite-iron—that is, materials containing less than 0.1 per cent. of phosphorus, which is a very injurious ingredient if present in much larger quantity. Using a mixture with poorer quality iron, with phosphorus running up to 0.15 per cent., good results may still be obtained—that is, the castings still compare favorably with ordinary malleable castings. In using scrap-steel, which is necessarily low in phosphorus, it was found that manganese interfered with the production of good castings, a result rather unexpected. Since almost every melter devises various mixtures of his own, as circumstances permit, it is but natural that we find the best features of the mitis process united with some other old-established practices. Thus, in one mitis plant in this country the mixture for melting was composed of—

Mitis scrap	35 per cent.
Hematite muck bar	35 "
Wrought-iron punchings	$12\frac{2}{3}$ "
Soft-steel scrap (0.1 per cent. carbon) . .	$12\frac{2}{3}$ "
White pig-iron	3 "
Ferro-silicon (10 per cent. silicon) . . .	1 "
Ferro-aluminium (6 per cent. aluminium) .	$\frac{2}{3}$ "

It is seen that in this charge the melter used a little white iron as a flux, which would probably introduce 0.1 per cent. of carbon; then the virtues of ferro-silicon for making sounder castings are utilized by adding 0.1 per cent. of silicon to the charge; lastly, 0.04 per cent. of aluminium was introduced.

29

In general, it may be said that if iron free from impurities is used, very good castings are obtained; if iron is used with a large percentage of phosphorus, proportionately brittle and unsatisfactory castings result.

The ferro-aluminium used should be, for similar reasons, free from any considerable amount of such impurities as generally injure wrought-iron.

Since the castings are almost identical in composition with the charge of iron melted, the following analyses of mitis metal, made by Mr. Edward Riley, will show the range of material or mixture to which the process has been successfully applied :—

Raw material.	Carbon.	Silicon.	Phosphorus.	Manganese.
Hematite bar . . .	0.067	0.161	0.068	0.022
Swedish scrap . .	0.053	0.044	0.077	0.027
Refined iron . . .	0.130	0.124	0.137	0.014
½ Staffordshire iron } ½ Swedish scrap }	0.130	0.035	0.150	0.026
⅔ Staffordshire iron } ⅓ Hematite bar }	0.070	0.093	0.194	0.014
Staffordshire iron . .	0.106	0.080	0.250	0.014

The above figures are percentages; sulphur was present in all as a trace. The first in the table, those low in phosphorus, gave the best castings, the last the poorest; with over ¼ per cent. of phosphorus, the castings were brittle. As already stated in considering steel castings, mixtures containing higher percentages of carbon have been treated, but there seems to be a limit to the increase of this element, above which the addition of aluminium is no longer helpful but even deleterious.

Method of treatment.—The charge of wrought-iron is placed in covered crucibles and brought to a temperature of about 2200° (Mr. Ostberg), at which heat it is just losing the solid and assuming the pasty condition. If it were desired to cast the iron without adding aluminium it would be necessary to superheat it several hundred degrees above this point, not only to give it the desired fluidity, but also to permit it being carried around the casting shop. It is during this superheating that a large part of the gases contained in the molten iron are absorbed. If, therefore, the charge is treated with aluminium immediately on reaching the melting point, the effect is such that this superheating with its accompanying deterioration of the iron is rendered unnecessary. This is possible for the reason that on adding ferro-aluminium sufficient to introduce 0.05 to 0.1 per cent. of aluminium the charge immediately liquefies, and is so far from its setting point that it can be removed from the furnace and poured into numerous moulds, retaining all the time its exceptional fluidity. The metal acts just as if it had been superheated several hundred degrees, but this has been accomplished without leaving it in the furnace for half an hour or so, thus attaining an economy in fuel which is not to be ignored. When the crucible is taken from the furnace the charge is perfectly *dead melted*, lies quiet in the crucible, evolves no gas and teems like molten silver. It is cast in either sand or iron moulds, and on account of its fluidity does not require large heads to bring the castings up sharp and show the finest impressions of the mould.

Several devices are used in connection with this process which it may be interesting to note. The furnace used is one designed and patented by Mr. Noble, and burns naptha or crude petroleum or petroleum residues. A full description with drawings may be seen in Engineering and Mining Journal, May 8, 1886. With this furnace are melted on an average 8 heats in 10 hours. Starting cold, the first charges are melted in $1\frac{1}{4}$ hours, when the furnace is fully up to heat only $\frac{3}{4}$ hour is necessary, so that the furnace is equal to 24 heats in 24 hours. Any one familiar with steel melting will recognize this as a great improvement in melting furnaces. The difficulty met with in this country has been to get oil of uniform quality. At the Chester Steel Casting Works they state that with one car of oil the furnace works splendidly, but with the next they may have difficulty in keeping up the heat. A supply of oil of uniformly good quality is necessary for the successful working of this furnace. A patent pouring ladle is also used in which the metal is kept up to its original heat as long as is needed in order that a number of castings can be all poured at the same temperature. The moulding material used is pure fire-clay, hard burnt, finely ground and mixed with sugar or molasses as a binding material. This is perfectly fire-proof at the temperature of the molten wrought-iron, and is said to answer well.

Properties of mitis castings.—The material being primarily wrought-iron, the castings do not have to be annealed before using. The thinnest or most complicated castings can be produced which it would be almost impossible to forge in wrought-iron, thus furnishing difficult forged pieces at not much greater expense than ordinary castings. When there is less than $\frac{1}{4}$ per cent. of phosphorus present, the castings can be welded and forged in all respects as wrought-iron. The castings come out of the above moulding material with a remarkably smooth surface and a peculiar bluish tint; there is no sand burnt into their surface. The castings are as ductile as the iron from which they are made, but when tested for elongation under stress it was found that they did not elongate so much. This is counterbalanced, however, by the fact that as mitis metal contains no intermingled slag and absolutely no fibre, it has the same strength and elongation in all directions. In general, the tensile strength of the iron is increased, Mr. Ostberg says 20 to 50 per cent., but the lower figure is probably nearer correct. Experiments at the Bethlehem Iron Works showed 10 per cent. increase in tensile strength with no change in the elongation. Mitis castings are, in short, objects cast on molten iron yet having all the desirable properties of wrought-iron. The uses to which they can be put are very numerous, including all purposes for which malleable castings are suitable and particularly to replace complicated or even impossible forgings in any shape which admits of casting.

Rationale of the process.—The following facts are to be explained : 1. On adding ferro-aluminium to wrought-iron brought into a pasty fusion, the charge immediately becomes very liquid. 2. The castings made of metal thus treated are almost entirely free from the blow-holes which render ordinary wrought-iron castings almost useless.

Mr. Ostberg's explanation of the first point is that the addition of the aluminium produces a sudden lowering of the fusing point of the wrought-iron by some 150° to 250°, thus leaving the metal superheated to that extent above its *new melting point* and consequently with greatly increased fluidity. That this view is erroneous has been shown by the fact that the aluminium added does not remain in the iron. Numerous analyses made abroad have failed to find any aluminium in the castings. Mr. R. W. Davenport, a trustworthy analyst, could find it in no instance. Mr. A. A. Blair, of Philadelphia, one of the greatest authorities on the analysis of iron and steel, has been unable to find any, and considers it very improbable that 0.03 per cent. could escape detection. Mr. Ostberg also admits that it has never been detected, and virtually abandons his explanation by saying, in a letter to Mr. Howe, " An iron may contain 1 or 2 per cent. of aluminium without any noticeable effect in the making of castings; it is not the presence of the aluminium but the act of adding it at a certain moment that produces the effect."

Mr. R. W. Davenport offers the following explanation: A rise of the temperature of the metal would explain the phenomena as satisfactorily as a fall in the melting point. Since the aluminium oxidizes and passes into the slag, probably according to the reaction,

$$2Al + 6FeO = 3FeO.Al^2O^3 + 3Fe,$$

the high calorific power of the aluminium would supply a considerable quantity of heat, in spite of its small amount. Mr. Davenport then assumes the calorific power of aluminium as 10,000, leaves out of consideration the heat absorbed by the reduction of FeO to Fe and the union of FeO with Al^2O^3 to form the aluminate, and from this calculates that the oxidation of 0.06 per cent. of aluminium would produce 640 calories of heat and raise the temperature of the bath about 40°. He further implies that a rise of 120° would thus call for the oxidation of only 0.18 per cent. of aluminium, which might easily have been added to Mr. Ostberg's castings.

This explanation is as untenable as Mr. Ostberg's, for the following reasons: 1. The calorific power of aluminium burning to alumina is very nearly 7500*; when producing hydrated alumina (the datum usually given in the tables) it is only 100 higher. 2. The heat required to reduce ferrous oxide is pretty accurately known, and there is no reason why this should not be taken into account. 3. The heat of combination of ferrous oxide and alumina is certainly not known, though probably quite small; but on inspecting slag from mitis metal, white patches or flakes

* Ann. de Chim. et de Phys. June, 1889, p. 250.

of alumina are to be seen in it, showing that some of the aluminium, if not the greater part, escapes as alumina uncombined. It would be erring on the safe side to leave this quantity out of consideration altogether. 4. As the ferro-aluminium is thrown into the crucible cold, the melting of it will very nearly absorb as much heat as is developed by its chemical reactions. Reconstructing the formula with these points in view we have—

	Heat developed.	Heat absorbed.
Oxidation of aluminium		
0.06 × 7250 . . . . . =	435	—
Reduction of ferrous oxide		
$0.24 \times \frac{7}{9} \times 1286$ . . . . =	—	240
Melting of ferro-aluminium		
1.00 × 200 (Gruner) . . . =	—	200
	435	440

Making all reasonable allowances, the increase of heat due to the addition of the alloy will be too small to be noticeable. I might say, finally, that if the charge were left in the crucible untreated and heated one or two hundred degrees hotter than the temperature at which the charges usually receive their ferro-aluminium, the wrought-iron would not flow with anything like the fluidity shown by mitis metal. It is probably safe to say that wrought-iron untreated could not, at any practicable temperature, be made as fluid as the aluminium-treated metal.

What then will explain the increased fluidity? The author asks a consideration of the following facts: Every metal melter who has tried to run down wrought scrap of any metal, zinc, copper, tin, etc., knows how the melt will become pasty, and in many cases resist every effort to run it together. Now zinc is zinc, and its melting point is somewhere about 420°, and yet the surface of the scrap metal will keep together while the interior is quite fluid. This is particularly noticeable with copper. It appears that the previous working has driven particles of foreign matter, particularly oxide, into the pores of the metal, and this less fusible skin keeps the melted particles from coming in contact and running together. Again, it is noticeable with many metals that the absorption or solution of a minute quantity of its oxide tends to make the metal pasty. Let any one blow air for a very short time through perfectly fluid zinc, and in an incredibly short space the metal will thicken up, and an amount of mush out of all proportion to the amount of oxide which could have been formed, will float on the surface. It can also be noticed that on heating up this mushy metal again it passes out of the solid state at nearly the same temperature as pure zinc; but instead of becoming fluid it remains pasty for many degrees' rise of temperature above that point. The case of wrought-iron appears to me to be similar to those just noted. When wrought-iron scrap is heated it becomes soft at a moderately high temperature, but on heating it further to get it to form a homogeneous bath the hard, wrought surface is a great hindrance to its running together; a very high temperature causes the separate pieces to unite imperfectly into one body, but because of the scale and oxide present from the first, together with that formed during heating, the fusion is thick and viscid. Here the second phenomenon pointed out above can be noticed. This metal, because of the oxide in it, requires a higher temperature to make it fluid than if the oxide were not there. What follows then? Remove the oxide by some means, and the bath becomes perfectly fluid, and is superheated with respect to its *proper* melting point, *i. e.*, the melting point of the metal uncontaminated with oxide. The aluminium added does this work, reduces the oxide to metallic iron, the infusible and unalterable alumina produced rises to the surface, and the bath attains its extraordinary fluidity for the reason just given.

With regard to the lessening of blow-holes, we will first note their cause. First, they are not shrinkage cavities, which are caused by the metal chilling too quickly after pouring in the mould, and the sink-head not remaining fluid long enough to feed the cavities made in the body of the casting as it cools. The metal which is most liquid and least likely to chill quickly will produce the least number of unfilled shrinkage cavities, and these advantages are possessed by the wrought-iron when converted into mitis metal. But, in considering blow-holes proper, we note three distinct causes for them in wrought-iron castings.

1. When molten metal is poured into a mould prepared with the greatest care there is always some ebullition caused by the expulsion of moisture from the mould or moulding material.

This boiling will continue as long as the metal is fluid enough to allow the vapor to escape, but as soon as it stops escaping, there will be some gas entangled in the solidifying metal, producing cavities. The gas thus entrapped is principally hydrogen with some oxygen, most of the oxygen being caught by the metal and forming a lining of oxide inside the cavity.

2. When a stream of liquid, molten metal or water, is poured, it draws with it into the bath a considerable quantity of air. Every one who has poured water from a pitcher into a goblet has had opportunity to see this phenomenon, which occurs just as certainly and perhaps to a still greater degree when molten metal falls six or eight inches through the air and down a pouring gate. Such an arrangement is an actual suction apparatus. The gas thus drawn into the metal will be principally air with whatever proportion of moisture it contains. This gas escapes as long as the metal remains fluid enough, but will be largely entangled in the solidifying metal.

3. Metals possess the property of dissolving or occluding gases while molten, just as water dissolves air. They can retain some gas even when solid, but when they melt their dissolving power is largely increased. It results from this, that if iron is kept molten for some time it will be able to dissolve a certain quantity of gas. As it cools toward its setting point its dissolving power may increase or decrease, I cannot say which, but it is certain that when very near to its setting point it suddenly loses this power of solution, and considerable quantities of gas are evolved. The corresponding phenomenon in aluminium is very marked and quite easy to observe (see p. 56). The gas being set free near to the setting point, much of it is entangled in the casting.

4. In castings made of cast-iron there is another cause of blow-holes; viz., the carbonic oxide produced by the carbon present reducing oxides; but since carbon is very low in mitis castings and the dissolved oxides are otherwise removed, this cause need not be taken into account.

Reviewing these causes of blow-holes we note that the first two will occur in casting any kind of metal, but that kind which is most fluid in the mould and remains fluid the longest time will permit most gases to escape and so set with the smallest number of blow-holes. The superiority of aluminium treated metal in these requirements gives it great advantages over these causes of blow-holes. For the same reason, when the third cause is considered, the fluidity of mitis metal down almost to its melting point, with a small range during which it is pasty, allows more of the dissolved gas to escape when once set at liberty. In the author's opinion, the increased fluidity of mitis metal and the closer definition of its melting point, are the chief causes of the comparative freedom of the castings from blow-holes. However, the point has been raised by Mr. Howe, that perhaps the aluminium imparts to the iron greater power of holding gases in solution, not directly by alloying with it, since none remains in the iron, but indirectly by removing the oxygen. In one of Mr. Davenport's experiments, wrought-iron was melted alone in a crucible, and while oxygenated and boiling gently, 1 per cent. of ferro-aluminium was added, introducing 0.04 per cent. of aluminium and 0.1 per cent. of silicon. It appeared to lessen the evolution of gas, and in $2\frac{1}{2}$ minutes the iron was perfectly still, and when poured 3 minutes later lay quiet in the mould like cast-iron. In all of Davenport's experiments with molten iron, the addition of aluminium seemed to check the evolution of gas. We cannot say that Mr. Howe's suggestion is impossible, yet it is very improbable, because one of the laws of solution of gases in liquids is that when a liquid has dissolved as much of one gas as it is able, it will yet take up as much of another gas as if the first were not present.* If, then, we have the case of solution of gases in molten iron, the removal of one gas would not affect the amount of any other gas which the iron might take into solution. When wrought-iron is boiling the cause is that carbon present (perhaps that still left in the metal, but more probably the graphite of the plumbago crucible) is reducing the dissolved ferrous oxide and forming carbonic oxide. To explain the action of the aluminium in stopping this ebullition it is not necessary to suppose that it gives the iron increased power of retaining in solution the gases which are escaping (Howe's explanation), nor yet that it reduces the carbonic oxide gas as it forms, which reaction might take place, but simply that it reduces

---

* Deschanel's Natural Philosophy, pp. 182, 183.

all the dissolved ferrous oxide at once and so leaves no available oxygen in the bath for the carbon to combine with. With regard to any carbonic oxide which might be held dissolved in the iron, and by being evolved near the setting point from blow-holes, it is quite probable that the aluminium takes the oxygen away from it also, and so lessens chances of blow-holes from this cause. If such gases as hydrogen or nitrogen are dissolved in the molten iron, the greater fluidity of the bath will have no tendency to cause their evolution, the aluminium cannot influence them chemically, and it is altogether probable that they remain, and must be evolved as the metal sets. But, since castings almost entirely free from blow-holes are obtained, it follows that the amount of such gases present is inconsiderable.

The arguments just presented may be summed up as follows :—

1. Treating with aluminium makes the bath fluid because it removes the dissolved oxide which made it pasty.

2. Treating with aluminium stops the evolution of gas because it combines with all the oxygen present and so removes the essential gaseous ingredient of the gas which was being evolved.·

3. Treating with aluminium lessens blow-holes in the castings principally because the greater fluidity of the metal allows the easier escape of the gases mechanically entangled in it during casting.

INFLUENCE OF ALUMINIUM IN PUDDLING IRON.

The Cowles Company state in one of their pamphlets that if a small percentage of aluminium is added to iron in the puddling furnace, the bath comes to nature quicker, and the wrought-iron produced is much stronger, equalling the best grades of mild steel.

An article in the "Iron Trade Review," September, 1887, stated that on adding 0.1 per cent. of aluminium to iron about to be puddled, the tensile strength was raised from 52,000 to 60,000 lbs. per square inch, an increase of 16 per cent., while the elongation was variously increased up to 20 per cent.

The only thoroughly reliable report on this subject is made by Mr. G. W. Thomson, a gentleman connected with the well-known firm, Messrs. P. & W. MacLellan, Glascow. He took a charge of 373 lbs. of No. 4 forge pig-iron, charged it into the usual type of puddling-furnace, and when nearly melted threw in an ingot of ferro-aluminium which weighed 13 lbs., contained 7.11 per cent. of aluminium, and so introduced 0.25 per cent. of aluminium into the bath. The operation of puddling then went on as usual, and with no noticeable change, except that when the charge was just getting pasty it suddenly swelled up considerably, slag flowed from it abundantly, and the charge was very soon ready for balling. In the shingler and rolls the balls worked decidedly stiffer than usual. The result was very satisfactory. The ordinary iron averaged 22 tons tensile strength, with 12 per cent. elongation. The aluminium-treated iron showed 31 tons tensile strength and 22 per cent. elongation, being gains of 40 and 80 per cent. respectively. These bars stood the bending test perfectly, and when polished and cut showed a remarkable fine surface and close grain. They also forged satisfactorily.

If the above reports are to be relied on, this subject deserves looking into by every puddling-mill manager desirous of improving the quality of his iron.

Influence of Aluminium on Cast-Iron.

Very early in the history of aluminium, away back in 1858, the Tissier Bros. suggested the possibility of this application of aluminium by saying: "When aluminium has become low in price, it will be interesting to see what qualities it can communicate to cast-iron, introduced in large or small quantities." This suggestion does not appear to have led to any experiments in this line until after 1885, when the discovery and publication of the mitis process turned many experimenters toward the determination of the effect of aluminium on cast-iron. In April, 1886, Mr. Sellers, of Philadelphia, remarked at the Washington meeting of the National Academy of Science that he had made a series of experiments on the use of aluminium with iron in casting, with the result that the castings produced were very sharp and without any flaws. In December, 1887, the Williams Aluminium Company, of Boston, began pushing the sale of an alloy called by them aluminium-ferro-silicon, which they recommended to founders for addition to cast-iron in the ladle, claiming increased fluidity of the iron and greater freedom from blow-holes. This company is now located in New York, with works in Newark, N. J., and manufacture and sell this alloy in tolerably large quantities. The claims of this company, as also of other companies selling ferro-aluminium for foundry practice, are certainly very broad, but we will discuss in how far they are probably true. These claims are, in general, that the addition of aluminium—

1st. Makes the iron more fluid.

2d. Makes hard iron softer.

3d. Frees castings from hard spots and blow-holes.

4th. Lessens the tendency of the metal to chill.

5th. Increases the resistance of the iron to chemical action.

It is also stated that while good, soft iron is made more fluid and benefited to some degree, yet the advantages of treating with aluminium are most evident with poor, hard, white iron. We will review these claims in the light of those trustworthy experiments which have been made and certified to.

It is now generally conceded that the addition of ferro-aluminium does affect the quality of the castings. The method of adding it which has been generally adopted is to put some pieces of broken ferro-aluminium into the bottom of a ladle, preferably a hot one, and tap the iron from the cupola directly on to the alloy. In this way the maximum benefit is obtained. A German experimenter states[*] that it is important that the iron be not too hot when the ferro-aluminium is added, for if it is white-hot, the aluminium burns with a greenish flame and a peculiar smell; a golden-yellow heat is recommended as the right heat for treatment. If the ferro-aluminium is thrown into the molten iron at this heat, the streaks playing on the surface of the metal disappear, and the bath becomes blistery looking. The same writer states that, in general, white iron is undoubtedly improved by this treatment, but that gray iron is made porous, the pores showing particularly in the lower parts of castings.

---

[*] Zeitschrift des Vereins Deutscher Ingenieure, 1889, p. 301.

While there have been many testimonials from practical men as to the benefits derived from the use of ferro-aluminium, testimonies so numerous that the fact of benefit has become indisputable, the only systematic investigation of this subject is that made by Mr. W. J. Keep, of the Michigan Stove Company, Detroit, with the co-operation of Prof. C. F. Mabery and L. D. Vorce. Their results are embodied in two quite lengthy papers, one read before the American Association for the Advancement of Science at their Cleveland meeting, August 17, 1888, the other published in the Transactions of the American Institute of Mining Engineers, December, 1889. As we shall quote many of the results given in these papers, we will first explain the methods employed in pursuing the investigation.

Two kinds of iron were used, having the following composition :—

	White iron.	Gray iron.
Silicon	0.186	1.249
Phosphorus	0.263	0.084
Sulphur	0.031	0.040
Manganese	0.092	0.187
Graphitic carbon . .	0.95	3.22
Combined " . .	2.03	0.33
Total " . . .	2.980	3.550

The ferro-aluminium used contained 11.42 per cent. of aluminium and 3.86 per cent. of silicon. The melting was done in a covered plumbago crucible, and the melt was run into test-bars one foot long, some having a section $\frac{1}{2}$ inch square, others 1 inch wide and $\frac{1}{10}$ inch thick. The ferro-aluminium was added to the molten iron, the smallest quantity first, and, after casting, part of this first cast was remelted with more ferro-aluminium, and so on. Another series of heats was made under exactly the same conditions but without adding aluminium, these tests serving for comparison and determination of the true effect of adding the ferro-aluminium. The general plan of the tests consisted in adding 0.25, 0.50, 0.75, and 1.00 per cent. of aluminium to the white iron, and 0.25, 0.50, 0.75, 1, 2, 3, and 4 per cent. to the gray iron, the test-bars being examined carefully as to strength, shrinkage, etc., and comparison made with the corresponding remelt of the iron alone.

The weak point of the first set of tests, recorded in the first paper, was the fact that many of the changes credited to the addition of the ferro-aluminium might probably have been accounted for by the silicon in the alloy added, and so the results could not be accepted as demonstrating the influence of the aluminium except where the change was in a direction contrary to that which the silicon could have produced. Mr. Keep recognized at once the necessity of differentiating the effect of these two elements, which was accomplished very ingeniously by finding an iron containing the same amounts of silicon, carbon, etc., as the ferro-aluminium, and making comparison tests with this iron in place of the aluminium alloy; also by adding pure metallic aluminium to the iron. Taking the second paper in connection with the first, the conclusions advanced may be regarded as final and beyond reasonable doubt. Since Mr. Keep's method of presenting his results is in some cases not easily understood, I have, from an inspection of his diagrams, re-cast the results into tabular shape.

Solidity of castings.—All Mr. Keep's tests bore on this point, but one particular test was made with white iron, adding only 0.1 per cent. of aluminium (0.03 of silicon). The castings were of slightly finer grain, but blow-holes and interstitial cavities were noticeably absent, this accounting for the largely increased strength. The resistance to dead weight was increased 44 per cent., and to impact 6 per cent. No check test was made to eliminate the effect of the silicon added, but the effect produced was much greater than can with any probability be ascribed to the silicon alone.

Does the aluminium remain in the iron?—To determine this question, enough ferro-aluminium was added to white iron to introduce 0.25 per cent. of aluminium (0.08 of silicon), and the resulting metal was re-melted five times. Samples were taken of each melt, and found to contain at first addition 0.23 per cent. of aluminium, and at the successive remeltings 0.20, 0.18, 0.15, 0.13 and 0.10 per cent. respectively. On comparison with white iron remelted alone the same number of times, the influence on the strength is also seen to endure through the remeltings; for instance—

No. of remelting.					Per cent. of aluminium.	INCREASE IN STRENGTH (per cent.)	
						Dead wt.	Impact.
					0.23	35	109
1	.	.	.	.	0.20	118	235
2	.	.	.	.	0.18	115	165
3	.	.	.	.	0.15	123	150
4	.	.	.	.	0.13	32	62
5	.	.	.	.	0.10	21	39

The analysis of Mr. Keep's other tests also answer the above question affirmatively, since, as before explained, the percentage of aluminium was increased gradually by adding ferro-aluminium to a previous melting, giving the aluminium several chances to escape if it tended to do so, before a large percentage was reached, but the calculated amounts agreed with those actually found as follows—

Percentage by calculation.	FOUND ON ANALYSIS.	
	White iron.	Gray iron.
0.25	0.25	0.10
0.50	0.54	0.14
0.75	0.89	0.32
1.00	1.28	0.75
2.00	—	1.50
3.00	—	2.23
4.00	—	3.84

There were, of course, unavoidable irregularities in the making of the tests, but the general conclusion from the above analyses is that all the aluminium remains in the white iron and almost all in the gray, the reason of the slight loss in the latter case not being apparent. On adding small amounts of aluminium to wrought-iron, none remains in the metal; the reason for the contrary phenomenon in the case of cast-iron is that the presence of carbon in large quantity prevents the presence of iron oxide dissolved in the iron, and the aluminium remains because there is no such oxidized compound present to slag it off.

Transverse strength.—The addition of aluminium as ferro-aluminium had the general effect of strengthening the iron, the white iron showing the greater improvement, and the resistance to impact being increased more than the resistance to dead weight. The following table gives the percentage increase in strength in each case, the minus quantities in parentheses meaning decreased strength :—

Percentage of aluminium	White Iron.		Gray Iron.	
	Dead wt.	Impact.	Dead wt.	Impact.
0.25	32.5	82.8	—(12.5)	—(15.5)
0.50	128.0	291.0	—(8.9)	44.0
0.75	113.6	240.0	5.8	23.0
1.00	117.6	350.0	2.4	30.0
2.00	—	—	—(10.4)	29.0
3.00	—	—	4.6	11.0
4.00	—	—	15.8	130.0

Since for every 1 per cent. of aluminium added, 0.34 of silicon was contained in the ferro-aluminium, the question very naturally occurred, how much of this benefit was due to the silicon. Tests were therefore made on this point, proving the part taken by the aluminium. The figures show the percentage increase in strength, as in the former tables.

Addition.	White Iron.	
	Dead wt.	Impact.
1 per cent. aluminium in ferro-aluminium . .	117.6	350.0
Cast-iron introducing the same quantity of silicon	126.8	94.6
1 per cent. of aluminium as pure aluminium .	141.7	156.5

The conclusions to be drawn are, therefore, that while the silicon in the ferro-aluminium is sufficient to explain the increased resistance to a dead weight, yet the increase in resistance to impact is clearly due in large part to the aluminium.

Elasticity.—The closing of the grain of the iron on treatment with ferro-aluminium caused the iron to be less brittle, or more elastic. The deflection of the different specimens for a fixed weight was measured, and the increase in deflection was found to be (in percentages)—

Percentage of aluminium.	White iron.	Gray iron.
0.25	31	125
0.50	89	116
0.75	100	147
1.00	153	133
2.00	—	133
3.00	—	194
4.00	—	193

To distinguish the effect due to the silicon added, tests made with silicon and aluminium alone showed increased deflections as follows :—

Addition.	White-iron.
1 per cent. of aluminium in ferro-aluminium	153
Cast-iron containing an equal quantity of silicon . . .	11
1 per cent. of aluminium as pure aluminium	100

It is thus proved that the increased elasticity is due to the aluminium, caused, as Mr. Keep believes, by a very uniform distribution of the graphitic carbon when aluminium is the element precipitating it, a phenomenon to be examined further on.

Effect on the grain.—Mr. Keep found that the addition of ferro-aluminium made the grain of the iron decidedly darker, caused by the separation of more carbon as graphite. It is well known that silicon acts in the same direction, but Keep's first impressions were that the separation of graphite took place much nearer to the setting point of the iron than he had ever observed to result from silicon acting alone. A check test, adding cast-iron without aluminium, confirmed this impression; for instance,

Addition.	White-iron. (Description of fracture.)
———————	White.
0.25 per cent. aluminium as ferro-aluminium .	A few gray specks.
0.50 " " " " .	Light gray.
0.75 " " " " .	Gray.
1.00 " " " " .	Dark gray.
Cast-iron containing an equal quantity of silicon	[specks.
to preceding	White—a few gray
1.00 per cent. aluminium as pure aluminium .	Dark gray.

The comparisons made show that aluminium is undoubtedly active in changing combined into graphitic carbon; from a comparison of the analyses of the above tests it appears that it is even more powerful than silicon in accomplishing this result, since 0.25 per cent. of aluminium (with 0.20 per cent. of silicon) seems to give a fracture identical with that produced by 0.62 per cent. of silicon alone.

Mr. Keep notices that the separation of the graphite seems to take place instantaneously just as the iron is about to set, and not before, the result of which is that there is very little opportunity for any gathering together of the graphite into soft spots in the casting, and also that no matter how quickly the iron sets the graphite will be mostly separated out, and thus the iron chills

30

less. This effect is very noticeable in gray-iron, where the first addition of 0.25 per cent. of aluminium as ferro-aluminium decreased the depth of chill fully one-half and slightly darkened the fracture, subsequent additions of two, three and four times as much reduced the chill to nearly nothing, while with 2 per cent. of aluminium added, the drop of the graphite was so nearly instantaneous that no chill was visible.

Fluidity of the iron.—The general conclusion from Mr. Keep's tests is that, with white-iron, small additions of aluminium, such as would be used in ordinary foundry practice, increase slightly the fluidity; one-half per cent. of aluminium and over decreases the fluidity. Gray-iron is rendered decidedly less fluid by any addition of aluminium. Mr. Keep notices a peculiarity of cast-iron containing aluminium which is similar to that we have remarked in pure aluminium (p. 350); viz., that as the metal flows it seem to have a skin in front of it, causing it to run with a very thick edge, and if two currents of this iron come together in a mould, they are apt not to unite but to simply chill without union.

Shrinkage.—Mr. Keep measured carefully the shrinkage of the different specimens of aluminized cast-iron. The general conclusion was that aluminium reduces the shrinkage if enough of it is added. The following table shows the reduction in the shrinkage, in percentage of the original shrinkage, under the different conditions—

	REDUCTION OF SHRINKAGE (per cent.)			
	WHITE-IRON.		GRAY-IRON.	
Percentage of aluminium	Square bar.	Thin bar.	Square bar.	Thin bar.
0.25	—(4)	—(4)	0	14
0.50	0	—(4)	—(3)	4
0.75	16	—(4)	4	14
1.00	21	—(4)	9	16
2.00	—	—	19	16
3.00	—	—	34	16
4.00	—	—	28	26

It will be noticed that, particularly with the square bar, the first two additions of aluminium have very little effect either way, but that with subsequent additions the amount of shrinkage is reduced 5 to 30 per cent. Mr. Keep thinks that this behavior is dependent on the action of the carbon; that the small amounts of aluminium are chiefly active in closing blow-holes and giving soundness to the casting, but the larger amounts have a noticeable influence on changing combined carbon into graphite, and that the increased deposition of graphite just as the metal sets increases its volume and decreases the amount of shrinkage.

Hardness.—The indications from Mr. Keep's tests are that aluminium of itself hardens cast-iron, but, by its influence in changing combined carbon into graphite it indirectly renders the iron softer. It was noticeable that if an iron cast with soft spots, the parts in between being hard, that the addition of aluminium caused the graphite to be dropped so near to the setting point that it had no opportunity to collect into spots, and was therefore uniformly distributed, rendering the iron uniformly softer.

Aside from the above determinations, many testimonials could be quoted from practical iron founders as to the practical benefit to poor iron gained by adding ferro-aluminium. Perhaps one of the most striking results is the increased time which the aluminium-treated iron will remain molten. For instance, Mr. Keep found that 0.02 per cent. of aluminium added to a ladle of iron caused it to keep fluid 5 minutes, while a similar ladleful of the same metal without aluminium became solid in $2\frac{1}{2}$ minutes. This property of keeping fluid longer is of direct usefulness in a foundry where it is necessary to run a large number of small castings, during which operation there is usually much trouble experienced in keeping the iron fluid unless it was very hot to start with. I am quite assured of the fact that the addition of a very small amount of aluminium does have the effect described above. A friend of the author's described an experiment in which a large ladleful of iron was tapped from a cupola and taken for pouring about 200 yards, partly through the open air. The iron was not hot enough to fill the moulds satisfactorily. Another ladleful, similar in all respects, was tapped immediately after, some ferro-aluminium being placed in the ladle. This iron was taken to the same place for casting, filled the moulds perfectly, and when brought back to the cupola the metal left in the ladle was still fluid enough to make good castings. I have heard similar reports from trustworthy sources, all stating that the judicious use of a small quantity of ferro-aluminium will result in nearly doubling the time during which a bath of iron will stay fluid.

The practical results observed by the foundry men are that they obtain cleaner, more solid, softer castings, with a large reduction in the percentage of defective castings. Mr. Adamson, President of the British Iron and Steel Institute, says that "since using ferro-aluminium in his foundry, 80 per cent. of the waste had been saved and all the work manufactured was improved in quality." The general testimony seems to be that the castings come out of the sand cleaner, are much more free from blow-holes, work more uniformly in the lathe or planer because of the absence of hard or soft spots, come up sharper in the mould and are generally stronger. There is not much improvement made on good gray foundry iron, in which case it is best to leave the good iron to itself, but when the quality of the iron is low, and difficulty met in getting good castings, then ferro-aluminium is of undoubted benefit. Very small additions are relatively of greatest effect; Mr. Keep has stated that with only 0.00067 per cent. of aluminium added to a poor quality iron it could be observed that the blow-holes were lessened and the transverse strength noticeably increased. It is probable that the first one-hundredth of a per cent. of aluminium added has more effect than the next five-hundredths, and it is fortunate that this is so, for it opens up an extremely large field for the use of aluminium alloys.

The *rationale* of the action of aluminium on cast-iron may be said to be an open question. That 0.25 to 0.50 per cent. begins to affect the carbon, has been proven by Mr. Keep, yet these are quantities which are not used in ordinary foundry practice, where 0.10 per cent. may be taken as the maximum amount which the founder can afford to use, while 0.01 to 0.05 per cent. may be said to be the usual additions. It would appear to me that such small proportions of aluminium can only exert the effects attributed to them by (expressing it figuratively) holding a sort of balance of power, by which it determines a much greater result than the aluminium alone could possibly bring about. This idea is alluded to by Mr. Keep when he says that to use ferro-aluminium to best advantage the cast-irons should be mixed so as to get mixtures most suitable for treatment, or, in other words, to get an iron more sensitive to the aluminium. What the conditions are which render an iron sensitive to the action of the aluminium has not been definitely determined. We might infer that since it acts in many respects similarly to silicon, the less the amount of silicon present the larger the scope of the aluminium, that is, the more room it has to act. Or, similarly, the more combined and less graphitic carbon present, the more opportunity is given the aluminium to benefit the iron. But, why the aluminium-treated iron should stay fluid so much longer, we cannot see. This seems to be one of those effects out of proportion to the quantity of aluminium present. For the present, we will rest the case here: practically, very small additions of ferro-aluminium can be made very advantageous; theoretically, we have, as yet, no satisfactory explanation of the facts observed.

CHAPTER XVII.

ANALYSIS OF ALUMINIUM AND ALUMINIUM ALLOYS.

COMMERCIAL aluminium may contain the following elements besides aluminium: Silicon, iron, lead, tin, zinc, copper, silver, carbon, sodium, chlorine, fluorine. The method of attack generally preferred is solution in pure caustic soda. Hydrochloric acid usually attacks aluminium very energetically, and more easily the larger the percentage of silicon it carries, for the purest aluminium is not attacked very violently. When much silicon is present, the odor of silicuretted hydrogen is plainly perceptible during the action of the acid, thus indicating some loss of silicon. In dissolving in caustic soda this loss does not take place. Solution in bromine or iodine solution also offers similar advantages as respects avoiding loss of silicon. When analyzing particularly for carbon, the solvent used in the determination of carbon in iron may be appropriately used. We will consider these more in detail further on.

A qualitative test may very appropriately precede the analysis, and will often save much time in the quantitative determinations. These qualitative tests may generally be made on the same lines as the others; though in some cases shorter methods are practicable. Thus, after solution in hydrochloric acid, neutralizing with and adding excess of ammonia will show the presence of copper. If the solution is made hot, a spot of sulphuric acid will show whether lead is present. Iron may be detected by potassium ferro-cyanide. Silver will remain as a white residue after the action of the acid. Chlorine is detected most easily by Berzelius' blowpipe test with oxide of copper. Lead and zinc can be detected to a certain extent before the blowpipe. A specimen of aluminium which gave an unmistakable test for lead on charcoal before the blowpipe, was found on subsequent analysis to contain nearly 7 per cent. of that metal. It appeared as though a much smaller proportion could have been thus detected. Another specimen similarly treated gave a good test for zinc, and the quantitative analysis gave 6.25 per cent. of that metal. Smaller amounts than this could probably be easily detected. On the other hand, however, it is not probable that the presence of tin would make itself evident in the charcoal test; for aluminium seems to protect the tin very strongly from oxidation. An alloy known to contain 90 parts of tin to 10 of aluminium was tested on charcoal, and would not, with the hottest flame at my command, give the usual white coat indicating tin. Such being the case, it appears highly improbable that aluminium containing a small percentage of tin would give the test in question. The better test would be the white residue left on solution in hot, concentrated nitric acid. The presence of iron or copper, or of both together, can usually be immediately recognized by dissolving a small piece of the aluminium in a borax bead, on a platinum wire, in the oxidizing flame. Copper thus detected by the author was found on analysis to be less than 1 per cent. of the weight of the aluminium tested. The presence of sodium is generally shown by the metal decomposing water heated nearly to boiling, setting free hydrogen. This test can be easily made in a test-tube.

The specific gravity of the metal, accurately taken, furnishes some intimation as to the presence of any of the heavy metals in any considerable quantity. Thus, 5 per cent. of silver increased the specific gravity from 2.65 to 2.8, 6 per cent. of lead from 2.75 to 2.9. This test is not, however, of much value, since, because of the very low specific gravity of aluminium, small amounts of heavy metals have only a small influence in increasing it, while silicon, an impurity most likely to be present, is lighter than aluminium (specific gravity 2.35), and therefore neutralizes to some extent the effect of the heavy metals. However, in commercially pure aluminium, containing only iron and silicon, the specific gravity can be made useful in indicating the amount of iron present within rather wide limits—say within 1 per cent.

Another test of somewhat similar utility would be that given by Fr. Schulze.* He proposes to dissolve the aluminium in caustic alkali and measure the volume of hydrogen set free. If the metal contains no zinc, this volume will be approximately proportional to the amount of aluminium in the metal. Thus, $\frac{1}{2}$ gramme of one specimen gave 648 cubic centimetres of hydrogen, a similar weight of another, 580 cubic centimetres. These figures are then taken as expressing the relative purity of the two samples. In an aluminium works where a quick, approximately accurate test is needed, which can be made, if need be, by a person not necessarily a skilful chemist, and which is applied to testing samples of nearly the same composition, this test would appear to be of practical utility.

Determination of silicon.—Deville recommended the following method : " Dissolve in pure hydrochloric acid and evaporate to dryness in a platinum dish. The evaporation to dryness is indispensable in order to render insoluble the quite important quantity of silica which is kept in solution by the presence of the acid. There remains an insoluble residue consisting of silicon, silicon protoxide and silica, which is washed by decantation with hot water and thrown on to a filter. This mixture of siliceous material is then calcined with the filter in a platinum dish at a low temperature. A little flame may often be seen coming from different parts of the mass, caused by the production (at the expense of

---

* Wagner's Jahresbericht, x. 23.

the silicon protoxide) of a little silicon hydride, according to the reaction—

$$3SiO + 2H^2O = SiH^4 + 3SiO^2$$

causing a slight loss of silicon. This may be altogether avoided by moistening the material with ammonia before calcining. The residue after ignition is a mixture of silicon and silica, the silicon protoxide having completely disappeared, and is carefully weighed. This done, it is put into a platinum crucible, and treated with a little dilute hydrofluoric acid, which dissolves the silica and leaves the silicon, which is washed with care. This residue is then dried and weighed, and by subtracting from the former weight, the weight of silica is known with which it was mixed. This silica is calculated to silicon, and when the silicon weighed directly is added in, the result is the total weight of silicon in the metal tested."

The above method probably does determine with accuracy the amount of silicon which remains after solution of the aluminium, but Rammelsberg observed (see p. 55) that when aluminium contains considerable silicon there is always some silicon hydride formed during its solution in hydrochloric acid, which escapes and so causes error in the analyses. By passing the gases produced during solution through a solution of caustic potash, the silicon hydride was intercepted and the amount of silicon thus escaping was determined. It was found to be in two instances, 0.74 and 0.58 per cent., being 7 per cent. and 22 per cent. respectively, of the total silicon in the metal (the first was very siliceous). It follows, therefore, that to make an accurate determination of the silicon in aluminium, hydrochloric acid cannot be used for attacking the metal unless care is taken to catch and determine the amount of silicon passing off as silicon hydride. Prof. Rammelsberg concluded from his study of the subject that silicon occurred in two forms in aluminium, a small amount free (like graphite in iron) the larger amount combined, and that on treatment with hydrochloric acid the free silicon remained as such while the combined silicon partly escaped as silicon hydride and the rest was converted into silica. Such being the case, we can readily see the superiority of caustic potash or soda solution for dissolving the aluminium. Graphitoidal or crystalline silicon is dissolved by hot potash solution, the combined silicon will be dissolved, and if the nascent hydrogen forms for an instant any silicon hydride—as it does when acid is used—this gas is at once decomposed by the alkali solution. The result is that if aluminium is attacked by hot potash solution, all the silicon present is oxidized and none lost. The solution of the aluminium should take place in a silver or platinum dish or crucible, a porcelain dish, however, is very slightly attacked, but glass should not be used. Care should be taken that the solution of caustic does not contain alumina or silica. After solution is complete, the liquor is filtered from any residue, hydrochloric acid is added until the reaction is acid, the bath evaporated to complete dryness until no smell of acid is perceptible, then moistened with a little hydrochloric acid to dissolve any alumina formed, water added and the whole brought to boiling. The silica is then filtered out, dried, ignited, weighed and calculated to silicon.

Determination of iron (and aluminium).—Deville's method of procedure was as follows: "The metal is dissolved in pure hydrochloric acid, evaporated to complete dryness in a platinum dish, and the insoluble, siliceous materials filtered out. The solution is mixed with a large excess of nitric acid, evaporated in a porcelain dish covered by a glass, thus converting the bases into nitrates, which are then transferred to a platinum dish. Here the solution is evaporated to dryness and calcined lightly on the sand-bath, the dish being covered, until abundant vapors of nitric acid rise from all parts of the mass. Cool, and moisten with a solution of ammonium nitrate containing free ammonia. Heat until all odor of ammonia has disappeared, take up with water, and separate by decantation all soluble matter. (Decanting for greater precaution on to a filter.) The solution obtained contains all the sodium which was in the aluminium (see Determination of Sodium), while the insoluble residue is a mixture of aluminium and ferric oxide. This is heated to redness in the platinum dish which contains it, and transferred in whole or part into a tared platinum boat, where it is weighed. (It is best to make all these weighings with the boat inclosed in a glass tube closed by the flame at one end, and at the other by a well-fitting cork. The tube and boat are then weighed together.) The boat is then placed inside a porcelain or platinum tube, heated up to redness, and a current of pure hydrogen passed over it. When the tube is bright-red, the hydrogen is replaced by hydrochloric acid gas, which transforms all the reduced iron into ferrous chloride without touching the alumina. At the end of the operation, when the tube is just below redness, the hydrochloric acid gas is replaced by hydrogen. When nearly cold, the boat is drawn out and weighed, the loss in weight being the ferric oxide removed, the portion still remaining being pure alumina. From these data the iron and aluminium are calculated. Very little trust can be placed on the alumina being perfectly white, to conclude that it is, therefore, free from iron; for experience has taught me that one may be very greatly deceived in making this conclusion. Experience will show about how long and at what heat the operation must be continued to remove all the iron, but to be absolutely certain, the operation should be repeated for a short time, when the alumina will remain constant in weight if it is perfectly pure."

The above operation is the most accurate method of determining iron and aluminium, and is more applicable to estimating small quantities of iron in aluminium rather than *vice versa*, since in the latter case the operation is much prolonged in reducing and volatilizing so much iron compounds. The estimation of small amounts of aluminium in presence of a large quantity of iron generally calls for special methods of separation, which will be detailed under the analysis of aluminium-iron alloys. Confining ourselves here to the determination of iron in commercial aluminium, we may suggest the following modifications of the above method. After solution in hydrochloric acid and separation of silica, the addition of a few drops of sulphuric acid will precipitate any lead which may be present, and then the iron and aluminium may be precipitated with ammonia in slight excess. Any copper present will remain in solution, and the precipitate may be washed well, filtered, ignited, and weighed. Instead, then, of removing the ferric oxide by the method given by Deville, the method of H. Rose may be used, which consists in fusing the two oxides with caustic potash (by alcohol) in a silver crucible, when potassium aluminate will be formed, and on boiling the mass with water and filtering, the alkaline fluid will contain the aluminium, while the residue will be ferric oxide containing some potash.

If a solution has been prepared containing only iron and aluminium (lead, zinc, copper, etc., having been separated out) many methods have been proposed for separating these two elements. The best known is to make the solution nearly neutral and then pour gradually into excess of pure caustic potash solution heated nearly to boiling in a platinum or silver dish. The iron is precipitated, while aluminium remains in solution. The details of this test can be found in any treatise on quantitative analysis. It is not to be relied on in many cases, especially for determining a small amount of aluminium in presence of much iron, since it is always probable that some aluminium is retained by the iron precipitate. Dissolving the iron hydryoxide and reprecipitating will partially correct this. Solution of caustic soda is more apt than caustic potash to contain alumina, and so give aluminium results too high. In either case, the alkaline solution of potassium or sodium aluminate is heated nearly to boiling, and mixed with a large excess of ammonium chloride, when the alumina is entirely precipitated.

The method most frequently used to separate ferric oxide from alumina is to weigh the two oxides together, then dissolve in concentrated hydrochloric or sulphuric acid, reduce the solution by any suitable reducing agent (zinc, sulphurous acid gas, etc.) and determine the amount of iron present by titration with potassium permanganate or bichromate solution. These results are sufficiently accurate if the proportion of aluminium is not small; when the latter is the case, as in ferro-aluminium, other methods give more satisfactory results, and are given further on in considering the analysis of ferro-aluminium.

The solution of the aluminium to be tested in caustic alkali offers the quickest method of separating out the iron, for the alkali dissolves out all the aluminium and leaves iron in the residue. The best way to conduct this method of analysis is to roll out the aluminium into a thin sheet, put in moderately concentrated alkali and allow it some time to dissolve. On filtering, the solution will contain all the aluminium while all the iron will remain, with various other metallic impurities, in the residue. This residue is then dissolved in acid and the iron easily determined in it without any interference from aluminium.

Determination of lead.—Dissolve the aluminium in hot hydrochloric acid, evaporate to dryness, moisten with hydrochloric acid, add hot water and boil. Filter hot to separate out silica. To the hot, boiling solution add a few drops of sulphuric acid and let stand an hour. The precipitated lead sulphate can then be filtered out and weighed.

Determination of copper.—Proceed as in determining iron, when the copper will pass into the filtrate on precipitating iron and aluminium by excess of aqua ammonia. If any amount of copper is present this solution will be blue, especially when concentrated by evaporation. The solution may be evaporated to dryness, taken up with a little acid and the copper determined by any ordinary method; or, the ammoniacal copper solution may be acidified with sulphuric acid, crystals of oxalic acid added, the liquid boiled and the copper thus deposited as oxalate. This is washed in boiling water, dried, calcined at a gentle heat and weighed as cupric oxide. The ammoniacal copper solution might also be acidified with acetic acid and the copper precipitated by lead foil.

Determination of zinc.—Dissolve the aluminium in hydrochloric acid or in caustic alkali and separate out silica, taking up, however, with strong acetic acid. On passing sulphuretted hydrogen through the solution, zinc will be precipitated free from aluminium or iron. The precipitate of zinc sulphide is washed carefully with distilled water saturated with sulphuretted hydrogen, dried, calcined very carefully in a muffle and afterwards very strongly over a blast lamp, and weighed as zinc oxide.

A very satisfactory separation is also obtained by nearly neutralizing the solution, adding a limited quantity of acetic acid and then sodium acetate, and thus precipitating the aluminium (and iron) as basic acetates, leaving the zinc in solution, from which it can be precipitated by a current of sulphuretted hydrogen.

Determination of tin.—Solution of the aluminium in hot nitric acid will leave the tin as insoluble metastannic acid. This residue is washed and treated with warm, dilute hydrochloric acid, which dissolves the tin compound and leaves the silica. The tin is then thrown down in the solution of stannic chloride by any of the ordinary methods of precipitation, preferably by sulphuretted hydrogen. The stannic sulphide is ignited gently, moistened with nitric acid, ignited more strongly, and the tin weighed as stannic oxide.

Determination of silver.—Dissolve the aluminium in weak aqua regia, dilute and filter out the siliceous residue which will contain also all the silver as chloride. Wash carefully, and then dissolve out the silver salt with concentrated ammonia. On neutralizing the ammoniacal solution with nitric acid, the silver chloride is again precipitated, washed by decantation, dried at 250° to 300°, and weighed.

If the aluminium is attacked with caustic alkali, the silver will remain in the residue. This is washed, filtered, and treated on the filter with dilute nitric acid, which dissolves the silver. The solution of silver nitrate is precipitated by hydrochloric acid or sodium chloride, and the operation finished as before.

Determination of sodium.—In the first steps of the determination of iron, as given by Deville (p. 473), a solution was obtained free from iron and aluminium, and containing all the sodium which was in the aluminium tested. Deville describes the estimation of the sodium in this solution as follows: "To the solution is added a drop of ammonium oxalate, which sometimes precipitates a trace of calcium, indicative of the presence of fluorspar in the slag with which the metal may be impregnated. After filtering (if necessary) evaporate to dryness in a weighed platinum dish, cover and heat to 200° or 300°, to decompose the ammonium nitrate. Nitrate of soda remains. This is moistened with water, and on it are placed several crystals of oxalic acid. Dry, calcine, and there remains sodium carbonate, which is often impregnated with a little carbon from the decomposition of the oxalate of soda. Dissolve the residue in water; if not clear, filter. Mix the solution with a little hydrochloric acid, evaporate to dryness, heat to 200°, and weigh as sodium chloride."

Commercial aluminium rarely contains metallic sodium, but when it does exist it can usually be detected from the fact that the amount of chlorine present is not sufficient to combine with the sodium found. The presence of fluorine would weaken this conclusion, but it is seldom present in any quantity.

Determination of chlorine.—Dissolve the aluminium in pure caustic soda, neutralize with nitric acid in very small excess, filter, and add several drops of nitrate of silver. The chloride of silver precipitated is washed well, dried at 300°, and weighed.

Determination of carbon.—I know of no accurate determinations of carbon in aluminium. If it does occur, it is probably all as combined carbon. A large proportion of this carbon would necessarily escape on treating the aluminium with acids or alkalies, for the nascent hydrogen developed during solution would form volatile hydrocarbons with it. To obviate this, I see no reason why the ammoniacal solution of copper chloride should not be used as a solvent, as in carbon determinations in iron. Solution in this, or, perhaps, in bromine, should give a means of estimating the total carbon present without much reason to doubt its accuracy.

Detection of fluorine.—Deville recommends the following procedure: "Dissolve in caustic soda (using no more than is necessary), filter, and nearly neutralize with pure sulphuric acid. It is necessary to take care to leave a very small amount of free alkali without separating out any alumina, which falls at the moment when neutralization is complete. Evaporate the whole in a platinum crucible, and heat, covering with a watch-glass coated with varnish through which regular lines have been traced with a copper point. A section of quartz prepared in the same way does still better. Vapor of water and hydrofluoric acid are disengaged on heating the crucible, and the latter etches the glass quite perceptibly. Sometimes, by breathing on the glass, the lines become more apparent.

Analysis of Ferro-Aluminiums.

It is not intended to give directions for the complete analysis of ferro-aluminium. The determinations of silicon, carbon, sulphur, phosphorus, etc., are made the same as in steel or pig-iron. The only point offering special difficulty is the accurate determination of a small amount of aluminium in presence of a large amount of iron. For doing this, the ordinary methods do not give satisfactory results. Thus, when determining the iron volumetrically and the aluminium by difference, it is almost impossible to get concordant results. When separating by caustic alkali, the results in aluminium are too low, for the iron precipitate being so abundant, carries much aluminium down with it. Also, a large amount of caustic alkali must be used, and since it sometimes contains alumina, considerable error may be thus introduced.

Supposing that the alloy has been dissolved in acid, silica separated out and a solution containing only iron and aluminium obtained to work with, we will take up the various methods proposed to determine accurately the small amount of aluminium.

Mr. H. N. Yates analyzed several hundred specimens of ferro-aluminium, and found that the caustic alkali separation gives aluminium too low, and is unreliable; the method of weighing the two oxides together and afterwards determining the iron volumetrically with potassium bichromate gives pretty fair results with aluminium from 1 to 19 per cent.; but, with less than 1 per cent., as in steels, the most satisfactory results were obtained by the sodium thiosulphate separation. This latter is known as Chancel's separation, and is operated as follows: The solution is neutralized with sodium carbonate, made dilute, solution of sodium thiosulphate (hyposulphite) added and the liquid boiled until no more sulphurous acid is disengaged. All the alumina is precipitated as hydrate, with free sulphur, which may be washed, dried, ignited and weighed as alumina. The boiling until all sulphur smell is gone is a tedious operation, and the following modification is said to give equally accurate results: To the slightly acid solution, sodium thiosulphate is added more than equivalent to the amount of free acid present. The liquid is then boiled in a flask 10 to 15 minutes. The precipitated alumina is in a fine, granular state, easy to wash. The liquor is rapidly filtered, the precipitate washed with boiling water, dried, ignited and weighed.

A great disadvantage of the foregoing method is that any phosphoric acid present in the solution will be precipitated with the alumina. With certain iron alloys this would materially affect the result. To overcome this disadvantage, Mr. Peters suggested converting the alumina entirely into phosphate, still keeping the same method of separation from iron. The process, as modified, is as follows: If less than 1 gramme of iron is in the solution, it is diluted to 400 or 500 c.c. with cold water, and ammonia added until the solution is dark-red in color but contains no precipitate. Now add 3 c.c. of hydrochloric acid (sp. gr. 1.2), and 2 grammes of sodium phosphate, dissolved in water and filtered. Stir till the precipitate formed is re-dissolved and the solution is clear. Add now 10 grammes of sodium hyposulphite dissolved in water and 15 c.c. of acetic acid (sp. gr. 1.04), heat to boiling, boil 15 minutes, filter as rapidly as possible, wash with hot water, dry, ignite in a porcelain crucible raising the heat very carefully until all the carbon has burnt off, and weigh as $AlPO^4$ ($Al^2O^3.P^2O^5$).*

Mr. R. T. Thompson, an English chemist, stated that he found Chancel's separation ineffectual in determining aluminium in presence of a large quantity of iron (probably mainly because of the phosphoric acid present), and devised the following method of separation: The iron is reduced to the ferrous state by a current of sulphurous acid gas, excess of this gas is boiled off, and when cool, phosphoric acid or sodium or ammonium phosphate is added in excess of that required to precipitate all the alumina, then aqua ammonia until a faint permanent cloudiness is formed, finally excess of ammonium acetate. The precipitate generally contains a little iron, but on washing it, re-dissolving in acid and repeating the precipitation, it is obtained free from iron. The precipitate is dissolved in hydrochloric acid, a little nitric acid added, boiled, and nearly neutralized with caustic soda, and then boiled with an excess of the latter. The aluminium is precipitated as phosphate, is washed with a 1 per cent. solution of ammonium nitrate containing about 0.1 gramme of ammonium di-hydric phosphate per litre, and ignited and weighed as aluminium phosphate.†

* The Chemical Analysis of Iron. A. A. Blair.
† Journal, Society of Chemical Industry, V, 152.

A method of separation which has given the author satisfactory results where the caustic alkali and even the sodium hyposulphite separations were unsatisfactory is the following :* To the cold, concentrated and slightly acid solution add an excess of solid sodium bi-carbonate in such quantity that after stirring a little remains undissolved and all the iron appears to be thrown down. Now add solution of potassium cyanide until the precipitate dissolves, then heat gently until the yellow color of potassium ferrocyanide is produced. Add a few drops of caustic potash to the somewhat turbid solution until it is perfectly clear; then add excess of ammonium chloride and boil. Aluminium hydrate is precipitated free from iron, nickel or cobalt. As an analytical operation this method works very satisfactorily, care must be taken, however, in handling such large quantities of the very poisonous potassium cyanide.

It has been stated that if an excess of tri-methylamine is added to a dilute solution containing iron and aluminium, and let stand twenty-four hours, all the iron is precipitated and all the aluminium remains in solution.† The accuracy of this separation has not been thoroughly tested.

A. A. Blair recommends the determination of aluminium in iron and steel (when it occurs in very small amount) by the direct separation of ferric oxide from alumina, the method used being as follows: Dissolve the iron in strong hydrochloric acid, in a flask provided with a valve, thus keeping out air and allowing the iron to dissolve in the ferrous state. Neutralize with sodium carbonate, cool, dilute, add "milk" of barium carbonate, let stand several hours, the flask being meanwhile well stoppered, and filter. The precipitate consists of all the alumina, ferric oxide, chromic oxide, phosphoric acid or titanic acid mixed with the graphite and insoluble silica of the alloy. Wash well, treat with dilute hydrochloric acid, boil the solution with a slight excess of sulphuric acid, to precipitate the barium in the solution. Settle, filter, wash with hot water and concentrate the solution by evaporation. To separate the iron from the aluminium, add citric acid to the

---

* Chemical News, March 29, 1888.
† Zeitschrift für Anal. Chemie, xxiv, part 5.

amount of about five times the weight of oxides present, and excess
of ammonia. If the solution does not stay clear, acidulate with
hydrochloric acid, add more citric acid and excess of ammonia.
Heat the clear solution to boiling and add fresh solution of ammonium sulphide until all the iron is precipitated. Let settle,
wash with water containing ammonium sulphide. The iron
sulphide can be dissolved in acid and the iron thrown down with
ammonia. The solution can be acidified with hydrochloric acid,
boiled, the sulphur filtered out, evaporated to dryness, ignited,
the residue fused with sodium carbonate, dissolved in water, filtered, acidulated with hydrochloric acid and the alumina precipitated by ammonia. It would be quicker to take the solution
containing iron and aluminium and divide it into two portions.
In one part the iron and aluminium are thrown down together
by ammonia and weighed as oxides; in the other the iron is
separated as above. The last part may then be omitted and the
aluminium found by difference. The imperfection of this method
is the fact that the precipitate of alumina finally obtained contains
any chromic oxide, titanic oxide or phosphoric acid that may be in
the original solution. The first two elements may be seldom
present, but some phosphorus is usually present, and the alumina
will be too high for this reason. If the phosphoric acid is determined in this precipitate, this source of error may be eliminated.
A more serious defect, however, is the fact that the iron is not
completely precipitated, as may be proved by small flakes of iron
sulphide being deposited if the solution is let stand several days,
showing that the ammonium sulphide has the power of holding
small quantities of iron in solution. The aluminium results are
thus apt to be too high. Tartaric acid might be used instead of
citric, but it is more liable to contain alumina and so give results
too high in aluminium.

For separating large quantities of iron from small quantities
of aluminium, the electrolytic method seems to be particularly
applicable, for aluminium cannot be deposited from aqueous solution by the battery (except under exceptional conditions as alumina), while the iron can be totally deposited in a form easy to
weigh. This method of separation is superior in accuracy to any

of the preceding chemical methods, and is not difficult of application.

Dr. Classen gives the following method of procedure :* The solution may contain iron, cobalt, nickel, zinc, and aluminium. The solution of their sulphates is made very nearly neutral with ammonia and an excess of ammonium oxalate added, so that there are 2 or 3 grammes of this salt present to every 0.1 gramme of oxides. When the solution is not above 40° C. (its volume is best 150 to 200 c.c.) it is electrolyzed with a current not exceeding 10 or 12 c.c. of oxyhydrogen gas per minute. If the current is stronger than this, alumina may be precipitated. If the amount of aluminium is not greater than that of the iron (and other metals), this method gives accurate results. The solution remaining is evaporated to dryness, heated gently to decompose the aluminium salts, and finally ignited strongly to alumina.

Prof. Edgar F. Smith gives an electrolytic method which for accuracy leaves nothing to be desired.† The solution of the sulphates is made dilute, and about 10 per cent. of sodium citrate and a few drops of citric acid added to it. This is electrolyzed with a current of about 12 c.c. oxyhydrogen gas per minute, using platinum electrodes. The deposit of iron is firm, and is washed successively with water, alcohol, and ether, and weighed. If the iron and aluminium have been already determined together, the aluminium can be calculated by difference. If it is desired to weigh the aluminium directly, the citric acid solution must be evaporated to dryness, ignited to drive off organic matter, dissolved in acid, and the alumina precipitated by ammonia or ammonium sulphide.

The author would suggest the following method of determining iron and aluminium, which can be quickly executed, and if done carefully, gives tolerably accurate results : Divide the solution into two equal parts. In one part, precipitate the iron and aluminium by ammonia, and weigh their oxides. In the other part, precipitate with ammonia and ammonium sulphide, well washing the precipitate, and igniting with sulphur in a Rose

---

* Quant. Chem. Analyse durch Electrölyse, p. 79.
† American Chemical Journal, 1888, p. 330.

crucible in a stream of hydrogen sulphide. This ignition only takes a few minutes, and the iron remains entirely as ferrous sulphide, while the alumina is unchanged. In the first case, we weigh alumina and ferric oxide; in the second case, alumina and ferrous sulphide. Since 160 parts of ferric oxide are equivalent to 176 parts of ferrous sulphide, the difference in the two weights is one-tenth the amount of ferric oxide present in the first weighing. It is true that any errors are multiplied by ten, but the method is quick and of unquestioned accuracy, so far as the weighings are concerned. It might be even modified so far as to weigh the oxides precipitated by ammonia, mix them with excess of sulphur, and ignite in a Rose crucible in a stream of hydrogen sulphide. This ignition could be done inside of fifteen minutes, and tolerably accurate results obtained.

Analysis of Aluminium Bronzes.

The ingredients usually present are copper, aluminium, silicon, iron, and sometimes zinc, tin, nickel or lead. Solution in nitro-hydrochloric acid and evaporation to dryness will serve to separate out the silicon as silica. If dissolved in hot nitric acid, the tin remains in the residue as metastannic acid, which can be ignited along with any silica remaining and weighed as stannic oxide. The silica is left on treating this residue with hydrochloric or sulphuric acid, and its weight being subtracted from the previous weighing gives the net weight of stannic oxide. Lead would be precipitated from the nitric acid solution filtered out above by nearly neutralizing the solution and adding a few drops of sulphuric acid. The precipitated lead sulphate is filtered out and weighed. Copper may be precipitated free from zinc, iron or aluminium by evaporating the last filtrate nearly to dryness, to drive off nitric acid, acidulating with hydrochloric acid and precipitating with sulphuretted hydrogen. The precipitate of cupric sulphide may be mixed with sulphur and ignited in a Rose crucible in a current of hydrogen and weighed as cupric sulphide. Or, it may be dissolved in a few drops of nitric acid (the least possible quantity), a few centimetres of sulphuric acid added and the solution electrolyzed. The electrolysis of the solution can also be made directly in the presence of iron and aluminium by using a sulphuric acid solution with only two or three drops of free nitric acid present. The precipitation of copper by the battery is advisible in many respects, since the copper is simply removed from the solution without leaving any reagent behind it, and the other metals can be easily separated out of the solution remaining.

If the copper has been removed, either by sulphuretted hydrogen or by the battery, the solution contains only iron, aluminium, zinc, nickel or manganese. In the first case, it must be oxidized by a little nitric acid and the precipitated sulphur separated out. The metals remaining can be separated in several ways, the best, however, is to precipitate the iron and aluminium as basic acetates. To do this, the solution is neutralized with carbonate of soda until a faint precipitate forms which redissolves only after two or three minutes' stirring. Dilute, add about 4 per cent. of acetic acid and excess of sodium acetate. Boil two or three minutes and then let the precipitate settle. Wash quickly with boiling water containing a little sodium acetate. The filtrate contains all the zinc, manganese, cobalt or nickel which were in the solution. The precipitate can be dissolved in dilute hydrochloric acid and the iron and aluminium separated by any of the methods already given. The filtrate may be evaporated to dryness, taken up with hydrochloric acid, sodium carbonate added till a permanent precipitate just forms, and then a drop or two of hydrochloric acid added to re-dissolve this precipitate. On passing sulphuretted hydrogen through the solution the zinc is precipitated as sulphide, while any manganese, nickel or cobalt present remain in solution. When all the zinc is precipitated, allow to stand twelve hours, filter, wash with sulphuretted hydrogen water, re-dissolve in hydrochloric acid and throw down the zinc as carbonate by sodium carbonate and ignite to oxide; or, the zinc sulphide may be mixed with sulphur, put in a Rose crucible and ignited in a stream of hydrogen sulphide. It is in this case weighed as sulphide.

If the solution from which the zinc has been precipitated and filtered out be made strongly acid with acetic acid, and excess of sodium acetate added, sulphuretted hydrogen may be passed through again and will precipitate nickel or cobalt sulphides.

These are filtered out and the filtrate concentrated, ammonium sulphide added to it and then acetic acid. The remaining nickel and cobalt will be precipitated. The two precipitates are united and the nickel (and cobalt if present) determined by any of the ordinary methods of precipitation. The filtrate is neutralized with ammonia, ammonium chloride added and let stand at least 24 hours in order to precipitate out the manganese as sulphide. Manganese might also be separated out of the filtrate from the basic acetate separation by adding hydrochloric acid and boiling with bromine water. The manganese is completely precipitated as dioxide.

If it is not wished to determine the copper, but only the aluminium present, the copper can be easily removed by adding a slight excess of ammonia to the hot hydrochloric acid solution. The solution is boiled a few minutes and filtered. The precipitate is apt to carry down and retain some copper. It is therefore necessary to re-dissolve it in acid and repeat the precipitation. All the iron and aluminium are thus obtained in the precipitate, along with manganese and possibly some zinc and nickel, if these are present. The precipitate can be dissolved in hydrochloric acid and the iron and aluminium precipitated alone by a basic acetate separation.

GENERAL REMARKS.

In analyzing aluminium-tin alloys, hot nitric acid will dissolve the aluminium and leave the tin as meta-stannic acid. Aluminium-silver alloys may be attacked by caustic alkali, leaving silver undissolved in the residue, or may be dissolved in hot nitric acid, nearly neutralized and the silver precipitated by hydrochloric acid or sodium chloride. Alloys of aluminium with either zinc, nickel or manganese are best analyzed by bringing the alloy into solution and separating aluminium from the other metal by a basic acetate precipitation. Aluminium-lead alloys may be dissolved in hot nitric acid and the lead precipitated by sulphuric acid after nearly neutralizing the solution and adding alcohol to it.

I would recommend that some qualitative tests such as are

suggested in the beginning of this chapter—blow-pipe tests, wet tests, etc.—be always made preparatory to the quantitative analysis; and then, knowing what is present and which elements it is desired to estimate and which to neglect, the method of attack and analysis should be decided on. Half an hour spent in making qualitative tests and ten minutes in reflection as to the best method of analysis to adopt, will often save several hours of unnecessary work and frequently prevent the exasperating necessity of having to stop an analysis and start over again.

INDEX.